David Haskell's work integrates scientific, literary, and contemplative studies of the natural world. He is a professor of biology and environmental studies at the University of the South and a Guggenheim Fellow. His 2017 book, *The Songs of Trees*, won the John Burroughs Medal for Outstanding Nature Writing. His 2012 book, *The Forest Unseen*, was a finalist for the Pulitzer Prize and the PEN/E.O. Wilson Literary Science Writing Award, and won the 2013 Best Book Award from the National Academies, the National Outdoor Book Award, and the Reed Environmental Writing Award.

* * *

Praise for *Sounds Wild and Broken*

"*Sounds Wild and Broken* is a symphony, filled with the music of life. It is fascinating, heartbreaking, and beautifully written."
　　　　　—Elizabeth Kolbert, author of *The Sixth Extinction*

"Earth sings and rings and warbles: a musical planet, maybe the only one in the universe. As David George Haskell tells it in his captivating new book, *Sounds Wild and Broken*, it is astonishing good fortune—and a fearsome responsibility—to be given this music and the ears to hear it with. . . . *Sounds Wild and Broken* offer[s] one delight after another."
　　　　　—Kathleen Dean Moore, *Scientific American*

"Listen to David Haskell: he will transform the way you hear the world. Haskell is one of those rare scientists who illuminates his topic—the magnificent natural sonic diversity of our planet, what we have to gain from its richness, what we have to lose from its diminishment—in lyrical, erudite prose that both informs and inspires."
　　　　　—Jennifer Ackerman, author of *The Genius of Birds*

"[Haskell] is something of an idiosyncratic genius. . . . [His] previous works leveraged two tools that established him as one of America's premier nature writers: his Zen-like ability to pay granular attention to what most people ignore and a lyrical writing style few scientists can muster. . . . Haskell enlivens the science by taking us on a journey, hopping from continent to continent. He wanders the mountains of southern France, treks Ecuador's Amazon jungle, and noses about eucalyptus forests in New South Wales, all to illustrate the connection between sound and place." —*Outside*

"A stunning call to reinhabit our ancient communion with sound. Gorgeous prose and deep research meld wonder with intellect, inspiring reverence, delight, and a sense of urgency in protecting aural diversity. The voice of the earth is singing with beauty and need—Haskell shows us the extraordinary gift and responsibility of being available to listen."

—Lyanda Lynn Haupt, author of *Rooted:*
Life at the Crossroads of Science, Nature, and Spirit

"A moving paean to Earth's fraying soundtrack . . . [Haskell] traces, beautifully and brilliantly . . . the infinite serial interactions between communication and reception. . . . [*Sounds Wild and Broken* is] a reminder that the narrow aural spectrum on which most of us operate, and the ways in which human life is led, blocks out the planet's great, orchestral richness."

—*The Guardian* (London)

"In *Sounds Wild and Broken*, David George Haskell once again expands our sensory universe, revealing not only the grand variety of earthly song, music, and speech but the astonishing ways in which sound originates, evolves, and binds us together. His careful listening will sharpen your ears."

—Michelle Nijhuis, author of *Beloved Beasts:*
Fighting for Life in an Age of Extinction

"A soaring panegyric not just to the human ear but also to the auditory equipment of every living being . . . It's beautiful, Haskell's devotion to his ears. . . . Haskell wants us, above all, to listen, to use our glorious ciliary hairs for good. Those twitching hairs delivered us from pond scum, after all. Maybe, if properly attuned, they can deliver us from catastrophe."

—*Los Angeles Review of Books*

"Haskell's voice is unique in contemporary nature writing. . . . [He] creates a pleasing poetry of nature, his carefully crafted sentences luring readers in for the long haul. . . . Glorious." —Chapter16.org

"This is how scientific writing should be, and almost never is: suffused with wonder and pathos, throbbing with the music of the wild. Haskell shows us— no, lets us hear—that we are resonant animals in a thrillingly resonant universe, and that our fulfilment depends on finding the frequency that will make us resonate with everything else. His superb book sent me on my way singing, and trying to join in with the songs I heard on the way."

—Charles Foster, author of *Being a Beast* and *Being a Human*

"Thoughtful, insightful . . . Haskell presents a clear-eyed thesis on the impact of worldwide environmental destruction and human noise on what we hear. . . . With persistent intelligence and understated wit, Haskell uncovers one subtle mystery after another, forming a gorgeous argument for protecting all we long to hear."

—*Booklist*

"This brilliant book will change the way you hear everything. Haskell takes us deep inside the music of human and nonhuman life, revealing one marvel after another, and makes a powerful case for conservation that not only preserves species but the sensory experience of life itself."

—Jonathan Meiburg, musician and author of *A Most Remarkable Creature*

"A joyous celebration of the music of life. . . . Seamlessly melding history, ecology, physiology, philosophy, and biology, Haskell exults in the delightful cacophony created by birds and insects, wind and sea, human voices and musical instruments. . . . Sparkling prose conveys an urgent message."

—*Kirkus Reviews* (starred review)

"In luminous prose, David Haskell teaches us to hear the beauty and tragedy of the whole history of life on Earth. *Sounds Wild and Broken* will change the way you listen to nature and to yourself, and may this help us heal our planet before it's too late."

—David Rothenberg, author of *Nightingales in Berlin* and *Why Birds Sing*

ALSO BY DAVID GEORGE HASKELL

The Forest Unseen:
A Year's Watch in Nature

The Songs of Trees:
Stories from Nature's Great Connectors

Sounds
Wild
and
Broken

SONIC MARVELS, EVOLUTION'S CREATIVITY,
AND THE CRISIS OF SENSORY EXTINCTION

David
George
Haskell

PENGUIN BOOKS

PENGUIN BOOKS

An imprint of Penguin Random House LLC

penguinrandomhouse.com

First published in the United States of America by Viking,
an imprint of Penguin Random House LLC, 2022
Published in Penguin Books 2023

ISBN 9781984881564 (paperback)

THE LIBRARY OF CONGRESS HAS CATALOGED THE HARDCOVER EDITION AS FOLLOWS:
Names: Haskell, David George, author.
Title: Sounds wild and broken: sonic marvels, evolution's creativity, and
the crisis of sensory extinction / David George Haskell.
Description: New York: Viking, [2022] |
Includes bibliographical references and index.
Identifiers: LCCN 2021028957 (print) | LCCN 2021028958 (ebook) |
ISBN 9781984881540 (hardcover) | ISBN 9781984881557 (ebook)
Subjects: LCSH: Bioacoustics—Environmental aspects. |
Nature sounds—Environmental aspects. | Acoustic phenomena in
nature—Environmental aspects. | Sound—Physiological effect.
Classification: LCC QH510.5 .H37 2022 (print) | LCC QH510.5 (ebook) |
DDC 591.59/4—dc23
LC record available at https://lccn.loc.gov/2021028957
LC ebook record available at https://lccn.loc.gov/2021028958

Printed in the United States of America
2nd Printing

Designed by Alexis Farabaugh

Dedicated to Katie Lehman,
who opens my ears to marvels

CONTENTS

PREFACE

On the sidewalk that runs along the edge of Prospect Park in Brooklyn, katydids and crickets spice the air with their late-summer songs. Sunset was hours ago, but the heat dallies, animating pulsing rasps and trills of insects hidden in tree branches. The pavement's light has its own rhythm, a regular pattern from widely spaced streetlights along the park's wall. The insects are drawn to the lights, gathering in the glowing orbs of leaves around each lamp. As I walk, sound and light rise and fall around me, a subtle swell.

The katydids sing with snappy, buzzing triplets—*ka-ty-did*—repeated in a steady pulse, one per second. A few singers abbreviate the song to doublets and slow the pace. Unlike nights when the performers unite in a park-wide beat, powerful enough that I feel it in my chest, tonight's katydids seem uncoordinated, each finding its own rhythm. These pulsations contrast with drawn-out, single-toned trills of tree crickets that twine their songs into a sweet and almost unvarying drone.

Security lamps behind a building in the park spill light upward into a cluster of oak trees. One hundred or more starlings gather in the branches. No sleep for these roosting birds, though. Stimulated by bright lights, they squeal, chitter, and whistle at one another, fluttering and jostling among twigs.

A large airplane passes low overhead, lined up along the western edge of the park as it completes its descent into LaGuardia Airport. The sound starts as a thread on the southern horizon, fattens to a heavy, rough rope as it smothers the insects' songs, then tapers to a frayed, rumbling tail as it leaves us. In the daytime, during peak landing hours, these planes pass every two minutes.

Other vehicles join: the whirring complaint of car tires on asphalt, the bark and rumble of accelerating engines, a distant clash of horns at the angry intersections of Grand Army Plaza, and the fizz of speeding e-bikes.

I walked here from a chamber music concert in the basement of the public library. Musicians merged their bodies with wood, nylon, and metal, a chimeric union of animal, oil, tree, and ore that reawakened sound from its slumber on printed sheet music. Afterward, I spoke with friends and our tremulous vocal folds imparted fugitive meaning to breath. In music and speech, nerves enlist the air as a neurotransmitter, erasing the physical distance between communicating bodies.

All these sounds draw their energy from the sun. Algae basked, grew, were entombed, then turned to dark oil. We hear the algae roaring now as their long-buried stores of sunlight are released from jet and car engines. The e-bike is juiced by electricity from a coal power plant, the snared light of old forests. This year's crop of sunlight, held in maple and oak leaves, feeds the katydids and crickets. Wheat and rice do the same for humans. It is night here, but the sun still shines, photons transmuted to sound waves.

An ordinary evening. A few insect sounds and some birds. Cars and planes on their rounds. Human music and voices. I take this for granted. A planet alive with music and speech.

Yet it was not always this way. The wonders of Earth's living voices are of recent origin. And they are fragile.

For more than nine-tenths of its history, Earth lacked any communicative sounds. No creatures sang when the seas first swarmed with animal life or when the oceans' reefs first rose. The land's primeval forests contained no calling insects or vertebrate animals. In those days, animals signaled and connected only by catching the eye of another, or through touch and chemicals. Hundreds of millions of years of animal evolution unfolded in communicative silence.

Once voices evolved, they knit animals into networks that allowed almost instantaneous conversation and connection, sometimes at great distances, as if by telepathy. Sound carries its messages through fog, turbidity, dense thickets, and night's dark. It passes through barriers that block aromas and light. Ears are omnidirectional and always open. Sound not only connects animals, its varied pitches, timbres, rhythms, and amplitudes carry nuanced messages.

When living beings connect, new possibilities appear. Animal voices are catalysts for innovation. This is paradoxical. Sound is ephemeral. Yet in its passage, sound links living beings and wakes the latent powers of biological and cultural evolution. These generative powers, acting over hundreds of millions of years, produced the astonishingly diverse sounds of the living Earth. The words on this page, inked stand-ins for human speech, are but one of the productions of the fruitful union of sound, evolution, and culture. Hundreds of thousands of other wonders ring out across the world. Every vocal species has a distinctive sound. Every place on the globe has an acoustic character made from the unique confluence of this multitude of voices.

The diverse sounds of the world are now in crisis. Our species is both an apogee of sonic creativity and the great destroyer of the world's acoustic riches. Habitat destruction and human noise are erasing sonic diversity worldwide. Never in the history of Earth have sounds been so rich and varied. Never has this diversity been so threatened. We live amid riches and despoliation.

"Environmental" problems are often presented in terms of atmospheric change, chemical pollution, or species extinction. These are essential perspectives and measures. But we also need a complementary frame: Our actions are bequeathing the future an impoverished sensory world. As wild sounds disappear forever and human noise smothers other voices, Earth becomes less vital, blander. This decline is not a mere loss of sensory ornament. Sound is generative, and so the erasure of sonic diversity makes the world less creative. The crisis exists within our own species too. The burdens of noise—ill health, poor learning, and increased mortality—are unjustly distributed. Racism, sexism, and power asymmetries create dire sonic inequities.

Listening opens us to the wonders of communication and creativity. Listening also teaches us that we live in an age of diminishment. Aesthetics— the appreciation and consideration of the perceptions of the senses— should therefore be central guides amid the convulsions of change and injustice that we live within. Yet we are increasingly disconnected from sensory, storied relationship to life's community. This rupture is part of the sensory crisis. We become estranged from both the beauty and brokenness of much of the living world. This destroys the necessary sensory foundation for human ethics. The crises in which we live, then, are not just "environmental," of the environs, but perceptual. When the most powerful species on Earth ceases to listen to the voices of others, calamity ensues. The vitality of the world depends, in part, on whether we turn our ears back to the living Earth.

To listen, then, is a delight, a window into life's creativity, and a political and moral act.

Sounds
Wild
and
Broken

PART I

Origins

Primal Sound and the Ancient Roots of Hearing

A t first, sound on Earth was only of stone, water, lightning, and wind.

An invitation: listen, and hear this primal Earth today. Wherever life's voices are hushed or absent we hear sounds largely unchanged since Earth cooled from its fiery start more than four billion years ago. Pressing against mountain peaks, wind yields a low and urgent roar, sometimes twisting into itself with a whip crack as it eddies. In deserts and ice fields, air hisses over sand and snow. On the ocean shore, waves slam and suck at pebbles, grit, and unyielding cliffs. Rain rattles and drums against rock and soil, and seethes into water. Rivers gurgle in their beds. Thunderstorms boom and the surface of the Earth echoes its reply. Sporadic tremors and eruptions of the underworld punctuate these voices of air and water, sounding with geologic growls and bellows.

These sounds are powered by the sun, gravity, and the heat of the Earth. Sun-warmed air stirs the wind. Waves rise as gales strafe the water. Solar rays lift vapor, then gravity tugs rain back to Earth. Rivers, too, flow under gravity's imperative. The ocean tides rise and fall from the pull of the moon. Tectonic plates slide over the hot liquid heart of the planet.

About three and a half billion years ago, sunlight found a new path to sound: life. Today all living voices, save for a few rock-eating bacteria, are animated by the sun. In the murmurs of cells and the voices of animals, we hear solar energy refracted into sound. Human language and music are part of this flow. We are acoustic conduits for plant-snared light as it escapes to air. Even the growl of machines is animated by the burn of long-buried sunlight.

The first living sounds came from bacteria that sent infinitesimally quiet murmurs, sighs, and purrs into their watery surroundings. Bacterial sounds are now discernible to us only with the most sensitive modern equipment. A microphone in a quiet laboratory can pick up sounds from colonies of *Bacillus subtilis*, a species of bacteria commonly found in soils and mammalian guts. Amplified, these vibrations sound like the hiss of steam escaping from a tight valve. When a loudspeaker plays similar sounds back into flasks of bacteria, the cells' growth rate surges, an effect whose biochemical mechanism is as yet unknown. We can also "hear" bacteria by balancing them on the tip of a microscopic arm. This bacteria-coated strut is so small that every shudder from their cell surfaces makes it quiver. A laser beam directed at the arm records and measures these motions. This procedure reveals that bacteria are in constant shimmering motion, producing tremulous sound waves. The crests and troughs of the waves—the extent of the cell's vibratory movement—are only about five nanometers, one-thousandth of the width of the bacterial cell, and half a million times smaller than the deflections in my vocal folds when I speak.

Cells make sound because they are in continuous motion. Their lives are sustained by thousands of inner streams and rhythms, each one tuned and shaped by cascades of chemical reactions and relationships. Given this dynamism, it is not surprising that vibrations emanate from their cell surfaces.

Our inattention to these sounds is puzzling, especially now that technologies allow our human senses to extend into the bacterial realm. Only a couple of dozen scientific papers have so far examined sound in bacteria. Likewise, although we know that bacterial membranes are studded with proteins that detect physical movement—shear, stretch, touch—how these sensors function with sounds is unknown. Perhaps there is a cultural bias at play here. As biologists, we're immersed in visual diagrams. In my own training, not once was I asked to use my ears in a lab experiment. The sounds of cells exist not only on the edge of our perception, but of our imagination, shaped as it is by habits and preconceptions.

Do bacteria speak? Do they use sound to communicate with one another just as they use chemicals to send information from one cell to another? Given that communication among cells is one of the fundamental activities of bacteria, sound would at first seem a likely means of communication. Bacteria are social beings. They live in films and clusters that are so tightly woven that they are often invulnerable to chemical and physical attacks that easily kill solitary cells. Bacterial success depends on networked teamwork and, at the genetic and biochemical levels, bacteria are constantly exchanging molecules. But to date, there are no documented examples of sonic signaling among bacteria, although their increased growth rates when exposed to the sounds of their own kind may be a form of eavesdropping. Sonic communication may be ill-suited to bacterial societies. They live at a scale so tiny that molecules can zip from one cell to another in a fraction of a second. Bacteria use tens of thousands of molecules within their cells, an extensive, complex, and ready-made language. For them, chemical communication may be cheaper, faster, and more nuanced than sound waves.

Bacteria, and their look-alike cousins the Archaea, were the only life on Earth for about two billion years. Larger cells—amoebas, ciliates, and their kin—evolved about 1.5 billion years ago. These larger cells, the eukaryotes, later gave rise to plants, fungi, and animals. Single eukaryote cells, like bacteria, are full of trembling motion. They, too, are not known to communicate by sound. No yeast cell sings to its mate. No amoeba shouts warnings to its neighbors.

Life's quiet continued with the first animals. These ocean dwellers had bodies shaped like disks and pleated ribbons made of cells held together by strands of protein fiber. If we could hold them now, they'd feel like filmy seaweed, thin and rubbery. Their fossil remains are lodged in rocks about 575 million years old. Collectively, they are known as the Ediacaran fauna, named for the Australian hills where some of their number were unearthed.

The bodily simplicity of the Ediacaran animals obscures their pedigree, leaving no telltale marks to assign them to groups we'd recognize today. No segmented body armor like arthropods. No stiff column down their backs like fish. No mouths, guts, or organs. And almost certainly, no sound-making devices. There is no hint on these animals of any body part that could make a coherent scrape, pop, thump, or twang. Contemporary animals with more complex bodies but superficially similar body shapes—sponges, jellyfish, and sea fans—are also voiceless, suggesting that these first animal communities were quiet places. To the hum of bacteria and other single-celled creatures, evolution added only the sloshes and swirls of water around soft disk- and fanlike animals.

For three billion years, life was nearly silent, its sounds confined to the tremors of cell walls and the eddies around simple animals. But during those long, quiet years, evolution built a structure that would later transform the sounds of Earth. This innovation—a tiny wiggly hair on the cell membrane—helped cells to swim, steer, and gather food. This hair, known as a cilium, protrudes into the fluid around the cell. Many cells deploy multiple cilia, gaining extra swimming power from clusters or pelts of the beating hairs. How cilia evolved is not fully understood, but they may have started as extensions of the protein scaffolding within the cell. Any motion in the water is transmitted into the weave of living proteins in the core of the cilium and then back into the cell. This transmission became the foundation for life's awareness of sound waves. By changing electrical charges in the cells' membranes and molecules, cilia translated motions exterior to the cell into the chemical language of the cells' interiors. Today all animals use cilia to sense sonic vibrations around them, using either specialized hearing organs or cilia scattered on the skin and in the body.

The rich animal sounds that we live among today, including our own voices, are a twofold legacy of the origin of cilia 1.5 billion years ago. First, evolution created diversity of sensory experience through the many ways that cilia are deployed on cells and on animal bodies. Our human ears are just one way of listening. Second, long after sensitivity to vibrations in water first appeared, some animals discovered how to use sound to communicate with one another. The interplay of these two legacies—sonic sensation and expression—fed evolution's creativity. When we marvel at springtime birdsong, an infant discovering human speech, or the vigor of chorusing insects and frogs on a summer evening, we are immersed in the wondrous legacy of the ciliary hair.

Unity and Diversity

In the moment of our birth, we are dragged across four hundred million years of evolutionary time. We turn from aquatic creatures to dwellers of air and land. We gasp, sucking the alien gas into lungs previously filled with warm, salty ocean. Our eyes are pulled from the dim, reddish glow of the deep into jabbing brightness. The chill of evaporation slaps our drying skin.

No wonder we wail. No wonder we forget, burying the memory in the soil of the subconscious.

Our earliest and only experience of sound before birth was the hum and throb of an aquatic cocoon. Our mother's voice found us, as did the sounds of her surging blood, breath flowing in lungs, and churning digestion. Fainter were the sounds of the world beyond our mother, from places then unimaginable to our mostly unformed brains. High tones were attenuated by the enclosing walls of flesh and fluid, and so our first sonic experiences were low and often rhythmic as her body pulsed and moved.

In the womb, hearing develops gradually. Before twenty weeks, our world is silent. At about twenty-four weeks, hair cells start to signal through nerves running to rudimentary auditory centers in the partly developed brain stem. Cells tuned to low-frequency tones mature first, and so our hearing starts with bass throbs and murmurs. Six weeks later, furious growth and differentiation of tissues result in a frequency range of hearing similar to that of an adult. Sound flows from mother's fluids into ours, directly stimulating the nerve cells in the innermost part of our ears, unmediated by ear canals, drums, or middle ear bones.

All of this gone, in a moment.

Birth removes us from our watery surrounds, but our final aural transition to air happens hours later. The fatty vernix that swaddles us at birth lingers in the ear canal, muffling airborne sound for a few minutes or, for some, days. Soft tissues and fluid likewise recede over hours from the bones of the middle ear. When these vestiges of our fetal selves finally dissolve, our ear canals and middle ears are filled with the dry air that is our inheritance as terrestrial mammals.

Yet even in adulthood the hair cells of our inner ears are bathed with fluid. We keep a memory of the primal ocean and womb inside the coils of our inner ear. The rest of the ear's apparatus—pinnae, middle ear chamber, and bones—delivers sound to this watery core. There, deep inside, we listen as aquatic beings.

I lie belly-down on the wooden dock. The splintery boards toast me with the stored heat of the summertime Georgia sun. In my nose, the sulfurous, ripe aroma of salt marsh. The flowing water under the dock is turbid, a mud soup sweeping past on a falling tide. I'm on Saint Catherines Island, a barrier island whose eastern shores face the Atlantic. Here, on the western side of the island, ten kilometers of salt marsh separate me from the flood-prone piney woods of the mainland. In the humid air, these woods are mere haze on the horizon. Salt marsh grasses, interrupted by narrow, twisting tidal creeks, cover the intervening distance. These grasses grow knee or waist high on all the mudflats, as thickly packed and as deep green as lush fields of young wheat.

The marshes seem monotone, their uniform verdure spiced only by snowy egrets stalking the creek edges and the pumping wing beats of glossy ibises passing overhead. But these are the most productive habitats known on Earth, capturing and turning into plant material more sunlight per hectare than the lushest of forests. Marsh grasses, algae, and plankton thrive in the happy confluence of fertile mud and strong sun. Such abundance supports a diverse animal community, especially of fish. More than seventy fish species live in these tidal marshes. Ocean-dwelling fish also swim here to spawn. Their larvae grow in the protection and plenty of the marshes, then catch a ride to adulthood on an outbound tide.

For all terrestrial vertebrate animals, rich salt water such as this was our original home, first as single-celled creatures, then as fish. About 90 percent of our ancestry was underwater. I clamp headphones over my ears and drop a hydrophone from the dock. I'm taking my ears back to where they came from.

The heavy capsule, a waterproof rubber and metal ball containing a microphone, sinks quickly, pulling the cable after it. I wedge a cable loop under my knee, holding the hydrophone above the creek bottom's mud and debris, about three meters down in the opaque water.

When I first release the hydrophone, all I hear is the high gurgle of streaming water. As it descends, the swirling sounds fall away. Suddenly I'm plunged into a pan of sizzling bacon fat. Sparkles surround me, a sonic shimmer. Every glistening fragment is a fleck of sunlit copper, warm and flashing. I've arrived in the acoustic domain of snapping shrimp.

This crackling is common in tropical and subtropical salt waters worldwide. Its sources are the hundreds of species of snapping shrimp that live in seagrass, mud, and reefs. Most of these animals are half the length of my finger or smaller, equipped with one hefty claw for snapping and a lighter one for grasping. I'm hearing a chorus of claws.

As the claw snaps shut, a plunger slams into a socket, shooting forward a jet of water. In the wake of this jet, water pressure drops, causing an air bubble to pop into existence, then collapse. This implosion sends a shock-wave through the water, the snap that I'm hearing. The sound pulse lasts less than a tenth of a millisecond, but it is strong enough to kill any small crustacean, worm, or fish larva within three millimeters of the claw tip. Shrimp use the sound as a territorial signal and jousting weapon. As long as they keep a centimeter away from their neighbors, they can spar un-harmed.

The combined racket of snapping shrimp is, in some tropical waters, loud enough to befuddle military sonar. In World War II, US submarines hid among the snapping shrimp beds off Japan. To this day, navy spies deploy-ing hydrophones must work around the sonic haze of shrimp claws.

My first lesson in this sonic immersion is that the underwater world can be a boisterous place. Before I donned the headphones, airborne sound came to me in bursts: squalls of whistles from boat-tailed grackles, pulses of cricket and cicada sound, occasional nasal caws from fish crows, and the melodies of distant songbirds. Underwater, the shrimp innervate their sur-roundings with unflagging sonic energy. There are no silent spaces be-tween song phrases or cries. Sound travels more than four times faster in salt water than in air, adding to the sense of brightness. This is especially true at close range, between the reflective surfaces of the muddy bottom and the upper water boundaries, where sounds have not been attenuated by the viscosity of water.

Into the cloud of shrimp sound come stammering bursts of knocks. Each batch lasts a second or two, a cluster of ten or more taps. Then a pause of five or so seconds, more regular taps, interrupted by occasional hesita-tions. The taps sound like an impatient fingernail drumming on a hardcover book, sharp and low, with a touch of resonance. The sounds come from silver perch close by. These finger-length fish come to the salt marsh to spawn before returning in late summer to the deeper waters of the estuary and offshore. Alongside these knocks come faster bursts of tapping, almost purrs, the calls of the Atlantic croaker, a bottom-feeding fish that grows as long as my forearm.

Waa! The bleat of a lamb, but quieter. These complaints occasionally poke into the background of shrimp, perch, and croaker, and come from an oyster toadfish, probably hiding in its lair on the bottom of the tidal creek. Like their namesake, toadfish are scaleless and warty, with huge gulping mouths. Their fist-sized heads and tapered bodies are also well endowed with spines. Males call to attract females to shallow burrows. After mating, males stay with the fertilized eggs for weeks, defending them and cleaning the nest. The one I hear now is muffled and soft. He must be at some distance from the hydrophone, perhaps burrowed into debris around the dock's pilings.

All three of the fish that I hear through my hydrophone make sound by vibrating their swim bladders. Each bladder is an air-filled sac running inside the fish, stretching for about one-third of the body length below the spine. Muscles pressed against the thin walls of these bladders shiver, and these motions evoke squeaking or grunting sounds from the air within. The muscles are among the fastest known in any animal, contracting hundreds of times per second. Sound waves from the swim bladder flow into the fish's tissues and then into the water. For these fish, the whole body is an underwater loudspeaker.

The acoustic realm of these shrimp and fish seems alien to me. I'm used to the melodies, timbres, and rhythms of humans, birds, and insects. Here, though, percussive sounds dominate: the sparkle of thousands of hammer blows by shrimp claws, the knocks of perch and croaker, and the unmodulated burr of the toadfish.

But unity undergirds these differences.

The shrimps' stony, articulated exoskeleton bristles with fine sensory hairs. Sound also stimulates clusters of stretch receptors in their joints, where cilia transmit motion to nerves. At the base of antennae, tiny sand grains enclosed within gelatinous balls of sensory cells are stirred into motion by sound. Hearing, for snapping shrimps, is an experience for the whole body. Unlike human ears that detect pressure waves on our eardrums, shrimp and other crustaceans hear by detecting the displacement of water molecules, especially low-frequency motions. Sound, for them, arrives not as the push and shove of a wave, but as the tickle of moving molecules.

Fish, too, hear through sensors spread all over their body surface. Cells capped with jelly-enclosed cilia line both the skin and watery canals just below the skin's surface, a network known as the lateral line system. Unlike the touch receptors deeply buried in our dry, keratinous skin, these fish sensory cells live in intimate contact with the water around them. The lateral line system is especially sensitive to low-frequency sounds and wafts of flowing water. Rudiments of the lateral line system appear on human skin as embryos, but we lose any trace as we mature, shedding this sensory embrace of our surroundings long before we are born.

Fish also hear using inner ears. These are the same structures that our ancestors brought with them when they came onto land. We humans hear with modified fish ears.

Like the lateral line system, the fish's inner ears unite sensation of sound and motion. Three looping semicircular canals detect body motion by sensing the flow of fluids in the canals over hair cells. Connected to these canals are two bulging sacs lined with sound-sensitive hair cells. In many fish species, tiny flat bones in the sacs overlay some of these hair cells. When the fish moves, the bone lags, dragging on the hair cells and magnifying the sense of motion. In many species, the swim bladder also gathers and transmits sound waves to the inner ears.

Among land vertebrates, the fish's flat ear bones and swim bladders are absent. The hearing sacs are elongated into canals, expanding the range of sound frequencies that the ear can perceive. In mammals, the canal is so long that it coils, forming what we now call the cochlea, from the Latin for "snail shell." Our language divides sensations of "sound," "body motion," and "balance," but they all emerge from hair cells in interconnected fluid-filled canals in our inner ears. The link in human cultures between music and dance, and between speech and gesture, is deeply rooted in both our bodies and in the evolutionary history of animals.

Ancient kinship among vertebrates is present in sound making too. Although vertebrate animals make sound in very different ways, these processes share an embryological origin. A small segment of nervous tissue at the intersection of the hindbrain and the spinal column develops into the nerve circuit that controls sound making in adult animals. This circuit acts as the pattern generator for vocalization across animals with very different forms of sound making: from the swim bladders of fish, to the larynges of terrestrial animals, to the unique syringes in the chests of birds, along with thousands of sonic variations made by croaking and booming vocal sacs, strumming pectoral fins, and drumming forearms.

The region of the spine that orchestrates vocalization also coordinates the actions of the pectoral region, the muscles of the front fins or limbs. This linkage reveals the need for fine control of timing in both vocalization and movement. All calls and songs have rhythmicity, from the steady hum of the toadfish to the layers of repetition in the song of a bird. The same is true for the coordinated movements of fins, legs, or wings. Just as hearing in vertebrate animals is closely allied with a sense of motion, sound production is linked to body movement. Rhythmicity of sensation and action shares an embryological root.

When we humans talk and gesticulate, or sing and play musical instruments, we evoke ancient connections. When my hands thump out rhythms on piano keys or strum a guitar, I'm enacting the same bodily relationships among voice, limbs, and sound that create the bleats of a toadfish or the melodies of a forest songbird. When Henry Wadsworth Longfellow wrote that "music is the universal language of mankind," he stated an embryo-logical and evolutionary truth that far transcends the bounds of "man-kind."

Lowering a hydrophone from the dock was a revelatory moment. The expansion of my awareness came from two intersecting directions. I understood that my unaided human senses utterly failed to convey to me the richness of the marshes. The water surface, especially when obscured by streams of opaque tidal mud, is a formidable barrier to human understand-ing. When I heard the lively below-water chatter, I pierced, for a moment, a sensory barrier. Now when I'm at the marshes, I imagine and feel their diversity and fecundity, despite the visual uniformity of their above-water plants. Listening below the water surface opened me to the previously hid-den life of the marsh.

Alongside this understanding of the nature of a particular place, my sense of self changed. Lying on the dock and, later, reading about animal voices and ears, my thoughts and feelings about identity shifted. Evolution has drastically reworked the mammalian body as it transformed us from fleshy-finned swimmers to four-legged land lumberers. But under these terrestrial bodily accretions is unity with our distant aquatic relatives, unity not just of pedigree but of lived sensory experience. I'm a fish talking in air, strutting and breathing on land, yet experiencing the sea through trembling hair cells in coiled watery tubes in my ears. My hydrophone and headphones created a curious loop. In listening to the subaquatic world, I used tubes of modified seawater buried in my inner ears.

But human ears are only one of the sound sensors present here. Earth's sonic diversity is not only present in the varied voices of animals. Part of the world's richness is the diversity of aural *experience*.

As mammals, we inherited triplet ear bones and a long tightly curled cochlea. Birds have a single middle ear bone and a comma-shaped cochlea. Lizards and snakes have a short cochlea whose sound-sensitive hair cells are arranged in patches, not in a single smooth gradient as in our ears.

These are three independently evolved mechanisms within the vertebrate clan for hearing in air, dating back about three hundred million years. Each lineage lives within its own construction of sound. Lab experiments on the behavior of captives give us a crude sense of what these differences might mean for perception. Compared with mammals, birds cannot hear as high. Birds are relatively unconcerned with the sequence of sounds but are highly attuned to rapid-fire acoustic details in each note in a song, picking out subtleties that human ears miss entirely. Birds are also especially adept at hearing how sound energy is layered into different frequencies, the overall "shape" of the sound, rather than attending to the relative pitches that are the particular focus of mammalian ears and brains. Where we discern a melody in bird or human song—shifting frequencies between notes—birds likely experience the rich nuances of the inner qualities of each note.

Fish and shrimp are immersed in sound as the movement of water molecules directly stimulates their surface hairs and as sound waves flow unimpeded into and through their bodies. Bacteria and free-living eukaryotes, too, feel the vibratory signal on their membranes and cilia. On land, insects hear airborne sounds with hairs on their body surface and modified stretch receptor organs in their skeletons, the same organs used by both insects and crustaceans to feel motion and vibration in their legs. Specialized hearing organs independently evolved at least twenty times in different groups of insects. Crickets have drumlike hearing organs in their front legs, but grasshoppers hear through membranes on their abdomens. Many flies hear with a sensor in their antennae. Among moths, hearing organs evolved at least nine different times, resulting in "ears" on wing bases, along the abdomen, or, in the case of the sphinx moths, on the mouthparts. We humans can feel vibrations on our skin and in our flesh as well as in our ears, but these are crude and blurry sensations compared with the nuanced whole-body hearing experience of these other beings.

It is a convenient shorthand to say that the shrimp, fish, bacteria, birds, insects, and I "hear" the same sound. *To hear* is a verb that reveals the narrowness of our sonic perceptions and imaginations. We have no such limitation when we describe how animals move: They lope, strut, crawl, sidle, wing, creep, sashay, slide, trot, flutter, and bounce. Here is a lexicon that recognizes the diversity of animal motion. But we have an impoverished vocabulary for hearing. Hear. Listen. Attend. These words do little to open our imagination to the multiplicities of sonic experience.

What is the verb for the sensation created by a snapping shrimp's foreleg joints or the direction-sensitive hairs on its claws? When the bony plate in a croaker's ears slides over a membrane covered in hair cells, what should we name the resulting experience? The ciliary hairs in the lateral lines of the fish are immersed in the water around them, surely yielding a different experience from the movement of a triplet of bones in our middle ears. We lack any word to convey the mystery of the sphinx moth's mouth palpus when sensing an approaching bat.

Without a diverse vocabulary for hearing, our minds lapse into inattention and our imagination is limited. Hobbled by weak verbs, language must draw on adjectives, adverbs, and analogies. A shrimp claw listens spikily, perhaps, through narrowly tuned hairs. A fish's low-frequency lateral line hearing is oozy, deep, and fluid. The birds' aural attention, fueled by high body temperatures, is fevered and has a narrower range of pitch perception than ours, trimmed off at its top by a stumpy, uncoiled cochlea. Is bacterial hearing like pressing a trembling thumb into jelly, viscous and enveloping?

Yet despite the limitations of language and human sense organs, our experiences of the world are encouragements to imagination. Listening opens our minds to other ways of being. At any place on Earth, thousands of parallel sensory worlds coexist, the diverse productions of evolution's creative hand. We cannot hear with the ears of others, but we can listen and wonder.

At the dock, in my headphones, a whir cuts into the fish and shrimp sounds. It builds in loudness over five seconds then abruptly ends. Cough.

Another sputter. An outboard engine has been lowered—the whir was its electric motor easing down the blades—and is now cranking. Two more turns of the starter and the engine comes alive.

The engine's voice clouds the water, a chug pitched at about the frequency of human speech. The shrimp keep on crackling and their sound joins the outboard in my ears, two textures, one growly, one sparkly, each holding steady. The outboard idles for a minute, then, in an instant, roars. The propellers are spinning, shredding the water. As the boat pulls away, the intensity of the sound wavers, perhaps as the propeller turns toward and away from my hydrophone. Over the next minute, through the hydrophone, I hear the noise climb in frequency, up three octaves from the start, as the engine's scream fades into the distance. The croaker keeps pulsing its thumping song every ten seconds or so. The silver perch and oyster toadfish fall silent.

Sensory Bargains and Biases

Like a painter applying a delicate brushstroke to a canvas, my audiologist extends her arm and slides a slender foam plug into my right ear. A thin tube runs from the plug to an electronic console and a laptop. A gurgle bursts into my ear. Then the room stills. In the quiet, my senses waken: Winter sun through dusty clinic windows. Odor of floor cleaner and latex. A metal cart clinks far down the hallway.

Suddenly a high-pitched tone darts into the foam-plugged ear. No, I'm wrong, not a single tone but a weird two-note chord. It pulses, repeats, and pulses again, quieter. Then more tones, lower pitched. We're running down a series. Every time a sound hits my ear, two spikes leap from a trembling horizontal line on a graph on the laptop screen.

Unlike the hearing test I took last month, squeezing a trigger whenever I heard a tone, I now sit empty-handed. This test directly probes the cilia-bearing hair cells of my inner ear, with no conscious involvement on my part. On the screen, I see the graph twitch with every burst of sound. Sometimes the graph kicks up, but I hear nothing.

My audiologist loops the tube and earplug to my left ear. She clicks the machine back on. Another gurgle. Silence. Then come the tones, working their way through the sequence. Now that I've figured out how to read the graph, I stare unblinking at the line, waiting. There it is: my ear answering back! Just to the left of the two big spikes is a third, a miniature, that pokes up whenever sound floods my ear. It is ankle high to its tall companions, but jabs up always in synchrony with them. Nearly always. For some sounds, even ones that I can hear, the junior spike is absent or merely flutters.

The small spike on the graph shows me the hair cells of my inner ear in action. When the incoming double tone hits them, they shoot out a pulse of sound in answer. This reply is too quiet for me to hear, but the microphone picks up its signal. My ears, then, are not passive receivers of sound. They are active participants in the process of hearing, making their own vibrations. This ability comes from the cilia-bearing cells in the inner ear, descendants of the oar-like hairs on the membranes of ancient free-living cells, now lodged in watery coils in my head.

As I sit in the sterile, white-walled examination room, thinking of the motions of these tiny hairs, my imagination turns to pond scum. One of my favorite exercises with students is to scoop up some slimy ditch or lake water and peer into the lively throng through a microscope. The unaided eye sees only slime. Glass lenses directed at microscope slides reveal dozens of species in every drop. Some species, especially the emerald cells of the larger algae, creep like cargo ships maneuvering in port. Others, tethered by slender tails to fragments of vegetation, pump globular heads back and forth, wafting bacteria into cuplike maws. Green globules zip past, leaving eddying wakes. Glassy needles glide. Slipper-shaped cells spiral, halt, reverse, then set off again in new directions.

The motion we see under the microscope is all driven by cilia. Some cells have hundreds, a beating pelt, others have just a single one, elongated into what we call a flagellum. The beating of each cilium is powered by ten paired protein columns. Each of these columns is made from a coil of thousands of tiny subunits. Cross-linking proteins connect the columns. Rapid changes in the links among these proteins slide the columns over one another, driving the hairs' motions. Shuttle proteins run alongside the columns, replenishing and repairing the lively, flexing meshwork. To call this dynamism a "hair" is a convenient shorthand, but belies the inner complexity of the cilium.

Cilia on free-living cells beat at rates from one to one hundred times per second. If we could hear them, the sound would be a hum at and below the lowest pitches that our ears can grasp. But like the shivers of bacteria, these motions disturb only a thin layer of fluid around each cell, too quiet for human ears to detect.

All the descendant lineages of the first eukaryotes possess cilia, although many fungi have lost theirs. We are one of the ciliated descendants. The beating hairs in the pond scum under the microscope seem exotic append-ages with little connection to our human bodies. But these unfamiliar mo-tions are a reminder of the hidden activities of our own bodies.

Cilia line the passageways to our lungs, wafting out impurities. Eggs are swept along Fallopian tubes by beating cilia, and sperm cells are powered by waggling flagella. Our brains and spinal columns are washed by fluid circulated by ciliary hairs, and cilia coordinate the embryonic development of our organs. The light receptors in our eyes are modified cilia, the tips of their hairs no longer moving but welcoming light on their protruding arms. News of odors travels to our nerves via cilia that grab aromatic molecules. Our kidneys use cilia to sense, without our conscious awareness, urine flow and to regulate the growth of the kidneys' network of tubes.

We also hear with cilia. Each of the fifteen thousand sound-sensitive cells in our inner ears is crowned with a cilium bundled with smaller hairs. As a sound wave flows through the inner ear, its motions deflect these bundles. This movement causes the cells to signal to the nervous system. Physical motion is thus alchemized by cilia into bodily sensation.

Outwardly, complex animals seem to have little in common with the cells that swarm through pond scum and ocean water. Yet the vitality of our bodies and the richness of our sensory experience are grounded in the very same cellular structures that power our single-celled relatives. When we perceive sound or light or aroma, we experience deep kinship, a shared cellular heritage.

The cilia in my ears, mounted atop hair cells, are arrayed along a membrane sandwiched between coiled tubes of fluid. These coils, one for each ear, form the cochleas. Each is the size of a fat pea, and they are lodged in the skull just beyond the eardrums. The cochlear membrane is narrow and stiff at the end closest to the eardrum, but wide and floppy at the apex of the coil. High-frequency sounds cause the narrow end to vibrate. Low sounds stimulate the wide part. Every frequency within the range of human hearing thus has a place along the membrane's gradient of sound sensitivity, as if we had coiled up piano keyboards in our inner ears. Complex patterns of sound, like music or speech, stimulate waves at multiple places along the membrane's length. Vibrations are picked up by hair cells on the inner part of the membrane, the edge closest to the center of the cochlea's coil. These signal via the cochlear nerve to the brain.

Vigorous sounds have enough energy to buck the cochlear membrane and stimulate inner hair cells. But quieter sounds are too weak. Alone, they cannot trigger nerve impulses. Hair cells on the outer part of the membrane give these softer sound waves a boost so that the inner hair cells can perceive them. Outer hair cells are three times more numerous than those on the inner part of the membrane, underscoring their importance.

When a sound wave of the right frequency hits the outer hair cells, a protein leaps into action, pumping the cells up and down. The protein, prestin, is the fastest-known force generator in living cells. The up and down motion of the outer hair cells amplifies the wave, turning an anemic shiver into a surge. The magnified wave triggers the waiting inner hair cells. The teamwork of outer and inner hair cells allows us to perceive sound across a millionfold difference in energy levels, from a snowflake falling into a drift in the quiet woods to the clap of thunder echoing in a canyon.

What I see on the audiologist's screen is the activity of my outer hair cells. Normally the cells would pulse with the same frequency as the incoming waves. But the test I'm undergoing throws them into confusion. The two incoming tones are precisely calibrated to hit the membrane very close together and, like two people shaking a rug at slightly different rates, the activated outer hair cells cause the membrane to judder with the weird collision of these two drivers. Part of this judder—a harmless distortion of the waves in my ear—then flowed back out of the cochlea. The third spike on the screen was the squeal of my outer hair cells.

At the end of the test, my audiologist clicks at her laptop and the spiking lines disappear, replaced with a graph that shows how my hair cells performed. At low sound frequencies, the cells did fine in both ears. In my right ear, those tuned to higher frequencies have stopped bouncing or have slowed their motion. In my left ear, it is those focused on the midranges that have quieted. These inactive cells are not resting or asleep, they're defunct. Unlike birds that can regrow damaged hair cells, human inner ear cells get one life only.

The crystal ball, my audiologist calls this test. For someone in their fifties, my results are unexceptional. In future years, more hair cells will bow out, especially in the higher frequencies.

Most of us are born with hale outer hair cells, full of vim all up and down the cochlear membrane. But from then on, it's all downhill, part of the cellular die-off that marks time in our bodies. We can hasten the decline with loud sounds—guns, power tools, amplified music, engine rooms—and with medications poisonous to hair cells, including common drugs like neomycin and high doses of aspirin. But even a life spent drug-free in quiet surrounds would not protect our ears from the erosive power of passing years.

Such is the cost of living in a body richly endowed with sense organs.

Our every sensory experience is mediated by cells. Aging is a cellular process. Over time, cells accumulate defects in their form and DNA, eventually slowing or ceasing their work. And so to experience the passage of time in an animal body is to experience sensory diminishment. This is the deal evolution has bequeathed us: we get to enjoy sensory experience, but in bodies where the scope of perception dwindles as we age. The only animals known to have broken this deal are freshwater-dwelling relatives of jellyfish called *Hydra*. Their body consists of a sac topped by tentacles. Nerves weave through the body in a net, with no brain or complex sense organs. This uncomplicated design, made from a handful of cell types, allows *Hydra* to regularly purge and replace any defective cells. They live without any signs of aging. But these eternally youthful, inverted jellyfish have only rudimentary senses: a hazy grasp of sound and light delivered by single cells buried in their skins. Our bodies are too complex to self-renew as *Hydra* does. But we therefore have more well-developed senses, mediated by complex organs. We can blame advancing deafness and the other diminishments of age on Faustian forebears. They exchanged ageless bodies for richly sensual lives. This evolutionary bargain was forced on them by one of life's seemingly unbreakable rules: all complex cells and bodies must age and die.

I mourn the progressive loss of my hearing. The voices and music of people, birds, and trees give me connection, meaning, and joy. But alongside the sadness, I try to accept and enjoy evolution's bequest. These diverse voices exist only because our bodies are complex and therefore ephemeral.

Our hearing cells and organs not only lock us into a trajectory of aging. They also bias sensory experience. It is not the case that in my youth I had perfect hearing and now I've lost some of this transparent connection to the world. Even before my hair cells started dying off, what I heard was highly mediated. Everything that I hear is an imperfect rendering. The inner and outer worlds converse and entangle in my ears.

My mind protests. Sound is sound, surely? Am I not just hearing what surrounds me, connected to the world by open ears? No. This is an illusion. What we perceive is a translation of the world and every translator has special talents, errors, and opinions. Sitting in the clinic, gazing at spikes on a graph, I'm seeing the chatter of my cochlear hair cells. I'm face-to-face with part of the hidden chain of interpretation. Along every step of the path from external sound to internal perception, our body edits and distorts.

The ear trumpets, pinnae, on either side of our heads, along with the ear canal, amplify sound by fifteen to twenty decibels. This boost is the equivalent to walking across a large room to stand next to someone who is talking. Sound waves also bounce around the cups and folds of the pinnae. This clash of waves cancels out some high frequencies. Push your ear flaps forward. You'll hear a change in brightness. As we move our heads, the sound reflections shift, cutting out slightly different frequencies. From these subtleties, our brain extracts information about where sound is located on the vertical plane. We edit sound even as it enters the ear canal.

The middle ear—the eardrum and three ear bones—has the task of converting sound vibrations in air to vibrations in the fluid inside the cochlea. The air-to-water transition faces a physical challenge. When a wave in air hits water, most of the energy bounces back. This is one reason why we can't hear poolside chatter when we swim underwater. To solve this problem, the tiny bones of the middle ear gather vibrations from the relatively large eardrum and, using the levering action of the longer "hammer" bone pivoting onto the shorter "anvil" and "stirrup" bones, these bones focus vibrations onto a much smaller window leading to the cochlea's watery tubes. This conversion both amplifies, increasing the pressure of sound waves by about twenty times, and puts a slight filter on the sound, trimming extremely high and low frequencies.

Then the cochlea imposes a more severe filter. The upper and lower ends of our hearing are set by the sensitivity of cochlea. The stiffness of the membrane, the responsiveness of outer hair cells, and the tuning of nerve sensitivities determine not only upper and lower bounds of our perception of pitch but also our ability to discriminate among sound frequencies. In general, we can discriminate among pitches of one-twentieth of a half step on a piano keyboard. Between the notes B and C, for example, we can potentially, if we concentrate, hear twenty additional microtones. But this is only true for quieter sounds. Our ears hear subtle differences in pitch in whispered or spoken words, but for shouts our discrimination of pitch is coarser. Intense sound bucks the cochlear membrane and overwhelms auditory nerves. We have finer discrimination at lower frequencies than high. The shrill sounds of high-pitched insect songs, for example, all sound about the same pitch to us, even for those that, when experienced with the objectivity of a graph of sound frequencies, differ significantly. But for the lower sounds of human speech, we perceive subtle differences among sound frequencies.

Nerve signals and the brain's processing add their own layers of interpretation. Nerves in the cochlea fire when inner hair cells are stimulated. Each of these cells responds to a particular range of sound frequencies corresponding to its place on the high-to-low scale of the cochlear membrane. The width of these ranges and their overlap set another limit on frequency discrimination. The nerve impulses from the cochlea then flow to the auditory nerve through a series of processing centers in the brain stem and then to the cerebral cortex. There, the brain interprets incoming signals in the context of expectations, memories, and beliefs. What passes into conscious perception is an interpretation, not a transcript. This is most vividly illustrated by auditory illusions. By playing different sounds into each ear or by looping sounds to create repetition, pioneering acoustic psychologist Diana Deutsch found that she could trick the brain into hearing phantom words and melodies. These illusions reveal that what we "hear" emerges from the brain's attempts to extract order from incoming signals, even when no such order exists. The words and melodies that we hear are partly a product of our background, each of us hearing words and music relevant to our culture.

Our brains do not just receive input from the ears, they send out signals to the ears, adjusting the cochlea to local conditions. In noisy environments, the brain suppresses the sensitivity of the outer hair cells, like a hand reaching out to crank down the volume on a loudspeaker. This reduces the masking effect of noise, allowing meaningful sounds to be more clearly distinguished. The hair cells in our ears are less jumpy in a noisy restaurant, for example, than they are in a quiet forest.

These layers of interpretation bias our perceptions of loudness. When we walk on pavement, for example, we perceive the sound as about twice as loud as footsteps on soft grass. This accords with the increase in sound intensity, the amount of energy hitting our eardrums. But in a carpentry workshop, our ears mislead us. The circular saw sounds about twice or three times as loud as the power drill. But the actual sound intensity, the rate at which energy pounds our ears, is about one hundred times higher. The extent of this biased perception depends, too, on sound frequency. For loud low-frequency sounds—the clap of thunder, for example—muscles tug on middle ear bones, dialing back the intensity of the sound that flows to the cochlea. But for loud high-frequency sounds such as power tools, this protective reflex is weaker.

The distorted scaling of subjective experience adapted us to subtle differences in the quiet sounds of the preindustrial world. The meanings in human speech, especially the textures of emotion, are conveyed through tiny changes in sound intensity. The same is true for information gleaned from the sounds of wind, rain, plants, and nonhuman animals. Our ears evolved to pay attention to quiet voices and are out of place in persistently loud environments. In an industrial society—surrounded by engines, power tools, and amplified music—it would be helpful to have a more nuanced experience of the upper end of the loudness scale. We'd better appreciate the sonic variegations of this new world and be empowered to protect our inner ears from permanent damage.

We also have biased perceptions of sound frequencies. Our sensitivity is like a one-humped dromedary, with highest sensitivities in the midranges and duller perception in the low and high extremes, tuning our ears to environmental sounds most relevant to human survival: the sounds of our prey and predators, and the movements of flowing water and wind in vegetation. As we age, the hump sags at the high-frequency end or splits into a two-humped beast. Our ears' specialization on the intermediate frequencies works well for hearing the speech of other humans and some of the voices of nonhuman animals. But although we can hear many low and high sounds, we have an erroneous sense of their vigor. What we hear as faint, high trills from insects or dull, low roars from waves on a shore are in fact as intense as a strong-voiced human talking next to us. It is the biases of our ears and nerves that have cranked down the perceived loudness of these high and low frequencies. We live embedded within sensory distortion.

There are also many sounds beyond the ken of our cochleas. We hear, at best, from about 20 to 20,000 hertz (sound waves per second). Some whales and elephants hear down to 14 hertz. Pigeons can hear as low as half a hertz. Porpoises hear well up to 140,000 hertz and some bats all the way to 200,000. Domestic dogs hear up to 40,000 hertz and cats up to 80,000. Mice and rats chatter and sing to one another up to 90,000 hertz. If my feet represent the lowest sounds heard by animals and the top of my head the highest, we humans hear from just above the skin of my feet to the top of my hiking boots. Compared with most mammals, humans and our primate cousins live within a restricted aural world.

Thunder clouds, ocean storms, earth tremors, and volcanoes all sing and moan, calling out with sound waves as low as one-tenth of a hertz, far too low for our ears to detect. These low sounds carry for hundreds of kilometers, revealing the dynamics of seas, skies, and Earth. But we cannot hear them and thus live in a sonic world unaware of what stirs over the horizon. A similar limitation exists at the other end of the frequency scale. High frequencies attenuate very quickly in air and travel only short distances. We miss close-range dynamics of high-pitched songs of insects, cries of bats, much of the creaking of tree wood, and the quiet sounds of water fizzing through plant veins. There's a poignancy in these limitations. The world is speaking, but our bodies are unable to hear much of what surrounds us.

Our culture falsely divides us into those who can "hear" and those who are "deaf." But there is no sharp biological divide between hearing and deafness. We are all insensitive to most of the world's vibrations and energies. And every human body, regardless of our ears, can feel some sound in our body tissues and skin. Yet from the small portion of the sound waves that the majority of humans can hear, we have erected a sharp cultural divide. The "hearing" population relies on spoken language to such an extent that those who communicate by sight and gesture are too often excluded. A thriving Deaf culture rightly rejects the prejudice and aspersion that often accompany this exclusion, and has built communities united by rich nonvocal visual and gestural languages.

The limitations of human hearing reveal a paradox. As biological evolution endowed living creatures with a sense of hearing, connecting them to others, it simultaneously built perceptual walls. The bodily mechanisms of hearing work only because they focus on specific tasks. In becoming sensitive to the vibrations of the world, cells must narrow their abilities. Middle ear bones amplify sound and translate from air to water, but only within a particular range of frequencies. Proteins in hair cells pump up and down, but only at a speed set by their looping structure in the cell membrane. Hair cells boost quiet sounds, but that limits their finesse for louder ones. The cochlear membrane is too short to pick out very low and high tones, and too stiff to allow any finer discrimination of frequencies.

Like any other superlative achievement of evolution, mastery comes through specialization, and specialization circumscribes power. Hearing, like other senses, both reveals and distorts. It opens us to the multifarious sonic waves of the world. It also, necessarily, conveys warped and edited perceptions of sound energy.

And so evolution has built hearing organs tuned to the ranges of frequency and loudness most relevant to the success of each species. The human range of hearing therefore reveals the sounds our ancestors found most useful. If our ancestors had been eaters of mice and moths, both of which communicate with ultrasound, we likely would have evolved to hear much higher frequencies, as do many smaller predatory mammals like cats. If these forebears had sung underwater across ocean basins, they would have evolved water-adapted ears tuned to low frequencies, as did the whales.

The richer the sensory experience, the more convincing the perceptual illusion. Before I faced my own attenuation of hearing, I lived within that illusion, giving little thought to the limits of my senses. I had no embodied experience to teach me that my ears convey active interpretations of sound energy. Seeing at the audiology clinic the liveliness of my hair cells taught me otherwise. I understood that the price of sensory experience is to live—always, from birth—in a perceptual box, a space much smaller than the diverse flows of energy of the world. The walls of the box bend and filter incoming sound, manufacturing the shape and texture of my sonic perception.

The stab of sadness that I experienced in seeing the marks of my dead and dying hair cells on the audiologist's graph jolted me into a better appreciation of both the limits and the precious value of my senses. Distortions and narrowing boundaries are the price of nuanced and rich sensation. My hearing connects me to sound, of course, but also to the bargains that evolution struck on its long path from the cilia-covered cells of the primal oceans to the aural wonders of animal inner ears.

PART II

The Flourishing
of Animal Sounds

Predators, Silence, Wings

Grasshoppers clatter away from me as I walk the verge of a country road. Crickets chirp from hiding places in the unkempt thatch of grass. A fritillary butterfly wings past. Every minute or two, I pass through a thin cloud of midges and I wave my hands to sweep away their mote-like bodies. The cicadas, loud and persistent yesterday afternoon, give only sporadic croaks and stuttering whines in the cool morning.

On one side of the road, exposed rock the color of raw liver angles up the valley slope. Entombed within this stone are the ancestors of the insects that fly and sing around me. One of this fossilized swarm bears the earliest known sound-making structure of any animal, a ridge on the wing of an ancient cricket. This fossil is the oldest direct physical evidence of sonic communication.

There should be a shrine here. A monument to honor the first known earthly voice. But pilgrimages instead lead away from these mountains in southern France to the chapels and cathedrals of the lowlands. The Camino de Santiago passes by; pilgrims tread the road unaware that the deepest known root of all song and speech lies in the stones under their feet.

I am on the southern edge of the Massif Central, a complex of mountains and steep riverine valleys that curves inland along the Mediterranean coast, then stretches north, covering nearly one-sixth of France's land area. Unlike the coastal plain, the geography here is rugged and the human population sparse. Volcanism, collisions with the Alps and Pyrenees, and the push of continental plates have wrought a complex mix of rocks across the Massif. Where I walk, the carmine color of the stone alongside the road was born hundreds of millions of years ago in the hot, dry interior of a continent. Iron, leached and oxidized in wind-blown soils, left its mark. These rocks, the Salagou Formation, named for a local river, are made from sediments laid down in a semiarid basin into which heavy rains sometimes carved lakes and rivulets. Scrubby ferns and conifers grew beside these wet areas, adding patches and corridors of green to an otherwise bare landscape. The formation dates from the Permian, 270 million years ago, a time when all Earth's landmasses were united in one giant continent, Pangaea.

Jean Lapeyrie, a local medical doctor, discovered in the 1990s that the colorful outcrops near his home were, in some places, richly peppered with fossilized insects. He made collections and, through collaborations with researchers across the world, opened a rare window into a time when the earliest members of modern families of insects mingled with now-extinct groups. Mayflies, lacewings, thrips, and dragonflies flew alongside ancient forms, including several relatives of modern crickets and grasshoppers.

Most of the insect fossils are of wings. Insect bodies decompose quickly, but their wings are made of dry, tough protein. Blown or washed into water channels or mud cracks, the wings became entombed in silty ooze. Later, unearthed from their funerary vaults by geologists' hammers, wing veins and contours are visible as impressions in stone. Every type of insect has its own wing shape and vein arrangement, so a fossilized wing can identify the taxonomic family of its long-dead owner.

In the Permian rocks of the Salagou Formation, one wing bears an unusual feature. Normally wing veins are arranged in a web, supporting a thin membrane. On one fossil specimen, though, a cluster of veins near the attachment point of the wing are thickened and raised. A slightly curved prominent central vein is buttressed by side veins. This convergence of embossed veins is just a couple of millimeters long, the length of a letter on this page, on a wing half the length of my thumb. Such a structure, a raised ridge, had no function in supporting the wing membrane. Instead, it was likely the insect's singing device. When the wings rubbed together, the raised central vein scraped over the base of the other wing, making a scratchy sound. The large flat surface of the wing may have acted like a loudspeaker, broadcasting the sound.

Modern crickets use a similar wing structure to make sound, although theirs is of a more refined design. Corrugated ridges on the right wing rub against a nub on the left wing. This action of a plectrum against a file is amplified and projected by an adjacent thin, membranous window in the wing. The shape of the file and window is unique to each species, as is the strumming rhythm, producing a great diversity of sound among modern crickets, from mellow chirping, to sustained trills, to whines so high that human ears cannot perceive them. The raised ridge of the fossilized insect lacks the precise series of bumps on the file, and there is no evidence in the wing of an amplifying window. It is likely, then, that the sound the animal made was a simple rasp, without the purity of tone achieved by the precisely tuned structures of today's crickets.

The scientists who described the fossil in 2003, French paleontologists led by Olivier Béthoux, working with the discoverer, Jean Lapeyrie, named the species *Permostridulus*—from Permian, the geologic age of the fossil, and stridulate, the zoological term for rubbing body parts together to make sound. The ridge in *Permostridulus* is made from the union of a different set of veins than those of modern crickets. The species belongs in its own taxonomic family, a clan now extinct, an early, distant relative of modern crickets.

When it was alive, *Permostridulus*'s arthropod companions were other insects, spiders, scorpions, and, in the temporary pools, swarms of small crustaceans. Our distant ancestors and their kin were there, too, their lizard-like bodies leaving footprints in the mud, preserved as tracks of fossilized prints. These reptiles, known as therapsids, ranged from iguana- to crocodile-sized and stalked the land on legs held vertically, unlike the sprawling gait of most reptiles and amphibians today. Some of their kind would, over the next fifty million years, shrink, gain a furry pelt, and evolve into what we now call mammals. But in the Permian, therapsids were reptilian-skinned browsers and predators, the dominant large animals in many land environments.

These forebears of the mammals most likely could not hear the insects' sounds. The eardrum and triplet middle ear bones that deliver high-frequency sounds to our mammalian ears had yet to evolve. The sonic world of therapsids comprised only low frequencies, delivered to the inner ear through external ear holes and the bones of the animals' bodies. The thud of footsteps and the boom of thunder were probably all they could hear. Perhaps, too, they heard the murmurs of other reptiles, although there is no fossil evidence that these animals were vocal. An ear adapted to higher sounds would come later in the evolution of these animals, when forest and plains were filled with edible singing insects and when the therapsid body transformed into the compact insect-hunting frame of the early mammals.

The arthropods of the time, though, could hear *Permostridulus*'s song. In their miniature world, sensitivity to higher-frequency sounds was a boon. For a spider or scorpion waiting in ambush for prey, feeling the patter of tiny footsteps in the soil, the scrape of an insect limb, the flutter of wings, or even a brush of a tiny body against vegetation can deliver information leading to the next meal. For prey, too, vibrations in air or through the ground are useful, serving as warnings of close-at-hand danger. Sonic awareness of the presence of others also helps in the intimate social negotiations of mating. These sounds of insect bodies and movement—whishes, sighs, and crinkles—are quiet and travel only a few centimeters or, for the heavy rustlings of the largest, a meter.

Ancestral crickets possessed well-developed hearing organs in their legs, arrays of cilia-bearing cells that detect minute vibrations in the ground and pressure waves in air. After *Permostridulus*'s time, these capabilities were further expanded when evolution added a thin-membraned eardrum to cricket forelegs. This innovation, dating to about two hundred million years ago, was surely precipitated by the evolution of sound-making wings. Once sonic communication started, natural selection favored refinements in hearing.

We don't know why *Permostridulus* made its sound. Living crickets sing to attract mates and defend territories. It is possible that the wing's sound gave the ancient insect an advantage in the breeding season, perhaps by garnering attention, bluffing away rivals, or revealing location to searching mates, just as the chirping of crickets does today. As long as the breeding advantage was larger than the increased risk of predation, the song would have been favored by natural selection.

Perhaps, though, the wing's sound-making ridge served a defensive purpose. A burst of sound can startle attacking predators, buying time for an escape. This sonic defense would have been especially effective in a world where such calls were rare. Imagine the shock that a pouncing spider experienced on feeling a buzz in its jaws or hearing an unexpected rasp at close quarters. To this day, vibratory startle responses are common. Pluck an arthropod from its home and you'll often get a short blast of sound. Animals as diverse as lobsters, spiders, millipedes, crickets, beetles, and pillbugs all give defensive vibrations. Experiments with predatory wasps, spiders, and mice show that these vibratory alarms do indeed offer protection, startling attackers enough to allow potential prey to escape.

This uncertainty about the function of sounds highlights a difficulty in human language. In describing the sounds of other species, we project human nouns onto nonhuman beings. *Song* is anything we judge to have an aesthetic root, a sound made to please or persuade. Most often, we reserve the term for sounds whose repetitions have timbres or melodies that are pleasing to our ears. We name shorter sounds *calls*: the chirps of begging nestling birds; the sharp high notes of flocking birds; the bell-like exclamations of frogs in breeding season; or the grunts, cries, and sighs of monkeys discovering and sharing food. Calls can unite a flock, communicate from offspring to parents, signal alarm, or mark territories. But the functions of animal sounds are more diverse than our simple classification allows. Often the division between song and call is arbitrary and usually reveals more about the effect of the sound on human aesthetics than its roles among nonhuman animals. I follow common usage, but where social functions are unknown, as in *Permostridulus*, or only partly known, as in most nonhuman animals, this terminology is a mere sketch.

Whatever its function, the wing ridge in *Permostridulus* presaged further developments in an insect group whose relatives would become some of the world's singing champions. *Permostridulus* is close kin to the taxonomic order named Orthoptera, from "straight wing," which today comprises more than twenty thousand species, most of which sing. Some, the crickets and katydids, make sound by rubbing files and plectra on their wings. Others, grasshoppers and giant flightless crickets called wetas, rasp hind legs onto ridges on their abdomens. Sound-making wings and legs are supplemented in some species by rasping mouthparts, wheezing air tubes, drumming abdomens, and wings shaped to crackle and snap as they fly.

Permostridulus is, for now, the earliest known singer in the fossil record. But it was surely not the first animal to make a communicative sound. The fossil record is incomplete and gives us only very conservative estimates of the antiquity of evolutionary innovations, especially innovations like tiny ridges on insect wings that do not preserve well in stone. To cast our ears further back in time than the testimony of fossils, we can infer the past indirectly, using evolutionary family trees reconstructed from genetic comparisons among modern species. These trees, when calibrated to the ages of known fossils, give estimates of when groups of species diverged from one another. It seems that the cricket clan appeared around 300 million years ago. Almost all the living descendants of these first crickets sing. It is likely, then, that their common ancestor did too. Other contenders for early singers are the ancestors of treehoppers, cicadas, and other hemipteran bugs. Their common ancestors may have communicated with sound waves transmitted from vibratory organs in their bodies through wood or leaves. Like the crickets, these ancestors date to about 300 million years ago. Stoneflies, common insects in many waterways whose adults breed on stream-side vegetation, communicate by tapping out duets on vegetation, yielding drumming rhythms unique to each species. Their origin dates to nearly 270 million years ago, and so their soft percussion likely was another early animal communicative sound.

Later, other members of the Orthopteran order left spectacular fossils. In the Triassic, the geologic period after the Permian, fossils of cricket-like wings possess stridulatory files and, perhaps, rudimentary "windows." These windows, flat panes of membranous tissue, have no known function for flight and appear to be smaller versions of the wing windows in living crickets that focus and amplify sound, giving their chirps a clear tonal quality. These Triassic crickets likely sounded sweet, not ragged and raspy as the coarse file of *Permostridulus* surely did. The most well preserved of all Orthopteran fossil sound-making devices are the wings of a katydid from 165-million-year-old Jurassic rocks in Inner Mongolia. The fossil is so exquisitely well preserved that broad dark bands are still visible across the forewings. A sound-making ridge lies across each wing, close to the attachment point to the body, and comprises a row of just over one hundred small teeth. The spaces between these teeth gradually increase, as they do in many modern katydids. As wings scissor closed, they accelerate. Evenly spaced teeth would produce a sound of increasing pitch, like running an accelerating fingernail across a comb: *brr-eee*! But the increasing spacing between the teeth exactly compensates for this acceleration, yielding a pure-toned sound: *eeee*! It seems likely that teeth in the extinct species did the same.

The team of scientists who described this fossil, led by Jun-Jie Gua and Fernando Montealegre-Z, described the morphology of the wing and made a speculative re-creation of its sound. Comparing the dimensions of the fossil to those of living species with known sounds, they estimated that the katydid made sixteen millisecond pulses of just over six kilohertz. To human ears, these are brief taps of pure tone, with a high, bell-like timbre. Fossil plant remains from the same rocks as the katydid suggest that this singer's home was an open woodland of ancient coniferous trees and giant ferns. The katydid's sound frequency would have traveled especially well in this habitat, and so the song and its ecological context seem well matched. Unlike *Permostridulus*, this katydid was also likely heard by vertebrate animals. By this time, amphibians, dinosaurs, and early mammals could hear higher frequencies. Like many modern katydids, this ancient insect may have sung at night, reducing the risk from predators.

Insect wings first evolved as stubby extensions of the external skeleton. Studies of wing development in modern species suggest that this evolutionary feat was accomplished by a merger of the actions of genes that control body armor with those that build legs. We have no fossils of the first flap-like wings, but evolutionary trees built using the genes of living species strongly suggest that the first wings evolved between 400 million and 350 million years ago. These first wings probably slowed the jumping descent of plant-climbing insects, a behavior still seen today in the bristle-tails, cousins of modern insects. Many insects at the time grazed on plant spores that were held in capsules at the tips of branches. Gliding would have been a useful skill in these forests of fern- and conifer-like plants.

Wings also allow easy access to food, rapid dispersal to new habitats, and more efficient searches for mates. The earliest fossil of a complete wing—veined, shaped with leading and trailing edges, and large enough to support flight—is 324 million years old. By about 300 million years ago, the fossil record contains dozens of winged insect species.

Insect wings also provide materials that readily make sound. Their flat, lightweight surfaces broadcast vibrations, animal versions of the papery interiors of electric loudspeakers. Flight muscles move with fast, repetitive motion and are well supplied with oxygen for sustained action. Any insect that developed a propensity to repeatedly rub wings without flying might make sound. Thickened or corrugated wing veins made the sound louder and more tonal.

Among animals like primitive crickets that lived in thick foliage or the jumble of debris on the ground, sound making was perhaps especially advantageous. Sound allows mates to find one another in the tangle of miniature jungles where sight lines are blocked.

After the long silence of Earth's first 3.5 billion years, insects gave the terrestrial world its first songs. Ancient forests of fern, cycad, club moss, and conifer were brightened with sounds that would be familiar to our ears. When we hear crickets chirping from the mulch in a city park, in a mountain meadow, or along a rural road, we are transported to the first days of song on Earth.

Why did communicative sound take so long to evolve? Bacterial and single-celled life existed for three billion years with no known sonic signals. Although all these cells could sense water motions and vibrations, none reached out to others with sound. The first three hundred million years of animal evolution, too, seem to have lacked any communicative signals. No known fossil from this time has a rasp or other sound-making structure. The expert paleontologists whose advice I sought all said that they knew of no physical evidence of sound-making structures from animals until the first cricket- and cicada-like insects evolved. Of course, the fossil record is incomplete and some sound-making structures, such as the swim bladders of fish, leave little or no trace in rock, and so we hear imperfectly across these great stretches of time.

This long silence is a puzzle. Sound is an effective and inexpensive way to send signals. Soon after the Ediacaran, the time period when disk- and ribbonlike animals first evolved, animal bodies evolved skeletons and other structures that could easily make sound. These bodies surely made incidental noises as they crawled across the ocean floor, swam, and chewed. Yet as far as we know, the early oceans had no communicative sound. Perhaps the right mutations did not happen, depriving evolution of raw material? This seems unlikely, given that evolution in the early days of animal diversification had enough creative power to produce all the known branches of the animal kingdom, equipped with sophisticated eyes, jointed limbs, and complex nervous systems.

We cannot know for sure, but it is likely that the waiting ears of predators put a brake on evolution's sonic creativity, one that would only be eased when animals got swift and nimble enough to escape the maws of listening enemies.

After the Ediacaran, the number and variety of fossilized animals exploded in a geologic era known as the Cambrian. Starting about 540 million years ago, Cambrian oceans filled with diverse new animal forms, including the ancestors of the major groups we know today: arthropods, molluscs, annelid worms, and tadpole-like creatures that later evolved into the vertebrates. The first skeletons, jointed limbs, complex mouthparts, nervous systems, eyes, heads, and brains all appear in the fossil record in the space of about thirty million years.

Cambrian oceans were full of listeners. Animals inherited the cilia of their single-celled ancestors, now attached to skin and spines, lodged into exoskeletons, and on the surfaces of organs within the body. The animal kingdom thus came into being with a preexisting sensitivity to the motions of water, including sound.

All the early animals of the oceans sensed pressure waves and vibration in water. Arthropods such as crustaceans and the now-extinct trilobites had bodies cloaked in arrays of sensors. The first predatory cephalopods and, later, jawed fish added to the dangers. Early cephalopods detected vibrations and motions of water through sensors on their skin and with statocysts, organs in their heads lined with sensitive hairs. Ancient fish sensed vibrations through their lateral line system and early rudiments of inner ears.

The fossil record reveals a pattern of increasing peril in the ocean, especially in the Ordovician, Silurian, and Devonian, the geologic eras after the Cambrian. Many fossils of shells and other prey show the marks of predatory attacks. Over time, animals lower down the food chain evolved more elaborate defenses—spines and thicker shells—and even took to burrowing in mud when it came time to molt, a behavior recorded in fossils of animals that died and were entombed as they shed their skeletons.

To make a sound in the early oceans, then, was to reveal one's presence to a community of predatory arthropods, fish, and molluscs. No aquatic animal can entirely avoid making some sounds as it moves and feeds. No doubt many perished when paddling and chewing revealed their locations. The penalty for early attempts at sonic communication would likely have been death.

Sound making was likely also dangerous for the first land animals. Fossilized footprints of small arthropods walking on land date to 488 million years ago. These colonists may have grazed on terrestrial algae and worms, or perhaps ventured onto land in search of sand in which to deposit their eggs, much as horseshoe crabs do today. Predaceous scorpions and spiders were on land 430 million years ago. By 400 million years ago, the land was inhabited by mites, millipedes, centipedes, daddy longlegs, scorpions, spider relatives, and the ancestors of insects. All these creatures could, through sensors in their legs, detect vibratory motions in soil or plants.

The early animal communities in the ocean and on land, then, seem to have been hostile places for sound making. In water, where sound creates fast and far-reaching molecular movements, the danger was especially acute. But even on land, the fact that many of the early colonists were predatory scorpions and spiders likely created a high cost for sound making. If the first animals in the oceans and land had been only vegetarians, the sonic diversity of the world might have bloomed much earlier.

But this is not only a tale from long ago. A survey of living animals lends support to the idea that predation is a powerful silencer. To this day, animals whose lives are sedentary or slow and whose bodies lack weaponry are voiceless. Among worms and snails, for example, only a couple of species are known to make sound. A marine worm that lives enclosed in glass sponges in deep waters off the coast of Japan makes popping sounds when it fights, drawing water into its mouth then expelling the fluid with a snap. The sharp strands of the worms' glassy home protect the fighters from passing predators. A land snail of tropical forests in Brazil makes quiet squeaks as it oozes bright, likely poisonous mucus when attacked by predators, the equivalent, perhaps of the warning buzz of disturbed bees. The other eighty-five thousand species of molluscs and eighteen thousand species of annelid worms are, as far as we know, mute except for the slither and bubble of their body movements. The same is true for nematodes, flatworms, sponges, and jellyfish. This silence is not the result of anatomical deficiency. The plate-like doors to snail shells would make excellent rasps. Soft, muscular flesh can make sound too, as popping worms, fish swim bladders, and our own vocal folds demonstrate.

Only two branches of the animal family tree account for almost all the voices and songs of our contemporary world: the vertebrate animals—fish and their terrestrial descendants, including us—and the arthropods—

crustaceans, insects, and their kin. Both are often swift and weaponized. Sound required a measure of fearless verve from its first animal makers.

The first half a billion years or more of Earth's sonic history comprised the voices of wind, water, and rock. Then came three billion years of hum from bacteria and the slosh, skitter, and chomp of early animals, a time with many incidental sounds of life but no known communicative voices. A long silence from the living world.

Then, a revolution. Terrestrial insects evolved wings. This likely broke the silencing power of predation. Wings on a tiny insect enabled escape. The costs of sound making plunged, allowing sonic communication to gain a foothold.

That sound-making insects evolved after they gained the power of flight does not prove that a release from predation caused the evolution of the first animal calls and songs. Cause and effect are hard to infer across such spans of time. If predation did act as a silencer, we can make a prediction, though. If examples of sound making are found in the fossil record from creatures older than *Permostridulus*, they will be from fierce, fast, or heavily protected animals. Perhaps an early insect with powerful hind legs or wings, an ancient prototype of a grasshopper. In the water, we'd expect sound from predaceous trilobites or crustaceans, and fish well suited to rapid escape or bristling with defensive spines.

As I walk the road verge in southern France, I am struck by the vigor of the insect sound around me. At any one spot on the road, I hear a dozen grasshoppers purring. The air is a haze of blended chirps from uncountable crickets. Jean-Henri Fabre, the great French scientist and poet of insects, wrote of crickets in this region filling the air with their "monotonous symphonies" in the late nineteenth and early twentieth centuries.

This soundscape contrasts with that of the cultivated areas in the lowlands, away from the unruly woodlands and verges along this mountain road. In fields and along country roads in areas where agriculture is more industrialized, insect song is muted. There is little natural vegetation left in fields made tidy by herbicides and the vigorous application of the plow. Diverse native grasslands and forests have been transformed into monocultures of annual crops. Insecticide arrives both from the nozzles of farm machinery and in the wind and rain, the stirred-up vapors and dusts of now-banned chemicals from decades past.

A 2016 report synthesizing the knowledge of sixty experts in insect biology found that Europe's grasshoppers, crickets, and their kin are in crisis. About 30 percent of species are threatened with extinction, and the majority of species for which we have good population data are in decline. In North America, grasshopper populations are dwindling even in areas away from the plow and the fog of insecticide. In two decades, grasshoppers decreased by 30 percent on the Konza Prairie in Kansas, a reduction that was associated with a sharp drop in nutrients—nitrogen and minerals—in prairie plants. Likely stimulated by extra carbon dioxide in the atmosphere, plants on the prairie have doubled their growth in twenty years, but nutrients in this rank vegetation have become diluted. The grasshoppers' food is now more like bulky, savorless straw than nutritious salad.

Not only are crickets and grasshoppers in trouble, but so, too, are many other insects. A recent compilation of 160 long-term studies of the abundance of insects of all kinds—bees, ants, beetles, grasshoppers, flies, crickets, butterflies, caddis flies, dragonflies, and more—found an average rate of decline of more than 10 percent per decade for terrestrial insects, but the reverse trend for the few insects that live in fresh water. These insects are the foundation of most ecosystems on land. By biomass, insects outweigh all mammals and birds combined by over twenty times. By number of species, they are at least four hundred times more numerous. On land, the sonic diversity produced by hundreds of millions of years of evolution is being sharply cut back. In the growing silence of insects in forests and meadows, we hear the decline of the animals whose lives sustain the vitality of all terrestrial ecosystems.

This extinction of sensory diversity has many causes: technologies that deliver poisons; ever-rising carbon dioxide levels; economies that force the costs of production onto other people and other species, the "externalities" of business; and ever-expanding human appetites and numbers that shoulder out other species. All these social and economic factors exist in a culture of inattention and lack of appreciation. There is a connection between the anonymity of this fossil site in southern France—one of the great mileposts along life's evolutionary epic—and the silencing of the living voices in the surrounding landscape. Our ears are directed inward, to the chatter of our own species. Introductions to the sounds of the thousands of species that live in our neighborhoods have no place in most school curricula. We generally regard human language and music as outside nature, disconnected from the voices of others. When a concert starts, we close the door to the outside world. Books and software that teach us "foreign" languages include only the voices of other humans. Public monuments to sound are rare and honor a handful of canonical human composers, not the sonic history of the living Earth. *Permostridulus*'s discovery passed without comment in the media.

Even within the community of environmentalist activism, we speak of crises in the lexicons of chemistry and statistics: concentrations of gases and estimates of extinction rates. These are essential ways of knowing and thereby healing the world, but they omit the lived experience of animal senses. Life is made not only of molecules and countable species but of relationships among living beings. These relationships—life-giving interconnections between the "self" and the "other"—are mediated through the senses. Diversity of sensory experience is a generative force, a catalyst for future biological innovation and expansion, not merely a product of evolution's creativity.

The Permian period ended 252 million years ago in a spasm of extinction. In the seas, more than 90 percent of species went extinct. On land, both animal and plant diversity was reduced by more than half, including the loss of most of the insects and vertebrates whose fossils dominate the Salagou Formation. The causes of this global cataclysm are much debated, but they likely involved a combination of massive volcanic activity, global heating and deoxygenation of the oceans, and the release from ocean sediments of poisonous levels of hydrogen sulfide. We're now in a rapid decline of our own making, albeit one that is, so far, much less severe than the decimation of the end of the Permian. One necessary part of our response to this swift decline must be to reawaken our senses and thereby human culture to the community of life.

Paying attention to sound offers us a pleasing and instructive invitation to this reawakening. Because so much of our human communication is aural, our ears and minds are primed to listen and understand. Sound is, of course, a complement to the other riches of life's community: the aromas of soil and trees; the colors of birds, fish, and arthropods; the varied shapes and motions of plants and animals; and the textures and tastes of plants in the hand and mouth. Our curiosity, care, and love are evoked by all these senses. But sound's special qualities—unlike light, it passes through barriers; unlike aroma and touch, it carries far—make listening an especially important, joyful, and sometimes heartbreaking practice in this time of crisis.

I sit down on a slab of bloody stone and close my eyes. Cricket music glows in the air around me. I smile, astonished.

Flowers, Oceans, Milk

We live surrounded by the many gifts of flowers. Their aromas, colors, and varied forms are a delight for the senses, of course. But their fruits, roots, and foliage also, and less obviously, give us the vitality and diversity of the living world as we know it. Except for the products of the ocean, almost every bite of human food comes from a flowering plant. Wheat and rice are the starchy products of wind-pollinated flowers. Pressed fruits give us olive, canola, and palm oil. The flesh of domesticated animals is made from grass, corn, and other flowering plants. Leafy greens, sugar, spices, coffee, and tea, too, all come from flowering plants.

What is true for the human diet is also true for nonagricultural ecosystems. Prairies, tropical forests, deserts, salt marshes, and deciduous woodlands are populated primarily by flowering plants. Only in the chill of the boreal forests or the dry soils of subtropical pine woods do the flowering plants' cousins, the pines and their relatives, take over. On the tundra and on mountain peaks, lichens and mosses dominate, but even there, flowering plants can be common and provide the principal source of food for many nectar-supping insects and seed-eating vertebrates.

Might flowers have also given us some of the diverse sounds of Earth? This seems an improbable link. Yet voiceless greenery yielded much of the modern acoustic exuberance of animals. The first stages of Earth's sonic evolution were a slow burn: 1 billion years of wind and water, 3 billion years of bacterial hum and quiet animal motion, and 100 million years of chirping crickets. Then, between 150 million and 100 million years ago, Earth's terrestrial sounds flared into the stunning variety we know today. The trigger for this explosion was likely the evolution of flowers. Literally, a flourishing of sound.

This was not the only time that plants lifted the acoustic vibrancy of the world. The first plants to reach up with trunks and branches—mostly ancient relatives of ferns and club mosses—precipitated the evolution of insect flight and, later, the sound making of wings. The first forests therefore offered a leg up for sound. The first flowers offered not structural support but energy and ecological richness. Compared with the fine dust of fern spores or the seeds of conifers, flowers and fruits are a bonanza for animals, rich in sugars, oils, and proteins.

This abundance created new ecological connections among plants and pollinating and seed-dispersing animals. Co-evolution between animals and flowering plants fed the diversification of both groups, a creative reciprocity. This action was fueled, in part, by new underground symbioses. Flowering plants united their roots with communities of soil bacteria, to the benefit of both. Roots protected and nurtured the bacteria within root nodules. Bacteria helped plants by giving them biologically usable nitrogen, the chemical foundation of all proteins and DNA. Nitrogen is in short supply in most ecosystems and so the union of roots and bacteria gave flowering plants an edge over their competition. Animals were the indirect beneficiaries of this below-ground revolution because well-fertilized plants produce abundant foliage and fruit.

Flowers, fruit, new ecological connections, and enriched soil: the origin of flowering plants transformed the terrestrial world and spurred animal evolution.

Studies of the DNA of modern plants suggest that the first flowering plants appeared in the Triassic, 200 million years ago. Flowering plants then slowly diversified through the Jurassic and exploded in diversity in the Cretaceous from about 130 million years ago. The below-ground partnership with nitrogen-harvesting bacteria started about 100 million years ago and presaged further surges in diversification. It is from this fanning out of plant lineages in the Cretaceous that we have the first unambiguous flower fossils.

For life on land, the Cretaceous period—from 145 to 66 million years ago—saw the remaking of ecosystems. Habitats where formerly only conifers, ferns, and their relatives grew were invaded by flowering plants that soon became the most common species, even in forests where giant ferns were still abundant as overstory trees. This time span, barely 3 percent of the entire timeline of life on Earth, also saw the origins or the diversification of many animal groups, including most of the animals that sing in modern ecosystems. Biologists refer to this period as the "terrestrial revolution," a burst of creativity unrivaled since the great Ediacaran and Cambrian evolutionary explosions of the early oceans. It was also a time of revolutionary expansion in sound making.

Insect diversity, especially, rapidly grew in concert with the rise of the flowering plants. Family trees of katydids, grasshoppers, moths, flies, beetles, ants, bees, and wasps, reconstructed from fossils and DNA, splay out in profusion coincident with the appearance and rise of flowering plants. This flourishing changed the sound of Earth, lifting old voices to prominence and catalyzing the origin of new groups of singing insects. For many of these singers, evolutionary history resembles a river flowing into a delta. A long single channel suddenly fans into a bush of rivulets that then further ramify. The channel is the ancestral lineage and the fan is the explosion of animal diversity that followed the ascendance of flowering plants on Earth.

Nighttime insect choruses worldwide are dominated by katydids (also called bush crickets), a group that now numbers over seven thousand species. Katydids sing by drawing a plectrum at the base of one wing over a ridged file on the other. The date of origin of the group is contested, with some DNA studies suggesting 155 million years, others closer to 100 million years. These first modern katydids descended from a lineage of ancestral crickets that stretches all the way back to *Permostridulus*'s time, at the dawn of cricket evolution nearly 300 million years ago. This long ancestry then burst into new forms starting after 100 million years ago, followed by another expansion in diversity after the asteroid impact and mass extinction 66 million years ago. Katydids are mostly foliage eaters. Many look just like the leaves of their flowering plant hosts, with emerald bodies and elegant leaf-like wings. A few feed on conifers and some species prey on other insects, but most are entirely dependent on flowering plants.

Crickets and their kin sang long before the advent of flowering plants, and their sounds were likely the principal element of the animal soundscape from 300 to 150 million years ago. These ancient sounds, too, received a boost when flowers evolved. The "true crickets," the Gryllidae, that now sing from meadows, forests, and lawns worldwide appeared 100 million years ago in the midst of the diversification of flowering plants.

Grasshoppers are a much later addition to Earth's soundscape. Unlike the wing-rubbing katydids and crickets, grasshoppers sing by drawing their hind legs over ridges in the abdomen. This sound-making talent has independently evolved at least ten times within the grasshopper clan, perhaps a result of the evolution of hind legs whose great length folds them right next to the abdomen, a pre-adaptation to song. Although the grasshopper branch of the insect family tree broke away from its cricket cousins 350 million years ago, it was not until the Cretaceous, when flowering plants became abundant, that they started to sing. Grasshoppers then continued to diversify and add new singing members of their family alongside the ongoing expansion of flowering plants.

The buzz, whine, and shriek of over three thousand modern cicada species come from the tymbal organ on the side of their abdomens. Within the organ, muscles pop fine corrugation back and forth, sometimes hundreds of times each second, yielding a crackling sound that is then filtered and amplified by resonant chambers in the abdomen. In warm climates worldwide, the unique structure of the tymbal defines the soundscapes of hot afternoons. The cicada clans that we hear today diversified after the rise of flowering plants, starting one hundred million years ago. But the lineage of sound-making cicada ancestors is much older, dating to at least three hundred million years ago. A descendant of this ancestral group still lives among the moss-covered branches of Antarctic beech forests in Queensland, Australia. The sound of this "moss bug," a sonic living fossil, travels as vibrations through vegetation, a repeated low buzz transmitted through the insect's legs. The ancestral lineages gave rise to the modern cicadas and moss bugs, but also to the spittlebugs, planthoppers, and treehoppers, a clan of over forty thousand species that feed like ticks on plants, using piercing mouthparts to draw nutritious juices from within. Almost all these species make sounds inaudible to us, usually by transmitting vibrations to the leaves or twigs on which they live. From ancient roots, the modern representatives of these groups ramified into their present diversity following the expansion of flowering plants.

When we hear crickets, katydids, grasshoppers, and cicadas—the insect species that dominate the range audible to human ears in many habitats—we receive plant energies turned to sound by insects. This relationship is both of the moment, fueled by the plants' sugars and amino acids, and ancient, the result of the stimulus that flowering plants gave for the evolutionary diversification of these insect groups.

Diversity in other major insect groups was also boosted by the rise of flowering plants. Ancestors of moths and butterflies lived three hundred million years ago, feeding on nonflowering plants. Their nectar-supping proboscis appeared in the Triassic and when flowers became common, the diversity of the group shot up, largely in synchrony with the expansion of host plants that provided nutritious foliage for larvae and nectar-rich flowers for adults. Tiny drumlike ears evolved at least nine different times among moths, mostly around one hundred million years ago, organs that are variously located on the abdomen, thorax, or proboscis, depending on the type of moth. These ears hear into the ultrasonic range and likely first evolved to avoid attacks from predaceous insects and birds. Such excellent hearing opened a new avenue for courtship, and many moths sing by softly rubbing their wings together, giving swishing, whispery songs too high for human ears to detect. But, unlike ours, moth ears can detect these sounds. Electrodes inserted into nerves running from these ears show that they can pick up sound up to sixty kilohertz, far above the twenty kilohertz that is the maximum for humans. When echolocating bats evolved, fifty million years ago, ultrasonic-sensitive ears also allowed moths to detect and evade the bats' sonar blasts. The tiger moths went further and evolved bumps in their exoskeletons that, when buckled, release ultrasonic clicks. These sounds startle and jam the echolocating signals of hunting bats, and also signal the distastefulness of poisonous tiger moth species. This aerial sonic warfare is founded on the flowering plants that both feed the moths in the present day and stimulated the flaring of their species diversity long ago.

Before flowering plants evolved, the soundscape of the terrestrial world comprised only a few insect voices, the crickets, stoneflies, and perhaps ancestors of the cicadas and treehoppers. By the late Cretaceous, the insect chorus was like that of our time, a diverse mix of katydids, crickets, grasshoppers, and cicadas. The Cretaceous climate was hot, what geologists call a "greenhouse world" of high carbon dioxide, and the land was cloaked with lush forests, even close to the poles. Likely this was the first time in the long history of Earth that the air thrummed and pounded worldwide with the communicative sounds of life. Like modern rain forests, late Cretaceous forests were animated night and day by the crepitations, drones, buzzes, shimmers, bleats, and whines of singing insects. Earth, finally, was wrapped in song.

Birds were part of this chorus, but not in the way that we hear them today. Modern birds vocalize with a unique organ, the syrinx. Buried deep in their chests, at the Y-shaped confluence of bronchi and trachea, membranes and lips attached to modified cartilage rings impart sound to flowing air. In many species, the sounds of this syrinx are refined by a dozen or more muscles, each smaller than a grain of rice. The fossil record is incomplete, but it seems that this syrinx evolved late in the history of birds.

The first birds took to the air in the Jurassic, right at the time when DNA evidence suggests that the major lineages of flowering plants were splitting from one another. These birds were mostly predatory, feeding on the new diversity of insect prey, a boon partly built on the ecological productivity of flowering plants. Birds then flourished in the Cretaceous, diversifying in the ancient forests and colonizing waters with diving, fish-hunting species. The dominant birds in forests in those days were the "enantiornithes" (named for the "opposite" articulation of the shoulder). Most were small and nimble, looking somewhat like modern jays and sparrows, with feathers and wings like those of modern birds, and feet adapted to perching in trees. They were good fliers and it seems from their beaks that they lived on diverse foods like insects, small vertebrates, and fruit. A few species resembled woodpeckers and others foraged for small invertebrates on muddy shores. These parallels with modern birds end with a closer look. Their beaks had teeth. Their wings were clawed. This was a parallel universe of bird evolution, one now entirely extinct. There is no fossil

evidence that any of these species possessed a syrinx. Are the known fossils too degraded and incomplete to reveal such a delicate structure? Or
did this diverse lineage of birds, a sister group to modern birds, break away
and take its own path before the origin of the syrinx? If so, they may have
hissed and grumbled with their throats, as do many other reptiles, but
produced nothing like the complex, tonal and harmonic sounds that we
now associate with birds.

This early bird diversity was almost entirely erased by an asteroid impact at the end of the Cretaceous, 66 million years ago, which not only
snuffed out all the nonbird dinosaurs, but also decimated the birds. The
asteroid struck at the northern tip of what is now the Yucatán Peninsula in
Mexico, leaving a crater 20 kilometers deep and over 150 kilometers wide.
This crater is now buried by younger sediments, but geologists have
mapped its extent using rock samples and magnetic analogies. The impact
caused a mega-tsunami, sent out a pressure wave strong enough to deform
rock hundreds of kilometers away, and ignited fires worldwide. The
ejected vapor and rock, along with smoke from the blazes, fogged the atmosphere with dust, sulfates, and soot, bringing on a dark, cold "impact
winter" that lasted at least two years. The world's forests were mostly destroyed. In their place, ferns, mosses, and weedy flowering plants grew
back. Forest-dwelling and larger bird species, especially, were scythed
down. The great branching tree of Cretaceous bird diversity was cut to
just a few twigs.

It was from shortly before the asteroid calamity that we have the first
fossil evidence of a syrinx. The fossil is of a relative to living ducks and
geese, named *Vegavis iaai* for Vega Island in Antarctica from which it
was disentombed. *Vegavis*'s syrinx looks like that of modern waterfowl, but
less complex than songbirds'. It could honk, but not warble. The close familial relationship of *Vegavis* to living birds demonstrates that the syrinx
was very likely present in the ancestors of modern birds. The few species
that made it through the end-Cretaceous bird-apocalypse arrived in the

post-asteroid world equipped with an ability to sing. The diverse sound-scapes of birdsong around the world today are built from this legacy as survivors expanded their ranges and split into new species.

It is likely, therefore, that bird sounds as we know them only arrived after the resurgence of forests following the calamity at the end of the Cretaceous. In birdsong, we hear the evolutionary legacy of renewal after great loss.

Land vertebrates other than birds—frogs, other reptiles, early mammals—followed a path of sound making only partly shaped by the rise of flower-ing plants. All modern vertebrate animals possess a larynx, a fleshy valve at the top of the windpipe enclosed in cartilage. The larynx first evolved in lungfish, where it stopped water from choking the air-filled lungs. The larynx retains this function in land vertebrates today, where it directs food and water to the esophagus, not the airways. Muscular tissues at the top of the windpipe can also make sound, and in many vertebrate land animals today the larynx now serves as an anti-choking valve and a sound maker. Curtain-like extensions from the sides of the larynx, the vocal folds, vi-brate as air flows out. These fleshy tremulations give voice to animals from frogs to humans.

Vocal folds do not fossilize and so we cannot exactly reconstruct the tim-ing of sonic evolution in these animals. But comparisons among modern species, combined with family trees built using DNA and dated fossils, give our ears a conduit to the past.

The common ancestor of all living singing frogs (a few modern species are voiceless, descendants of ancient pre-vocal lineages) dates to about two hundred million years ago. From then on, the wetlands of Earth rang with frog chirps and trills. It was likely at about this time that reptiles also be-came more vocal. Until about two hundred million years ago, ancestral rep-tiles lacked eardrums and could hear only low-frequency sounds, mostly transmitted through their jaw and leg bones to the inner ear. But once higher-frequency hearing evolved, the possibilities for acoustic communi-cation opened up. Modern turtles call with tonal or wheezy pulses during breeding, crocodile youngsters chirp at their mothers and mating adults bellow, geckos chatter with calls richly layered with harmonics, and many other reptiles hiss when threatened. Early reptiles likely used some or all of this palette of vocalizations, supplemented by nonvocal sounds like scale rubbing, jaw snapping, and whip-cracking long tails.

For a few large dinosaurs from the Cretaceous we can make more precise reconstructions. The head of the nine-meter-long herbivorous *Parasaurolophus* bore a long, backward-extending crest. The tubes of the nasal cavity looped within this crest, giving the vocal tract a length of over three meters. Like a head-mounted tuba, the crest amplified and projected low-frequency sounds produced by the larynx. The skulls of *Parasaurolophus*'s relatives, the hadrosaurs, also possess cavities, suggesting that low bugling sounds may have been common among these giants.

Living alligators and larger birds use their windpipes and air sacs in the neck as inflatable horns, broadcasting low-frequency sounds. Given the widespread use of this sound-making technique, it is likely that the birds' close cousins, the extinct dinosaurs, made similar sounds. If so, alongside the subwoofers of hadrosaurs, other dinosaur species may have called with sounds like those of some modern birds that sing partly with air sacs: dove and pigeon coos, booming sounds like the *basso profundo* thump of modern bitterns, or the strangled belch of ruddy ducks.

The dinosaur sounds that we hear in films do not reliably evoke ancient sounds. They are built to evoke emotional responses in humans using manipulated recordings of modern animals. The roar of *Tyrannosaurus rex* is a slowed-down infant elephant trumpeting sound, merged in the studio with lion roars, whale blow-hole blasts, and crocodile rumbles. Gentoo penguins give voice to velociraptors.

And what of the mammals in this era? The mammals of the Jurassic and Cretaceous periods were formerly thought to be mousy creatures living in the shadows of the dinosaurs, precursors to the mammal diversity that bloomed after the nonavian dinosaurs went extinct. New fossil discoveries, especially from China, have overturned this view. Early mammalian evolution produced an explosion of ecological forms, species that resembled modern shrews, rats, water voles, moles, weasels, marmots, badgers, and even flying squirrels. Flowering plants were likely partly responsible for this explosion, albeit indirectly. A few early mammals fed on sap, seeds, and fruit, but many were insectivorous. The newly abundant and diverse insect fauna provided ready food for vertebrate animals swift enough to catch them. Good hearing helped. The evolution about 160 million years ago among early mammals of the three middle ear bones, followed by an elongated cochlea, opened a new perceptual world: the high-frequency rustles and songs of insect prey. We do not know what any of these early mammals sounded like. It is possible that they squeaked, purred, roared, barked, and bellowed, like modern mammals. Unlike other land vertebrates, mammals have a diaphragm, adding both fine control and force to the breath, and a band of muscle within the vocal folds, allowing more precise tuning of vibrations.

Listening to Cretaceous forests would be a disconcerting mix of familiarity and oddness. I imagine stepping into this world: Insect choruses like those of modern rain forests, a soundscape filled with cicadas and katydids and others. Frogs peep and trill from pond edges and the water holes in large trees. Squirrel-like mammals chatter and grunt. Large herbivorous dinosaurs groan like subwoofers. Others hoot and trumpet like modern primates. Birds hop among the trees, gleaning insects and plucking fruit, as they do today. One of the birds opens its beak, revealing rows of sharp teeth. Instead of sweet whistles or ornamented trills, the feathered animal looses a sibilant cry or harsh grunt. At dawn, no surge of birdsong meets the rising sun. The melodies that birds today stitch into the air are absent from this Cretaceous soundscape.

The great explosion of sonic expression in the Cretaceous was rooted in the ecological and evolution revolution brought about by the flowering plants. For many animals, the catalyzing effect was direct: flowering plants nourished animals and then co-evolved with them as pollinators, herbivores, and fruit dispersers. For other species, the boost was indirect, largely through the new varieties and abundance of insect foods made possible by flowering plants. If flowering plants had not evolved, if the land's food web was still entirely based on ferns and conifers, the world would sound less diverse and less exuberant. Many of our most familiar singers—katydids, cicadas, birds, and others—would either not exist as songsters or be muted and monotone.

In our present biodiversity crisis, this history offers a warning. We cannot destroy botanical diversity without also silencing the animals that give voice to the living Earth. Ninety percent of the half million plant species on Earth are flowering plants. Although we lack population data on most species, the current best estimate is that at least 20 percent of the world's plant species are threatened with extinction.

There are two significant exceptions to the tight association between floral diversity and the expansion of sonic expression. One is in the sounds of the oceans. The other is in the words you read on this page, the human voice captured in ink.

In 1956, when French explorer and filmmaker Jacques-Yves Cousteau released one of the first color documentaries about the ocean, a film that received both the Cannes Film Festival Palme D'Or and a US Academy Award, he called the work *Le Monde du Silence*, the world of silence. But the oceans are not silent. Human physiology was the first barrier to listening in the sea. Inattention was the second.

Our ears are adapted to air, not water. Submerged, we can only hear a few loud sounds. And so the aquatic world's many sonic textures and nuances are mostly lost on our unaided ears. Although hydrophone technologies were developed in the early twentieth century, they were deployed mostly by the military for listening to shipping and submarines. Adding to the problem, before the 1960s biologists mostly studied the ocean by kill-ing or otherwise silencing their subjects. In Cousteau's film, lobsters are tugged from their holes, fish hauled on board, sharks slaughtered, and coral reefs dynamited, methods that reflect the crude scientific tools of the time. Early scuba explorations brought scientists into more intimate and less de-structive contact with sea life, but listening was hampered by the constant drone of boat noise and the roar of bubbles streaming over divers' ears.

We now know that the oceans are full of sound. Biologists and sound recordists, including later work by Cousteau and his team, have deployed hydrophones from the Arctic to the tropical coral reefs, finding waters alive, always, with sound. A pioneer in this work was Marie Poland Fish, a biologist at the University of Rhode Island, whose navy-supported stud-ies of underwater sounds, starting in the 1940s, revealed "sea sounds and languages" among fish and crustaceans. She wrote, in the same year that Cousteau released his film, that "the din of animal life pervades the under-water world as it does our forests, countryside and cities." We now know that, far from being silent, waters crackle and glow with choruses of snap-ping shrimp and other crustaceans. Fish, sometimes gathered by the tens of thousands on their breeding grounds, drum, twang, and purr. Marine mammals—seals, sea lions, walruses, dolphins, whales—click, boing, moan, and ring like bells. These sounds of life combine with the seethe of wind-stirred froth, the boom of colliding waves, and the groan and crack of ice sheets. Sound in water travels fast and far. Unlike on land, its ener-gies flow unimpeded into animal bodies. Sound in oceans is ubiquitous and deeply felt by its creatures.

As was true on land, these sonic marvels of animals came late in evolu-tion. Even after trilobites, fishes, and other complex animals evolved, communicative sound was absent, or so it presently seems from the fossil record. Toothed jaws clicked, fins swished, and body armors grated and snapped. Most sea creatures could hear, both finding food and avoiding predation by eavesdropping on the sonic clues evoked by the motions of other creatures. But in ancient oceans, no known animals called out to mates, cried warnings at predators, or whispered to offspring.

The first ocean creatures to break the long communicative silence that marked the first 300 million years of animal evolution were likely the spiny lobsters. Recognizable today by their long, often prickly antennae and lack of large front claws, these distant relatives of "true" lobsters live in warm waters worldwide. They can grow to over a meter in length and are an important human food source, with an annual global catch of over eighty thousand metric tons. Next time you see one staring dead-eyed from a pile of ice chips in the supermarket, take a close look at the details of the face: you are in the presence of the ocean's earliest known communicative sound. A nub on the base of the antenna rubs against a smooth track running down from the eye. In life, this makes a yelping sound loud enough to scare away predatory fish or crustaceans. Today, in productive habitats such as the coasts of Japan and Western Europe, a hydrophone can detect dozens of spiny lobster calls per hour and the sounds of the largest animals can reach up to three kilometers.

The spiny lobster makes its defensive squeal with a unique sound-producing mechanism. Although the nub and track seem smooth, their microscopic structures create a "stick-and-slip" motion as the rubbery nub slides over a sheet of microscopic shingles on the track. As the antenna slides toward the eye, the nub jerks forward, sticks, then repeats, creating a juddering motion and a sound wave. A violin bow acting on a string works the same way. Its motion seems fluid, but the rosin-covered horse-hair of the bow goes through a rapid series of sticks and slips as it moves over the string, a jerky motion that drives the string into vibration.

The nub and shingled track can squeak even when the spiny lobsters' skeletons are soft after molting, the most vulnerable time in the life cycle of most crustaceans. Sound therefore not only gives these animals a defense mechanism, startling potential predators, it protects them when their other defenses are down.

Evolutionary family trees reconstructed from DNA sequences suggest that spiny lobsters first evolved in the Jurassic, about 220 million years ago, followed by diversification from 200 to 160 million years ago. The first definitive fossil specimens are 100 million years old.

Fossil evidence suggests that other sound-making crustaceans evolved after the spiny lobsters, around 95 to 70 million years ago. Crabs and lobsters with ridges along their thorax and claws first appeared at this time, and these structures were similar to those used by living animals to purr and growl. Like those of the spiny lobsters, these sounds are used as defenses against attacks, but some species also use them as mating or territorial displays.

The timing of the origin of snapping shrimp sounds—one of the loudest and most widespread animal sounds in the oceans—is uncertain. Genetic evidence suggests that the group may have split from other crustaceans in the Jurassic, 148 million years ago. But, the first fossil evidence of a claw that snaps dates to less than 30 million years ago, and much of the modern diversity of the group is less than 10 million years old. It seems likely, then, that although these animals or their ancestors may have been present in the Jurassic, their sparkling clouds of sound appeared much later.

One thousand species of modern fish are known to make sounds. This is likely a vast underestimate given that most fish species have yet to be studied in any detail. The known mechanisms of sound production are diverse, reflecting at least thirty different evolutionary inventions across the fish family tree. Catfish, piranhas, squirrelfish, and drums use high-speed muscles attached on and near the swim bladder to evoke purrs, taps, or squeaks from the gas-filled chamber. Butterflyfish and cichlids vibrate their ribs and limb girdles, causing the swim bladder to vibrate. Seahorses click their head and neck bones. Damselfish slam their teeth so hard that their swim bladders croak. Grinding teeth supplement the swim-bladder cries of grunts. Catfish strum their pectoral fins.

These are modern groups of fishes, evolved in the last 100 million years.

It is possible that fishes were calling out to one another with their swim bladders long before this time, but the thin-walled bladder and its muscles do not fossilize, and leave no evidence. Bichir and sturgeon, living descendants of lineages that split from other fish 350 million years ago, make knocks, moans, and rumbling sounds when they are close to others or spawning. Perhaps their ancestors did, too, although it is also possible that their sound making evolved during the hundreds of millions of years after the lineages split. The sounds of fish in deep time are hard to discern. We can, though, conclude that the many fish voices that now animate waters worldwide come almost entirely from groups of recent origin.

It seems that for hundreds of millions of years, the fish, crustaceans, and other animals of the oceans made few, if any, communicative sounds. Then, starting around two hundred million years ago and accelerating around one hundred million years ago, most of the voices of the ocean arrived.

Three factors seem to have driven the rise of sonic diversity in the oceans: the breakup of a supercontinent, a hothouse climate, and a sexual revolution.

Starting 180 million years ago, the supercontinent Pangaea fragmented, a process that continued for another 120 million years. These fractures created the major continents and oceans as we know them today. New shorelines and coastal habitats opened worldwide, increasing the extent and diversity of ocean habitat and opening new opportunities for colonization and adaptation. The sound-making animals of the seas diversified during this time when new oceanic habitats were expanding.

A long period of greenhouse climate also increased sonic diversity. Temperatures were so hot during most of the Cretaceous that ocean waters were tropical almost from pole to pole. There were no permanent ice sheets, and the sea level was up to two hundred meters higher than it is today, further broadening marine habitat as Pangaea broke up. North America was bisected by a large sea. Most of northern Europe and North

Africa were underwater. Life abounded in these spacious and hospitable waters. Photosynthetic plankton, the base of the ocean food web, was abundant and evolved a burst of new forms. Fish, crustaceans, snails, and echinoderms also multiplied. The sound-making animals that evolved and diversified during this time were almost all predators, most of them also formidably defended by tough skeletons or speedy bodies: spiny lobsters, lobsters, snapping shrimp, and fish. Sound making was a luxury enjoyed only by those at the top of a rich food chain. Prey animals at this time stayed silent and evolved thicker shells, and many took to living buried in mud and sand.

Mating behaviors also seem to have driven the origins and diversification of sound-making ocean animals. Many sea creatures, unlike any land animal, shed sperm and eggs into the water without ever coming near another member of their own species. Clams, many snails, corals, and others breed without intimate contact. These species are also generally silent. With no nearby mates, why sing? During the breakup of Pangaea, species that bred this way showed no increase in diversity. But animals that breed by coming into close physical contact, rubbing bodies together or grasping one another, tripled in diversity during this time. These animals often make sounds to attract mates or repel sexual rivals. Crabs and lobsters both woo partners and spar with rivals by stridulating their exoskeletons. The diverse arrays of thumps, squeaks, growls, and pulsed tones among fish are mostly breeding signals.

Why should intimate mating behaviors increase species diversity? Animals that breed by copulating do so only with mates that live nearby. This keeps gene exchange local, allowing species to break into regional variants and, eventually, new species. But species that broadcast eggs and sperm in water currents have widespread and homogenous gene pools. They are like large, monolithic human corporations. These giants may be good at what they do, but they cannot break into specialized, innovative subgroups. Species with behaviors that enforce local mating are more like swarms of start-up companies, each one able to pursue its own regional opportunities without being swamped by gene flow from far away. This likely yielded many new species during the time when Pangaea's breakup created new habitats.

There is one significant group of latecomers to the sonic blossoming of the seas: whales, seals, and other marine mammals. In a deliciously convoluted evolutionary path, the structure that stopped water flowing into the lungs of the lungfish and first land vertebrates, the larynx, returned to the water and sang. By blocking their blowholes or nostrils, these marine mammals use the vibrations of vocal folds in their larynx to send sounds through their body tissues and out into the water. Among the toothed whales, the larynx is supplemented by whistling air sacs and a sound-focusing "melon" in the forehead, sending focused beams of sound forward, like a sonic headlamp. When a solid object reflects the beam, the whale uses the echoes to home in on prey, avoid obstacles, or "see" its companions. Because sound penetrates tissues, this echolocating vision also reveals the inner form of other creatures. Sound, for toothed whales, gives a living MRI scan of the surrounding world.

Whales descend from pig- or deerlike ungulates, and their transition from land to water took ten million years, starting fifty million years ago. Seals and their kin are carnivores and arrived in the water later, twenty million years ago. The teeth and limbs of the transitional ancestors suggest that both groups were drawn to the water in search of the abundant food in nearshore habitats, just as polar bears and sea otters today spend much of their time foraging in the water or at its edge.

To the creative forces of climate, biogeography, and mating that brought forth the sounds of fish and crustaceans, we can add the later opportunistic colonization of the seas by hungry mammals. The hot blood, large brains, specialized teeth, and communicative vocal networks of these pioneers— all qualities that first evolved on land—gave them an advantage when they turned their attention to the seas. We hear the result in whale cries loud enough to traverse entire ocean basins or in the squealing of seals where fish abound in nearshore habitats.

Today ocean waters are a tumult of engine noise, sonar, and seismic blasts. Sediments from human activities on land cloud the water. Industrial chemicals befuddle the sense of smell of aquatic animals. We are severing the sensory links that gave the world its animal diversity: whales cannot hear the echolocating pulses that locate their prey, breeding fish cannot find one another amid the noise and turbidity, and the social connections among crustaceans are weakened as their chemical messages and sonic thrums are lost in a haze of human pollution. Combined with overfishing and climate change, these assaults produce what biologists call the defaunation of the oceans: 90 percent declines among large fish, ongoing losses of marine mammals, calamitous reductions in coral reefs, and although data on many species are scarce, sharp population and range contractions among many other ocean-dwelling animals. The best current estimate is that about one-quarter of marine species face an imminent risk of extinction and many more are in decline.

Sound is one of animal life's ancient creative processes. The title of Jacques Cousteau's film *Le Monde du Silence* was a manifestation of our ignorance about aquatic sounds. It was also an unintended warning about the consequences of our actions for other species. As we get louder and more voracious, we silence other living voices, cutting back both the diversity and the evolutionary creativity of the oceans.

In the long view, we owe our human voices to milk. Specifically, the milk that ancient protomammalian mothers fed to their young. Before the evolution of lactation, protomammalian youngsters nourished themselves on whatever the environment supplied, sometimes brought to them by parents but often foraged for themselves. This diet of seeds, plant material, and small animal prey demanded guts able to digest complex and sometimes hard foods. Energy and nutrients were often in short supply, limiting the growth rate of the young. The invention of nutritive skin secretions broke these constraints and supercharged infancy. Mothers did the hard work of catching and digesting prey, then offered rich and easy-to-assimilate food. Nursing offspring connected directly to the strength and generosity of their mothers' bodies. Although the earliest stages of the evolution of lactation are still unclear, studies of the DNA of modern animals show that by two hundred million years ago, female mammals possessed mammary glands and specialized milk proteins. In addition to changes in the physiology and behavior of mothers, this new method of feeding demanded a reworking of the throats of infants. Much later, these innovations would allow humans to speak. Our languages are bequests from these ancient mothers.

No reptile can suck. Their mouths, tongues, and throats are weak and lack skeletal support for complex muscles. This changed early in the evolution of mammals. The thin V-shaped hyoid bone in reptilian necks transformed into a stout four-fingered saddle. Muscles attached to these fingers, strengthening and stabilizing the tongue, mouth, larynx, and esophagus. Judging by fossil evidence, by 165 million years ago the mammalian hyoid and its muscles had turned the slack, open maw of reptiles into a powerful and coordinated sucking device.

The diversification of the mammalian clan was built on the unique nutritive bond between mothers and offspring, a connection made possible by both mammary glands and throat anatomy. To this day, young mammals are born with fully developed hyoid bones, even when other bones are still partly grown. Adult mammals, too, benefit, masticating and manipulating food with their mouths in ways impossible for reptiles.

Although the primary function of the hyoid is to support feeding, evolution has also put it to use in the shaping of sound. The larynx imparts sound to air flowing up the windpipe from the lungs. These sonic vibrations then stream into the upper part of the windpipe, the mouth, and the nasal cavities, before flying free to find listeners. The mammalian hyoid and its muscles allow animals to change the shape and resonance of throat and mouth, giving sound its timbre and nuance, squelching some frequencies and lifting others. The hyoid both supports the mouth and tongue and anchors the larynx.

When we call the knobby larynx in our throats the voice box, we do a disservice to the complex architecture within our upper throats and heads, places where voice finds its shape and character. Open your mouth wide, flatten your tongue, keep your head immobile, and then try to speak: most of your vocal capacity disappears. The mammalian vocal system, then, acts like many musical instruments. The larynx is the reed in an oboe. The upper vocal tract is the oboe's body and finger keys.

Evolution has crafted many variations of the mammalian vocal tract, each suited to the ecological or social context of the species. In echolocating bats, part of the hyoid connects the larynx to a bony plate at the base of the middle ear. This connection allows the nervous system to compare the outgoing pulse of sound from the larynx with the returning echo in the ear. Toothed whales use their giant vocal folds to make whistles, but their echolocating pulses come from nasal air sacs below the blowhole. These whales feed not only by biting and grasping but by sucking large prey like squid out of the water then swallowing them whole. To support this predatory sucking, their hyoid bones are massive and have flattened surfaces for the attachment of muscles. Ultrasonic sounds in some rodents come from a larynx that directs a narrow stream of sound at a sharp ridge of tissue, somewhat like the air-to-edge sounds evoked in pipe organs or flutes. Among some roaring mammals—red deer, Mongolian gazelles, and lions and their relatives—deep sounds are achieved by lowering the larynx within the windpipe, elongating the vocal tract. This descent happens seasonally, dropping during the breeding season, and during the roar itself when the larynx falls then springs back up. The hyoid and its muscles and ligaments support this trombone-like slide. Because low sounds come from big bodies, the larynx's movement presumably serves to make an impression on listeners, the equivalent perhaps of human motorcyclists modifying their exhaust pipes to give the sonic impression of large, powerful engines.

The vocal tracts of primates seem especially amenable to evolution's creative powers. Compared with those of carnivores, for example, the larynges of primates are larger, have evolved faster, and are more variable in relation to body size. Many primates have large air sacs connected to the larynx that act as bellows and resonators. The most extreme of these modifications is among the howler monkeys, famed in the American tropics for their low, far-reaching roars and rumbles. In addition to large, paired air sacs on their necks, the howler monkey hyoid bone is expanded into a large air-sac containing cup that acts as an amplifier and broadcaster.

Strangely, we humans have no extraordinary elaborations of our vocal equipment. Our larynx and hyoid are about the size we would expect for an animal of our weight. Somehow, we've achieved the great complexity and nuance of spoken language with tweaks to basic mammalian gear. Losing the laryngeal sacs was likely a key early step. The bulbous sacs of our close cousins, the other great apes, are fabulous for making screams and moans that carry through the forest, but not so good for subtlety. We do not know why our ancestors lost these throat balloons. Perhaps early hominins benefited from quieter, more nuanced vocalizations or the sacs may have impeded them when they became bipedal runners and stalkers on the savannah. Whatever the reason, the loss of these encumbrances likely cleared the way for the neck and mouth to take on their modern human form.

Gently press your fingertip in the soft space under your chin, behind your lower jawbone. Now extend your chin a little and run your finger backward. At the junction of neck and jaw underside, your fingertip will find the front of the hyoid bone that wraps back into your neck. The ancestral mammalian four-fingered design of the bone remains, although two fingers dominate, giving it a horseshoe shape. This is the only bone in the body not attached to any other bone. Instead, it is suspended from the skull and jaw by strong straps of tissue. Keep moving your fingertip back and down. The next hard lump is the larynx, a thickened part of the windpipe. Inside, inaccessible to probing fingers, are the vocal folds. The larynx is suspended from the hyoid.

When we are born, the hyoid and larynx are pressed up against the back of the palate, as they are in many other mammals. As we grow, they both drop down. In adulthood, the hyoid sits just below the level of the lower jaw with the larynx suspended below, in the neck. The "Adam's apple" visible in many men results from rapid growth of the larynx and its cartilage during male puberty, resulting in lower-pitched voices.

Sound waves from vocal folds in the human larynx flow upward into a vertical stretch of windpipe leading to the back of the mouth. From there, sound moves forward, from the back of the throat to the lips. Say *aah* into the mirror and you'll see the horizontal space of the mouth take an abrupt downward turn behind your tonsils. Each space, throat and mouth, has its own resonance, adjustable through muscular action. The tongue is the ever-active mediator between these two resonant passages. No sound passes from one to the other without its involvement.

Articulate human speech starts with fine control of breath from the lungs. In the larynx, the vocal folds are drawn into the flow of breath and start vibrating, just as the mouth of a balloon vibrates when air rushes out. In most mammals, these folds are entrained in the flow of air, and their elasticity causes them to move back and forth, creating sound waves in the air. In the purr of a cat, these vibrations are boosted by rapidly pulsing muscles, but other mammals lack this enhancement. Sounds from the larynx then pass to the upper part of the throat and into the mouth. There the shape of airway and mouth enhances some frequencies and suppresses others. The tongue further filters the sounds as they flow into the mouth, where tongue, cheeks, jaw, and teeth also sculpt the sound. After departing the oral cavity, the lips impart plosive emphasis or hiss and, finally, the sound wings free into the air. Every part of this web of interacting muscles, bones, and soft tissues plays an essential role. Try speaking without air from the lungs, squirming tongue, or dancing lips. Impossible. The foundation stone for the whole edifice is the hyoid, the legacy of the first milk-producing mammalian mothers and their suckling offspring.

Attending to the differences between vowels and consonants gives us a sense of the importance of each part of the vocal tract. We hiss, spit, growl, and squeeze consonants from our mouths by constricting the air flow with our throat, lips, or teeth: *shh*, *buh*, *grr*, *kah*. Air flows freely from the larynx for vowels, shaped only by the tongue: *eee*, *ooo*, *aaa*. In each case, the larynx provides raw sounds that the mouth then sculpts. Khoomei singers, known in the West as Tuvan throat singers, take this to an extreme, using constrictions created by their tongues to filter out all but a few overtones while their tightened larynx drones. Theirs is a sophisticated vocal art that builds on the interplay of the larynx and mouth we all use as we speak or sing. The same is true for other mammals. When dogs or wolves throw back their heads to howl or squirrels lower their jaw and pull in their cheeks to chitter, they are shaping sound with their vocal tracts.

None of the structures that we use to speak are unique to our species. Our chests are more amply supplied with nerves for the fine control of breath than most primate species, but this is an elaboration, not an innovation. Our chimpanzee relatives also drop their hyoid bones and larynges. The descent is lower, though, in humans, opening a more voluminous resonant space in our throats. This, combined with protuberant faces of chimpanzees, means that the chimpanzee vocal tract is dominated by the mouth, with very little resonance in the throat. In humans, the resonant spaces in mouth and throat are about of equal size. Human and chimpanzee tongues are similar, although ours is more domed and larger relative to the size of our mouths. Anatomically, human speech is based on subtle changes in the proportions of structures present in other species. Contrast this with birdsong, which flows from a syrinx unique to modern birds. The evolution of both birdsong and human speech was a striking and novel expansion of the sonic diversity of the world. Theirs is the product of radical anatomical innovation, ours of tinkering.

Evolution used a heavier hand in our brains, creating new linkages that allow us to speak. These, too, build on talents and predispositions present in our close relatives. All great apes are keen learners. Infants take years to learn all they need in order to thrive within the social and ecological environment. This social transmission of behavior and tradition constitutes culture. But unlike in humans, the cultures of other great apes are founded almost entirely on close visual observation and tactile participation. Although other great apes are vocal, they do not, as far as we know, convey complex knowledge through sound. Our human ancestors connected vocal expression to culture. This union of two preexisting great ape abilities, vocalization and social learning, is the foundation of human language. We do not know exactly when this revolution took place. The hyoid bone was in modern form and position in ancestral humans, including Neanderthals, about five hundred thousand years ago. But there is nothing magical about the exact shape and position of this bone. Ancestors with higher hyoids and larynges might not have been quite as articulate as we are, but they had the anatomical capacity needed for complex sound making, just as other great apes do.

The conjunction of vocal production, learning, and culture left its mark in our brains and genes. Unlike in other primates, the nerves that control the larynx in humans thread directly into the "motor cortex," the part of our brain that controls voluntary movements. These connections give us finer control and, most important, bring vocal production into the realm of learning. We also have substantial and complex brain connections among the laryngeal nerves and those involved in vocal interpretation, sonic memory, and the control of body movements involved in speech such as those in the tongue and face. The richness of these links seems at least partly controlled by a gene, *FOXP2*, whose sequence in humans diverges greatly from that in other primates. The gene acts as a regulatory hub, stimulating and suppressing the actions of other genes that guide the growth and interconnections of nerve cells that coordinate muscular action, sensory input, memory, and interpretation. Like the hyoid, the human form of the *FOXP2* gene dates to at least five hundred thousand years ago and was shared with our relatives in the *Homo* genus, Neanderthals and Denisovans. Neanderthal ears were similar to those of modern humans. Reconstructions suggest that the middle and inner ears were, like ours, tuned to the frequencies of human speech. It is likely, then, that these now-extinct cousins could speak.

Brain networks, greatly elaborated in humans compared with other primates, allow humans to draw together vocal production, interpretation, and memory in ways that other species cannot. When we speak, we evince our human ability to comprehend: *com* "together," *prehendere* "to grasp hold of." Human speech is an achievement not only of tinkering, but of unification and interconnection. We are not alone in this talent. Many birds, and perhaps other vocal learners such as whales and bats, also have direct connections from the vocal organ to the motor portions of the brain, along with elaborated connections among regions of the brain concerned with memory, perception, analysis, and production of sound.

In reading these words, you take this human talent for integration one step further. Black-and-white glyphs are crystallizations of what was, until the invention of written language, ephemeral. Breath turned to ink. Vibrations in air frozen onto the page. Three hundred milliseconds after you gaze on a word, electrical energy courses through the visual cortex of the brain. Four hundred milliseconds after that, the auditory cortex fires, swiftly followed by brain regions that interpret sound and language. Within less than a second of attention to the written word, silent reading provokes a frenzy of activity in the "listening" portion of the brain. Silent reading thus opens us to apparitions, the ghosts of writers' voices. Movements of fingers on keyboards and pens cast these sonic wraiths out of the body and onto the page.

As your eye moves over these clusters of letters, sound no longer travels through air but in waves of electrical activation along wet, fatty cell membranes in a mammalian brain. Now speak these words aloud. The wave leaps from flesh to air. Just as it always has, sound moves from one being to another, from one medium to another, connecting and transforming.

PART III

Evolution's
Creative Powers

Air, Water, Wood

Listen! In the animal sounds around us, we hear the diverse physicality of the world. The songs of birds contain within them the acoustic qualities of vegetation and the voices of the wind. Mammal calls reveal how predators and prey hear one another in the varied terrains of forests and plains. Water's many moods are expressed in the forms of whale and fish songs. The inner structure of plant material is manifest in the vibratory signals of insects. Even the words on this page, voiced silently as you read, have living within them signatures of the air and vegetation in which human language blossomed.

I stand in a pine and spruce forest on the eastern slope of the Rocky Mountains of Colorado, on the upper reaches of North Boulder Creek where it descends from the continental divide. It is spring, but at this high elevation snow still covers the ground. All is quiet, except for the rich voice of a red crossbill. The bird's song is a slender watercolor brush flitting across paper. Strokes of warm color dash and sweep on a smooth, open surface. Each note rings with extraordinary clarity in the snowy hush and still air.

I rummage in my waist pack for a sound recorder and microphone, the zipper and fabric sounding obnoxiously loud, then I hold still, pointing the microphone toward the tip of the ponderosa pine tree where the bird perches. For a few minutes, I rest in the presence of the song.

Then a hiss and a bellow. The wind gusts from the northeast, passing unobstructed across the wide valley between mountains. The trees' sounds reveal the inner life of the air. Ropes of forceful flow draw roaring surges from the canopy, bands of sound that snake and leap. Eddies punch down from air into the trees, then dissipate. Pools of quiet move through this tumult like blown leaves on a lake surface, skimming, pausing, then veering in new directions. The volume indicator on my sound recorder lurches into the red, and I twist down the gain knob. Suddenly the forest is shouting.

But the bird keeps singing, somehow penetrating the fog of noise. The song's fine brushwork stands out, strokes of luminous pigment against the gray wash of wind.

The character of the mountain is contained within this song. When this male red crossbill offers his springtime melodies, the combined experience of thousands of ancestors flows to the air. Only those predecessors whose songs accommodated the particular challenges of the wind in these trees passed on their genes. Evolution shaped the song to the place.

Red crossbills live always among evergreen trees, wandering in search of seed-filled cones of pine, spruce, Douglas-fir, and hemlocks. Their relationship with these trees is so long-standing that evolution has carved the birds' beaks to fit conifer cones. The tips of the stout, hooked beaks cross, the sharp point of the lower mandible twisted to one side and the end of the upper mandible curving the other. The birds slide these beak tips between cone scales and, with a sideways slide of the lower mandible and a turn of the head, pop the cones apart, giving their long tongues access to the seed hidden at the base of the scales.

The birds' love of conifers has marked their songs too. These trees are vociferous when stirred by the wind, roaring even in modest gusts. And except in the calmer days of summer, the wind is a frequent presence. A map of North American average wind speeds at ten meters above the ground, the height of tall trees, shows a band of strong winds running down the spine of the Rocky Mountains. Houses here shake under its power for days. Trail walking often feels like wrestling with an inexhaustible adversary, especially in the season of strong winds in late winter and spring when crossbills are singing. In Europe and eastern North America, the closest comparison is the uncompromising force of a gale rising from a sea cliff: exhilarating to walk in at first, then draining.

My body feels out of place, but the trees are at home. Their springy branches accommodate the flow, flexing and shedding the wind's power. Unlike lowland pine trees, needles of upland conifers are like wires or spikes, toughened to resist the fraying, tearing action of the wind. An oak or maple here would have its branches snapped and leaves shredded. The tough needles and flexible branches of the mountain conifers produce wind-evoked noise unique to these forests, a sound that has likely shaped the crossbills' song. From wind, to trees, to birdsong.

Later, I pull up the sound recording on my laptop. A graph scrolls as the sound plays, showing how the sound frequencies change through time. Fine lines scribbled on a clear background reveal the structure of the crossbill's phrases. *Tee-tup-tup*, a sharp upward exclamation, then two shorter notes. He throws in a lower, raspy *bree-bree*. After a minute, he delivers a string of shorter, sweeter *tup* notes, ending with a very high *see*. Then galloping variants, in clusters of three or four. *Chik-a-eee* pops in, a snatch of song very similar to that of the mountain chickadee. In all, the song has a dozen elements and the bird seems to remix them as it goes, grouping, rearranging, and adding little flourishes and inflections. The result is sprightly and nimble, full of bright motion.

Suddenly a smear darkens the screen. The wind has arrived. The bottom half of the graph, the zone showing the lower frequencies, is fogged with the sound of the trees. The crossbill's song dances over this cloud. The bird's notes are all higher-pitched than the great whoosh of the pines and Douglas-firs.

When wind hits these forests, the resulting sound is almost all below one or two kilohertz. This is quite different from the sound of wind in other forests. When strong gusts hit oaks and maples, or pass through the canopy of the tropical forest, they evoke hissing noises that extend to much higher frequencies, five or six kilohertz. In the mountains, then, wind is a low roar that can last for hours or days. But in most other forests the wind is less frequent, and when it does come, it is high and sibilant. There's a human quality to the voice of these conifers. In them, the wind produces sounds in the frequency range of human speech, unlike the higher sighs and rustles of other tree species.

The red crossbill's song is higher than we would expect for an animal its size. Like musical instruments, the pitch of an animal's song is usually a product of its size. Ravens croak low, hummingbirds squeak high. But the red crossbill bends this rule, singing higher than other species of its size.

The forests are present in the red crossbills' songs not only in their relation to the wind but in the effects that tree cones have on the evolution of crossbill beaks. Red crossbills of the Rocky Mountains have stout beaks, suited to ponderosa and lodgepole pine. In the Pacific Northwest, birds of the same species have smaller beaks, adapted for opening Sitka spruce and western hemlock cones. Small beaks, by virtue of their nimbleness, can sing fast, with high-pitched trills. The variants of the songs of the red crossbill and its slimmer-beaked cousin, the white-winged crossbill, therefore, are partly shaped by the diversity of cone shapes among local trees.

It is not just crossbills that are high-pitched in the coniferous mountains.

In the autumn, elk fill the mountain valleys with mating calls that carry for kilometers as they echo from slopes and cliffs. Zoologists call the elk's song bugling, but its timbre is more like a flute hitting weird harmonics. The elk tips back its head and releases a nearly pure tone that slides up, holds steady for a second or two, then slips down, often edged with rough grunts. The first time I heard the sound, in a spruce forest in the Rocky Mountains, I could not believe that such a high sound could come from so massive an animal. A bull elk weighs more than three hundred kilograms. The steady central note of the elk's bugle is between one and two thousand hertz, a little higher than the squeak of a rabbit.

The closely related red deer of northern Europe makes a much deeper sound, two hundred hertz, a throaty roar at the frequency we would expect from an animal this size. Studies of vocal folds taken from hunter-killed carcasses have yet to resolve the mystery of how elk make their sounds. As expected for a large animal, their vocal folds are long, three times as long as ours, yet they force a high-pitched sound from this large instrument. Their throat bones and ligaments are shorter than those of red deer, though, suggesting that they may clamp or otherwise constrain part of their vocal folds, shortening them so that they can vibrate fast enough to make the extraordinary song.

In the autumn rutting season, male elk sometimes lunge at each other and butt heads in a clatter of antlers. But most of their jousting is at a distance, through sound. I've sat on mountain slopes above the tree line and heard males calling back and forth from a distance of five kilometers. Only airplane noise travels farther in the high country. Males usually call from open meadows along meandering streams or from the adjoining coniferous forests. To be effective, bugling calls have to travel hundreds of meters through conifers. Males are signaling to one another, but also to the female elk who live year-round in matrilineal herds. These close-knit groups gather in the autumn in mountain valleys, where rutting males compete for the privilege of attaching themselves to each herd. These aggregations are often completely out of sight of one another, but linked by bugling sounds from the males.

Just as the crossbill's song seems well matched to the particular sound of Rocky Mountain forests, so does the bugle of the elk. A bellowing, low call would be smothered by the wind. This is an unusual situation. In most habitats, low sounds are more effective than high sounds for long-distance communication because low, long-wavelength sounds flow around obstacles and are less degraded than high sounds by wind turbulence. But in coniferous forests with strong, persistent wind and stiff-needled trees, the masking noise of trees seems to override these advantages, pushing animal signals into higher frequencies.

Two examples from the mountains do not prove that environmental noise has shaped the voices of the animals here. The high voices of the crossbill and elk might have their origin in sexual competition and choice, sonic versions of colored plumes and exaggerated antlers. Or the ears of both species may be especially sensitive in the high registers, tuned to informative sounds from predators, competitors, and kin that are not masked by the roar of wind. Such a fit of hearing to habitat would then favor social communication at these higher frequencies. These hypotheses are, without more information about the history and society of each species, impossible to disentangle. Every time I visit these mountains, though, I am struck that forests with the loudest trees I have ever encountered are inhabited by animals whose voices are unusually high, leaping over the roar.

A wider survey of animal vocal communication reveals the effects of the physical environment on sound. Birds living on rocky coasts have cries loud and strident enough to be heard through the tumult of waves and the shearing action of wind. Gulls, oystercatchers, and shorebirds avoid soft murmurs or subtle inflections. Instead, they slice through the wind noise and the sound of pounding waves with emphatic strokes. Birds and frogs that live near rushing waterways vocalize with loud and high-frequency calls, vaulting over the masking rush of water.

In forests, vegetation attenuates and degrades animal voices. Leaves, stems, and tree trunks absorb and reflect sound, muffling and adding reverberation. At a distance, every note is hushed and blurred. Most forest birds therefore sing with slower, simpler whistles and slurs than their cousins of open country. The North American scarlet tanager's rich up-and-down warble, *chirru-cheery-chirru-cheery*, for example, is well suited to breeding territories thick with maple, oak, and hickory leaves, as are the singsong notes and inflected fluting of many Eurasian thrushes, Australasian whistlers, and songbirds of dense tropical forests worldwide.

In contrast, on the open prairies and plains, it is not vegetation that degrades sounds but wind shear and turbulence. Here, subtle inflections of pitch are erased by the wind. Many birds of grasslands and open rocky areas therefore buzz and trill, delivering repeated staccato notes that cut through the wind. The trilling songs of the dusky grass-wren in Australia, grasshopper sparrow of North American grasslands, and calandra lark of the Mediterranean and western Asia are all examples of whirring, rapid notes among birds of open country.

Unlike birds, mammals living in dense vegetation have higher voices than those from open country. This seems to be caused by differences in hearing. A survey of fifty species found that the average peak of hearing sensitivity for forest-dwelling mammals was nine and a half thousand hertz, three thousand hertz higher than that for species that live in the open. This difference is likely caused by the pressing need to hear quiet, high rustles and the soft whish of other animals brushing against leaves. With no wings with which to make rapid escape or arrival, forest-dwelling mammalian prey and predators rely on their ears to hear approaching danger and opportunity. These sounds of animal movement in vegetation are mostly high frequency, favoring animals with ears tuned to this range. This, in turn, favors high-pitched communicative sounds, hitting the most receptive spot in the ears of mates and competitors. The sounds of forest mammals therefore tend to be higher than those of their cousins of the plains and savannahs. The growls, chirrups, and meows of cats that live in forests, such as the Asian golden cat or the lynx, for example, are pitched higher relative to body size than the vocalizations of open country cats like the caracal of Africa and Asia or the manul of Asia. The same is true for the barks, *chips*, and chatters of forest-dwelling tree squirrels and chipmunks compared with their kin, the ground squirrels and other rodents of open meadows and deserts.

Human speech and hearing reveal our nature as a large mammal of open grassland and savannah. The peak of our hearing sensitivity falls between two and four kilohertz, and our speech is low, from eighty to five hundred hertz, spiced with sibilant sounds that reach up to five kilohertz or more. The peak of hearing sensitivity of our closest relatives, chimpanzees, is eight kilohertz, and they hear much higher sounds than we can, up to nearly thirty kilohertz. Chimpanzees have a diverse vocal repertoire, much of it high. Their long-distance call, the pant hoot, starts with quiet low grunts then climaxes with a scream akin to a piercing squeal from a human child, given at fifteen hundred hertz, much higher than an adult human scream, about four hundred hertz. Any pairwise comparison like this can be confounded by differences in body size—we're a little heavier than chimpanzees—and the quirks of each species' ecology. But in this case, the differences between humans and our chimpanzee relatives accord with the trends measured across broad surveys of hearing and vocalization in mammals.

Our voices are therefore ill suited to long-distance communication in forests. Words are swiftly muddied. Instead, human cultures that seek to connect through the forest use loud drums or whistles. There are dozens of whistled languages around the world, and most come from areas with dense forests. Whistles not only carry well through vegetation but, with practice, can be much louder than any human vocal sound, carrying messages for a kilometer or more.

Diet has also shaped the diversity of animal sounds. Birds with large beaks tend to sing slower songs, with narrower frequency ranges, because large beaks impose a physical constraint on sound making. This trend is especially evident in the woodcreepers, birds of Central and South American tropical forests. Beaks within the woodcreeper family range from stubby on the spot-throated woodcreeper to astonishingly pole-like on the long-billed woodcreeper. The longer the beak, the slower and narrower in frequency is the song: trills in the short-beaked species and drawn-out whistles for those with elongate beaks. The many species of Darwin's finches on the Galápagos Islands show similar patterns, as do different geographic variants of the red crossbill.

Diet seems to affect the acoustic form of human speech, as measured by comparisons across the six to seven thousand languages that we speak worldwide. The speech of hunter-gatherers tends to lack labiodental sounds, the *F*s and *V*s made by pressing lips to teeth. These sounds are three times more common in the languages of agriculturalists, whose softer foods cause the dental overbite of childhood to persist into adulthood. In hunter-gatherers and in our Paleolithic ancestors, the overbite disappears as teeth encounter tough foods and thus develop a strong edge-to-edge bite. *Form, vivid, fulvous, favorite*: in the sound of these English words we hear how domesticated foods have shaped our mouths and thus our language.

We might also hear the effects of climate and vegetation in human linguistic diversity. Languages from warm, humid, and densely vegetated areas such as tropical forests tend to use fewer consonants than languages from cool and open habitats, although this relationship is contested on statistical grounds by some linguists. The intelligibility of consonants depends on high frequencies and rapid changes in amplitude, features that are degraded by dense vegetation. Sonorous *oo* and *aa* may be more comprehensible in the forest than *pr* and *sk*. In addition, tonal vowels are more taxing for the larynx in dry air, further tipping languages in arid climates toward the use of consonants. I write these words in English, a descendant of languages from relatively open landscapes and dry air—Eurasia has many arid plains and savannahs, and, even in the wetter climes, winters are cold enough to depress humidity. My abundant English consonants and sparse vowels differ from vowel-rich languages that developed in tropical forests.

The environment also seems to shape human linguistic diversity at a regional scale. Lush environments with stable, year-round productivity of plants have higher densities of human languages than places with high seasonality or unpredictability. Productive areas support human cultural groups with smaller geographic ranges, favoring differentiation and high regional diversity of speech. From syllables to large-scale patterns of diversity, human sound making, like that of other animal species, is partly sculpted by the habitats in which we live and gain our sustenance.

What is true of the air is also true for water and solid matter. Each medium has its acoustic properties. Animals that live underwater or that communicate through wood or soil each find their own voices within the material properties of their homes.

For ocean animals that spend much of their lives in coastal waters, reflections from the sea surface and bottom conspire to diminish or mask lower-pitched sounds. Marine mammals like humpback, bowhead, and right whales that feed in coastal waters therefore tend to vocalize at a higher pitch than those of open oceans such as blue and fin whales.

Water over reefs, along wave-pummeled coasts, or in lively freshwater streams can be bedlam. Wind-stirred waves, breaking surf, or the agitations of tumbling water raise a din, blocking much of the acoustic space. Fish in these habitats sing to one another using highly repetitious pulses of knocks, buzzes, or whines, often pitched at the frequency least likely to be masked by the hiss and rush of water noise. Each pulse contains many frequencies and a distinct start and stop. The wide spectrum and repeated onset and offset of the calls increase the chance that the sound will be detected by mates and rivals in a challenging acoustic environment. Sonic communication in these species often happens only at very close range, after a mate or rival has been seen.

Levels of background noise also seem to have shaped the hearing abilities of fish. All fish species use their lateral line systems and inner ears to detect low-frequency movements of water molecules. Some have expanded the range of hearing into higher frequencies and evolved finer frequency discrimination. These species with excellent hearing, such as catfish, carp, and freshwater elephant fish, are mostly found in calm waters like sluggish rivers and ponds. The absence of background noise in the slow-moving waters of their homes may have opened a pathway toward better hearing. Species such as salmon, trout, perch, and darters that live in the tumult of watery noise in streams and seacoasts have little to gain from better hearing and retain only the low-frequency hearing of their ancestors.

To human eyes, the open ocean seems uniform. We might imagine this sameness penetrating all the way to the ocean bottom. Yet for sound, the ocean contains an invisible conduit, a passageway through which sounds can travel for thousands of kilometers. This "deep sound channel" is about eight hundred meters below the surface. Gradients of water temperature and density—cooler and denser in the depths—trap sound within the channel. When sound waves veer up or down, they are bent back into the channel by either warmer water above or denser water below. This watery lens transmits sounds across entire ocean basins, especially low sounds whose passage in water is unhindered by water's viscosity. Whales take advantage of this channel, and their rumbling, moaning, throbbing calls were, until humans invented the telegraph, the only animal signals capable of crossing oceans.

Sound also travels through solid matter, zipping through wood or rock ten or more times faster than in air. We use these waves in all our musical instruments, but these vibrating sheets and strings of wood, skin, and metal are designed to send their sounds to the air. For many other species, though, solid matter is the primary or only acoustic medium.

All land invertebrates like insects and spiders sense vibrations through nerves in their external skeletons and, especially, in the soft tissues of their leg joints. Imagine if every human toe, foot sole, and finger were an ear. This is the insects' world. They hear the vibratory energies around them through receptors on their body surfaces and inside their appendages. Most also use this ability to communicate. Spiders tap the ground with the feet, signaling to mates and competitors. Many hemipteran bugs—treehoppers and their kin—use buzzing organs in their abdomens to send complex trains of sound waves down their legs into leaves or twigs. These signals are usually inaudible in air but transmit with speed and clarity to the listening feet and limb joints of companions. Legs are, for these species, the organs of speech and hearing.

Insects live in a parallel world of sound, running alongside the aerial sounds that we humans hear. Only recently has the magnitude and diversity of this soundscape of solids become known. By attaching electronic sensors to vegetation, scientists have discovered that up to 90 percent of insects communicate using some form of vibration through vegetation or the ground. My own initiation into this strange world of insect buzzes, squeaks, and clicks came when I was gathering recordings for an exhibition of tree sounds. I hooked a tiny sensor to a cottonwood twig, capturing the many tremors and bangs that flow inside the windblown tree. Interspersed among the clatters of the tree itself were second-long high buzzes, regularly spaced like the ring of a cell phone set to vibration mode. I sent the sound file to Rex Cocroft at the University of Missouri, a pioneer in the exploration and study of insect communication, and he confirmed that the sounds were of an insect, likely a leafhopper. More precise identification is not possible because, unlike the well-known songs of birds, our knowledge of the diversity of these sounds is so rudimentary that we lack a comprehensive catalog matching sounds to species. For naturalists with an exploratory bent, the insect "vibroscape" offers fertile ground for discovery.

Every plant species and part of a plant has a different physical character. Young leaves are soft and spongy. Mature twigs are brittle and stiff. Bark is a wide sheet, but a leaf petiole, the fine stem that holds the leaf, is a tube of dense material around a more open core. Each of these materials transmits vibrations in a different way, favoring some frequencies over others. We get a crude sense of this when we hear our neighbors in an apartment building. The hardwood floor of the people living upstairs filters out nearly all high frequencies, but acts as an excellent transmitter of midrange footsteps. If our neighbors install cork—a form of tree bark—on the kitchen floor, only the lowest thuds come through. These varied properties of plant matter are the sonic world in which insects live. Such differences have created sonic diversity in insect sounds, just as differences in vegetation have done for airborne bird and mammal sounds.

Treehoppers in eastern North America offer a clear example of how physical differences in vegetation shape vibratory sounds. These diminutive relatives of cicadas suck fluids from tree leaves and stems using piercing mouthparts. A crest on their heads makes them look like little thorns. In the breeding season, male treehoppers whine and click, and females reply with lower grunts. This duet plays out entirely through tremors sent through leaves and stems.

Two-marked treehoppers, named for the yellow dots on their backs, are a group of closely related species that each specializes on a different plant species. This diversity arose when ancestral species expanded their ranges to new host plants. Colonist treehopper species not only encountered new food when they switched to novel hosts, but their sonic environment changed.

Two-marked treehoppers on eastern redbud, a common tree species of forest edges, give a low whine, about 150 hertz, the pitch of a throaty human hum. Treehoppers on wafer ash, another small woodland tree species, call much higher, about 350 hertz. The two varieties of the treehopper are the same size, and they stick with their song type even when plucked from one tree species and put on another. Each tree species has its own sonic qualities, transmitting some sounds better than others. Each tree-hopper species sings at the frequency that works best for its preferred plant species. Like human luthiers who know and use the subtle differences of woods, these insects have diversified their songs to match the material properties of their homes.

Insects that use many host plants have more widely transmittable songs than do the treehoppers. Harlequin stinkbugs, for example, feed on more than fifty plant species. They call with multifrequency buzzes whose sounds will travel through leaves and stems regardless of plant species. They are wandering troubadors whose songs work in any space, unlike the specialized two-marked treehoppers.

Wolf and jumping spiders attract mates with vibrations whose frequencies match the sound-transmission properties of the leaf litter on which they hunt. Elephants call to one another across great distances by making rumbles that then flow through the ground. They hear these sounds using dense patches of sensory cells in their feet, supplemented with transmission of the sound through their leg bones to the neck and then the inner ear. These rumbles are very low, too deep to be heard by humans, a frequency that transmits especially well over long distances in the soil.

The great diversity of sonic expression across the animal kingdom has its origin, in part, in the varied physical properties of Earth. When we hear a song or cry, we hear the material context in which it evolved. We are also surrounded by sounds inaccessible to our unaided ears, each one tuned to its environment. Our senses live confined in a small part of the whole. Yet we can imagine that under the river's surface are fish drumming to one another. Off the seacoast, whales sing into the deep sound channel and listen to answers from half a world away. In the trees and on the stems of grasses and flowers, insects duet. In human language, whether actively voiced or transmitted through the page, we hear the legacy of habitat, diet, and the physical nature of air and vegetation on our ancestors' speech.

In the Clamor

I t is two in the morning and I lie awake, listening to the rain forest. The cabin is in a small clearing, the top half of its walls open to the forest save for a shield of mosquito netting. My companions, scientists working at the Tiputini Biodiversity Station in the Ecuadorian Amazon, are asleep, worn out by treks on muddy trails. I woke from deep sleep into a glory of sound, an exultation born in the voices of hundreds of species.

A crested owl growls a sonorous *oor*, repeating every five seconds. This is the deepest sound in the forest tonight, delivered with the slowest tempo, a languorous bass. In the daytime, a pair of these crow-sized owls roost with their fledgling in low branches near our cabin. Twin white plumes crown the head of each adult, contrasting with their chocolate plumage. The youngster is all white. In the rain forest, we rarely see the animals whose sounds surround us, and so this family group is much photographed by visitors.

It rained earlier in the night, and drips from the soaked vegetation that overarches the cabin enliven our tin roof with snaps and spatters. In the forest, tree frogs yelp from low vegetation. Their call is tight and nasal, *yup! yup!*, and each singer has a slightly different pitch, perhaps reflecting differences in body size. I hear them all around the cabin, answering one another. I feel caught in the middle of a ball game among half a dozen frogs. On my left, a call smacks the rubbery projectile into the forest, then another frog on my right whacks it in a different direction, to a singer near my head, back and forth, the sound vaulting over me.

The songs of insects are not as easily localized by my ears as owl and frog sounds. I can pinpoint the direction of only a few crickets and katydids, but mostly I'm wrapped in their sonic mists. The clouds of sound are not homogenous, though. Dozens, perhaps more, of pitches, timbres, and rhythms coexist. My ears are used to the relative uniformity of the temperate world: quiet, singleton cicadas in the summer forests of the Rocky Mountains or Maine; the liveliness of field crickets in grassy meadows, a chorus of a handful of species at best. Even the relentless, ear-ringing pounding of katydids in late-summer forests of Tennessee and Georgia is dominated by one species and spiced with occasional bursts from half a dozen others. Here in the Amazon, species diversity is ten or more times higher, a magnificent convergence of sounds.

In the lower registers, a katydid gives short fibrillating bursts. This is overlain with higher, shimmering songs, like dry rice cascading into a steel bowl. Alongside, a hacksaw delivers regular strokes, the harsh bite of teeth on metal. A sweet trill floats over, pulsing once every second. At a faster tempo, another trill comes, higher-pitched and drier. Alongside, three species give continual buzzes, quite close in pitch, one ringing clear and bright, another slightly fuzzy, and the third very arid, like a stick dragging through sand. An irregular sound like the tinkle of metal shavings skips over the buzzes and whirs, so clear and bright that I see silver flashing. Pitched even higher are more pulses, some pumping every second or so, others coming in streams.

There is yet more sound here at higher frequencies, but the human ear cuts it out, a space we call ultrasonic but is, in fact, not "beyond sound" but merely beyond our perceptual abilities. Also evading my ears are many hemiptera—planthoppers, treehoppers, shield bugs, and others—that send songs made of chirps, trills, and pure tones through the solid material in leaves and stems. At least thirty genera of treehoppers live here, comprising an unknown number of species, as do more than four hundred species of planthoppers.

In the audible range, the insect sounds seem to occupy two bands. One is about the frequency of high birdsong. This is where most of the insects sing, a range familiar to anyone who has heard chirping crickets and katydids in parks or forests outside of the tropics. The other is much higher, a fine, crystalline gleam of sound. The lowest frequencies and the midrange seem sparser, save for the lowest insect trills, the owl, and tree frogs.

As I lie in the humid cabin air, sweat easing down my face and neck, pooling in my clavicles, I am befuddled by the experience of listening. I can attend to the insects in one of only two ways. Either let the sound wash over me as a whole or pick out one single species and focus on its shape and qualities. There is too much richness here to hold multiple species in close attention, as I do in temperate forests. In forests in northern Europe or the North American mountains, I can revel in the combination of several singing species, like enjoying the convergence of several spices in a meal. In the tropical forest, hundreds of flavors and aromas coexist at once, an extreme blast of sensory diversity that stuns my auditory palate.

This wonderful but unsettling experience is also radically unlike listening to human music. Whether in a folk song, a jazz improvisation, or a symphony, the human mind crafts sonic layers, each in close relation to the others, all emerging from instruments designed to complement one another.

One or, sometimes, a small number of people compose the music. Human music contains complex, divergent, and sometimes discordant narratives but emerges from a narrow generative source, the minds of its composers and the proclivities of the human ear. In the rain forest, there is no single composer and no agreed-upon collection of tonal or melodic rules. Many aesthetics and narratives coexist here. Listening in the rain forest is challenging and delightful because we hear many stories at once, each expressed with a voice suited to the aesthetic of its own species. These stories are connected through bonds of ecology and evolutionary kinship, but each is propelled and shaped by its own history, needs, and context. The anarchic equality of evolution—a process with no controlling central hierarchy—delivers sound that, to my ears, is joyful in its profusion, humbling me when I try to find its inner patterns. Listening here is a liberation from the tight control that we humans like to impose on the flow of sound.

From my cabin, I hear only the sounds of one spot in the forest, a single moment in the rhythm of the seasons and the night or day cycle. Last night, I walked with a small group of researchers to the riverside and then on a trail through the wet forest. The cloud of sound changed every ten meters or so, revealing new insects and, near the water, the varied crepitations, twangs, and tremolos of frogs. As dawn approached, night-calling species dropped out, one by one, replaced by voices of the predawn, then the day. Blue-gray spread into the sky's black, and howler monkeys suffused the forest with a low rumble and growl. A few birds joined at the first gleam of light, a chorus that peaked just after dawn. As light spread over the rain forest canopy and seeped to the understory, the soundscape filled with *krak* cries of pairs of macaws flying overhead and sneezy exclamations from flycatchers. As they did at night, insects dominated the new morning with dozens of tempi and pitches.

The cycle of day and night is marked here by shifting combinations of sound as each species calls at its preferred time. Rain and sun modify the shapes of this acoustic cycle. A downpour silences many of the birds, canopy-dwelling insects, and primates, but frogs and ground-dwelling insects persist or have their voices quickened by the rain. The sun-filled hour after a deluge evokes a burst of song, even from species that usually confine their vocal liveliness to the dawn. Midafternoon on a sunny day is the quietest moment for vertebrate animals and even for many crickets, but it is a rousing time for cicadas.

The soundscape varies greatly over the rain forest's terrain. As we walk the trails or climb on ladders from the ground into the canopy, we move through patches and layers. No two places sound alike. This is radically different from temperate or boreal forests. I can walk for hours in the spruce and fir forests of the Rocky Mountains in summer and hear combinations of the same half dozen bird, two squirrel, and two cicada species. No one knows exactly how many insect species live in the forests around Tiputini, but the count may be near 100 thousand, many of which are sound makers. Frogs and birds are better known. Nearly 600 bird species and 140 frog species live here. The same number of species as inhabit the varied terrain of North America are crammed here into the space of a few square kilometers. The sonic community is thus crowded and richly variegated.

The power and diversity of the rain forest's animal voices reveal sound's communicative power. Every species here is advertising presence, revealing identity, and conveying meaning to distant others without the danger of being seen. At night, darkness conceals. In the day, the dense profusion of rain forest foliage is almost as effective as a cloak. This is one of the most visually occluded habitats on Earth, perhaps rivaled only by the impenetrably dense thickets of young boreal forest or turbid seawater near a river mouth. No wonder sound blossoms here. Individuals can communicate through the crowds of leaves, all while remaining hidden from predators that hunt by sight. Hundreds of plants in every hectare, smothered in mosses and algae, create habitats of great visual complexity. This, combined with the cryptic color patterns of many insects and other species, makes seeing animals in the rain forest very challenging, even for dedicated and experienced naturalists. But we hear them.

What started on the arid plains of the late Paleozoic, 270 million years ago or more, with the thin rasp of *Permostridulus* and its kin, has now diversified into a thick weave of thousands of sounds in a single place. The sonic grandeur of these forests, though, presents challenges. The costs of vigorous sound making are borne by individual singers and also threaten the viability of sonic communication for the whole community. These dangers drive the diversification of sound in the rain forest, spurring evolution's creativity.

The first cost of singing is the same one that likely silenced ancient animals: making sound risks advertising your presence and location to predators. The risk increases with sustained sound, like the hours-long trilling of crickets or the repeated melodies of songbirds. The solution to this problem in *Permostridulus*'s time was a swift escape. The same is true now. Immobile or slow animals rarely vocalize. The rain forest's sounds come mostly from animals with wings, powerful jumping legs, or both: birds, frogs, monkeys, crickets, katydids, leafhoppers, cicadas, and their flighty, springy kin. But predators and parasites have honed their skills since the Paleozoic. Bounding escape is sometimes insufficient.

Singing insects in the tropics, for example, are plagued by tachinid flies. These hunters have paired eardrums on their undersides, just behind the head, that allow mother tachinid flies to home in on victims. Guided by ears tuned to the particular frequency and tempo of her preferred singing insect host, she alights and spills tiny larvae from her abdomen. These wrigglers swarm the victim and burrow through its exoskeleton. Lodged inside, the larvae grow for a couple of weeks, then burst out, killing their host.

Each tachinid fly species has its own sonic preferences, some preferring short trills, others rapidly delivered chirps, and each is sensitive to a particular range of frequencies. For prey, this specificity means that there is an advantage to sounding different from other species. Natural selection therefore favors sonic diversity. By sounding different from the crowd, singers avoid being assaulted by hordes of parasitic maggots, a strong incentive. Specialized hearing by parasites can generate regional variants in both the songs of hosts and the hearing preferences of the parasites. The diversity of rain forest trees here is explained, in part, by a similar process. Any tree species that becomes too common is cut back by fungi, herbivorous insect mouths, or viruses. Rarity buys a measure of safety. Over time, this results in more diverse communities.

Tachinid flies seek the sonic signature of a small number of species. Most other predators that hunt by ear, though, have ears and palates with more catholic tastes. Any large night-singing insect here advertises its location to listening crested owls. Calling frogs are picked off by slate-colored hawks that lurk in the vegetation along rivulets. Wolf spiders feel the tremors of singing insects both in the air and through vibrations in their legs. When an ornate hawk-eagle soars over the canopy, its ears, eyes, and talons seek mammals and birds alike, from doves to macaws, squirrel monkeys to spiny wood-rats.

These generalist predators also shape the sounds of their prey. If you've ever tried to creep up on a singing tree frog or katydid, you've experienced the sudden silence that a passing shadow or rustle of vegetation can elicit. But prey do not only fall silent when danger swoops in, they often give alarm calls, a seemingly paradoxical response. But by calling, prey signal to the predator that it has been seen. With no possibility of a stealthy, unexpected attack, the predator is often better off leaving the area to seek less wary prey. Alarm calls are also part of the cooperative networks that bind animal societies. Calling animals warn others, benefiting their own progeny and kin, storing up social capital with neighbors, and helping their group thrive while others perish.

The function of alarm calls is embedded in their acoustic structures. When a bird-hunting hawk lances through the forest, small songbirds often give high, thin *see* calls. Other birds dive for cover, responding to the alarm in one-tenth of a second. Hawks swoop at prey at up to fifty meters per second, and so vocal alacrity and split-second responses are essential if prey are to dodge the attack. The structure of the *see* call conveys alarm to others while minimizing risk to the caller. High, pure tones with tapered starts and stops are cryptic, like camouflage for sound, giving the hunter little information about where the caller is located. The calls are hard to find because they lack abrupt onsets that provide binaural cues about position and are shrill enough to be at the edge of the hawk's hearing. The high sounds also attenuate quickly in vegetation.

If a predator lingers, the element of surprise gone, songbirds switch to repeated, lower *pshht! pshht!* calls, vigorous harsh sounds that travel far and plainly advertise the birds' presence. These calls draw other songbirds within earshot into mixed-species crowds that mob the predator. Birds often dive-bomb the hawk or owl from behind, swooping through branches, then veer away on nimble wings. The recipient of such mobbing behavior often moves on.

Alarm calls are not generic. They do not merely convey the presence of danger. Some birds recognize the voices of mates and kin, responding more vigorously to alarm from these familiars than to strangers' calls that, to human ears, sound identical. Alarm calls of birds and mammals can also contain information about the species and proximity of the predator. Snakes, small owls, small bird-hunting hawks, and large hawks or eagles all elicit different calls from their prey. The signal for a distant predator is different from one whose strike is potentially imminent. Animals with highly developed social networks—crows, ravens, prairie dogs, monkeys—also use alarm calls to communicate the individual identity of a predator and the threat posed by this individual. Representation in sound of individual identity of predators demonstrates sophisticated cognitive capabilities. These animals recognize individuals, remember salient characteristics of each, then communicate this knowledge to others using sounds that carry information within their forms. In cleaving humans from all other animals, Descartes believed that *non loquitur ergo non cogitat*, "he does not speak therefore he does not think." If the philosopher had opened his ears and imagination to the alarm calls of the birds outside his window, his logic might have been reversed, *loquitur ergo cogitat*.

The information encoded within calls is a language that crosses species boundaries. By listening to what other species are signaling, birds and mammals assess the presence and identity of predators. Species that are preyed upon are knit together in a communicative network rich with nuanced representation of danger and identity. By giving the calls of other species our attention, we humans can join this network. If a bird stabs the air with a *see*, look up and see the hawk slicing low through the trees, hoping to surprise its prey. A cluster of scolding songbirds likely has a small owl in its center. A loud, repeated alarm indicates more immediate danger than the same calls given just once. A squirrel or bird giving a repeated harsh complaint and moving slowly through low branches is likely following the progress of a fox or other mammal. As I've worked over the years to open my ears, I've found that tuning in to the sonic network reveals previously unseen animals: a coyote in the brush at the edge of the park, a pygmy owl deep in the branches of a fir tree, or a hawk sweeping through gaps in the forest understory, in view for only a second.

Alarm-calling behaviors create opportunities for deception. An alarm given when no danger is present can distract and deflect a competitor or predator. If a male swallow suspects that his mate is in a tryst with a neighbor, his chittering call breaks up the assignation. Male lyrebirds in Australia sometimes mimic the sounds of a flock of alarm-calling birds. This causes females to pause as they peruse the territory, giving the males a little more time with potential mates. Some caterpillars give birdlike *see* alarms when pecked, surprising their attackers and allowing time to escape. Primates and dozens of species of birds give deceptive alarms during moments of intense competition for food. They shriek, then grab the food of fleeing competitors. The champion of this ruse is the African fork-tailed drongo, a bird species that mimics the alarm calls of forty-five other species. Drongos match the sound of their calls to the alarm call typically used by their victims. But this only works on the first heist. To avoid habituation in the victim, the drongos switch to different alarm calls on their second attempt with the same species.

Alarm calls give us a window into the complexities of nonhuman animal sounds. Unlike the many sounds made when animals are feeding, breeding, and interacting with their offspring, alarms are given in relatively simple contexts and thus are easy to study. Carry to the woods a taxidermied owl hidden under a sheet, then yank off the veil. The squirrels call out in fright. String a wire across a field and dangle a stuffed hawk from a pulley. The fake predator swoops, and songbirds shout their *see* alarms as they scuttle for cover. Prop a loudspeaker in a tree and watch the birds and monkeys as you blast out a prerecorded alarm. Compared with the many social and spatial subtleties of feeding and pair-bonding, these are straightforward encounters, easily manipulated in experiments. It is only in the last two decades that the inner complexities of alarm calls have been documented in the scientific literature, following a small number of pioneering studies in the twentieth century. If alarm calls contain within them such expansive meanings, what might the next decades reveal about the far richer sounds contained within other social signals? There is ample evidence from birds and mammals that songs carry information about the body size, health, and identity of the singer. Whether these sounds contain information that transcends the body of the singer, referring to objects outside the body as alarm calls do, is presently unknown. Might we also expand the breadth of our studies of subtle meanings in sound to include the insects, fish, and frogs? We know that some of these other species have individually recognizable sounds, but we have little idea whether these variations encode more information.

The diversity of sounds that I hear in the rain forest is partly the result of the grisly attentions of predators and parasites. Without them, trilling insects would be more monotone, and the vocalizations of birds and mammals would lack range and subtlety. Another threat to singing animals is acoustic competition. In such a loud and crowded place, the sounds of other species singing over one another are a potentially serious problem. These acoustic competitors may not dump hordes of writhing parasitic larvae or tear off your head with their hooked beaks, but if your sound cannot be heard over the din, your genes are likely to face oblivion nonetheless.

With hundreds, sometimes thousands, of species making sound in the rain forest, the masking effects of noise are severe. Animals here face challenges unlike those of other climes. In the Rocky Mountains, insects are silent for most of the year and give only weak chirps and clicks in the height of summer. They and the other mountain animals almost never overwhelm one another with their voices. The wind is the primary acoustic foe in the mountains. Even in the lush forests of the southeastern United States, in the most biodiverse temperate forests in the world, most of the year passes without intense sonic competition. Springtime birds can be voluble, but they don't make a clamor that blocks the sounds of others. Only in the hot days of midsummer do the cicadas fill the air at amplitudes that make my ears ring, surpassing, if they were in a factory, the legal threshold for required hearing protection. In the same forests on late summer nights, katydids unite in a low pulsation loud enough to make humans raise their voices as they converse. These choruses come after the birds' and frogs' breeding seasons, but any other insect species trying to be heard faces a barrier of sound. A challenge that exists, at best, for only a few weeks in the temperate zone is omnipresent in the rain forest. Evolution's response to this difficulty takes many forms, most of which promote diversification of sound.

Yelling louder is one solution to communication in noisy environments. This adjustment can happen in the moment or over evolutionary time. Birds, mammals, and frogs all sing louder in clamorous places and match their vociferousness to the amplitude of the background. Whether any insects do the same is presently unknown. Animals that live in persistently noisy places have evolved to make louder sounds at all times. Contrast the soft clicking sounds of the Putnam's cicada that sings from solitary perches in the uncrowded quiet of Colorado mountain pine forests, with the blast of the *Magicicada* periodic cicada that calls from dense aggregations of thousands in Tennessee woodlands. The former is gentle, sounding like a fingernail tapping on a dry twig; the latter is almost intolerable at close range, leaving my ears with the ringing, plugged-up feeling reminiscent of the aftereffects of a rock concert. The rain forest is so loud partly because the animals are shouting over one another, often pushing the physiological limits of amplitude.

As human aficionados of quiet dining can attest, you can sometimes avoid the fracas by changing your timing. A five or ten o'clock dinner reservation yields quieter surrounds than one at seven p.m. With only twenty-four hours in the day and hundreds of species jostling for space, this strategy has its limits in ecological communities, but several species are known to rework their schedules to avoid noise. A conehead katydid from Panama usually sings at night, but shifts to daytime chirping in places occupied by another species with a similar song. Experimentally removing the competitor causes the conehead to switch back to nocturnal song. But this is an unusual example. Most insect communities have extensive overlap in the timing of the daily cycle of their songs. A finer-grained division of time is possible, though, even when animals are singing at the same time of day. Some birds and frogs space their songs to avoid overlap. By timing their song phrases to fall in the silent gaps between the phrases of other species, singers avoid masking. This strategy, though, depends on all parties singing at about the same tempo. Birds with similar sounds, therefore, sometimes listen to one another's timing to squeeze in their songs, but other animals in the boisterous rain forest, especially insects at dusk, sing not in discrete phrases but in overlapping choruses or near-continuous trills.

Time is one way to slice the acoustic pie. Frequency is another. The lowness of the grunt of the crested owl makes it distinct from the higher squeak of tree frogs and the shrill whines of insects. By singing at different frequencies, animals might escape acoustic competition.

As I listen to the night sounds of the Amazon, it seems at first that animals have indeed divided up the frequency spectrum, making room for each species' voice. From owls to frogs to katydids to crickets, I'm hearing sounds arrayed across a wide spectrum, seeming evidence that evolution has produced a coherent whole, minimizing competition. This idea, proposed by pioneering sound recordist Bernie Krause, is hard to test. The song frequencies of animals may differ for many reasons, not just because competition has led to divergence and, in many cases, species do in fact overlap with one another. The frequencies of animal sounds depend largely on body size, for example. The wide spectrum of calls in the forest may reflect not acoustic competition, but the evolution of different body sizes for different ecological roles. Owls call lower than hummingbirds because they have larger sound-making membranes. The sonic differences between these two groups reflect the ecologies of each—owls hunt large insects, hummingbirds sup on flower nectar—and not a competition-induced division of the sound spectrum. A red howler monkey giving a deep roar at dawn from its treetop perch weighs about six kilograms. Its body is adapted to a diet of leaves and fruit plucked from the rain forest canopy. In the wet forest at the river's edge, pygmy marmosets trill to one another in high-pitched voices. These are the world's smallest monkeys, weighing only a tenth of a kilogram. They feed by gouging small holes in tree bark, then lapping at the oozing sap. The marmosets chitter and purr back and forth as they work their trees. Just as a violin cannot make as low a sound as a bass, a pygmy marmoset is physically unable to make as deep a sound as the howler monkey.

The observation that a species-rich place like a rain forest is filled with diverse sounds does not, therefore, amount to evidence for the idea that acoustic competition caused this diversity. A more rigorous test is to ask whether species that sing in the same places have more divergent song frequencies than we'd expect by chance.

A study of the dawn chorus of birds in the Amazon made such a test and rejected the idea that competition has caused acoustic differences among songs. The researchers analyzed sounds from samples of the dawn chorus recorded in more than ninety locations, comprising the songs of more than three hundred bird species. Amazonian birds, the study found, tend to sing at frequencies and speeds that allow for the best transmission of their songs through dense vegetation, a somewhat lower frequency and slower pace than those of temperate birds. With so many species singing in this fairly tight range of sounds, we might expect intense competition for acoustic space, perhaps driving apart the frequencies of species that sing together. If so, birds singing at the same time and place should have lower overlap in their songs than stimulated "communities" of birds randomly picked by the experimenters from their database. But the birds' songs showed the opposite. Species that sing together have songs that are *more* similar to one another than we'd expect by chance. This was true for all measures of acoustic structure: pace of delivery, highest frequency in the call, and range of frequencies or bandwidth of each call.

Individual birds singing in the Amazon sometimes adjust the second-by-second timing of their songs to avoid overlap, but at a larger scale there is no evidence that competition has caused species to diverge in the structure of their calls. Instead, the forms of the songs seem drawn together into clusters. Two factors may cause this sonic grouping. First, closely related species often share both habitat preferences and the structure of their songs. Small flycatchers are drawn to forest patches rich in flying insects. Large parrots gather in fruit-rich forests and antwrens forage where insects are abundant. Closely related hummingbirds feed from flowers growing in the same trees. Shared pedigree, and thus tastes for food and habitat, brings the sounds of closely related species into the same places.

Second, birds of different species may be linked in a communicative web. When competing species share and understand one another's sounds, they can communicate swiftly and unambiguously. This allows them to efficiently mediate their jostling for food and space, and quickly alerts them to incursions from outsiders. Thus, shared song characteristics paradoxically link competitors in a cooperative network. In the Amazon it seems that the more intense the territorial competition among bird species, the more similar are the structures and timings of the birds' acoustic signals. The need for mutually shared communication channels among competitors is not unique to birds. Governments in Moscow and Washington are connected by hotline, commercial competitors agree on aesthetic conventions for branding and the form of retail spaces, and competition within professions is mediated through the use of shared jargon.

We have mixed results from studies of vocal competition in animal species other than Amazonian birds. The eighteen most abundant cricket species in a forest in Panama do seem to have divided up sound frequencies to avoid overlap. A survey of eleven frog choruses found three in which competition appeared to have led to divergence of frequencies, but the others showed no such evidence. Birds in the temperate forests have widely overlapping song frequencies, although they separate their songs through their timing and spacing. Acoustic competition, then, seems at best only an occasional factor in the diversity of sound frequencies that we hear in natural settings. And in what is arguably the most acoustically crowded place on Earth, the dawn chorus of birds in the Amazon rain forest, sounds of species that sing together have converged, not diverged.

Singing is only one part of communication, though. Listening is the other. Evolution has addressed the challenge of a noisy environment by honing the ears and brains of listeners. Animals that live in noisy places are very good at focusing on the sounds of their kind and ignoring others. Their ears cut through sonic confusion to find what they need.

A study of poison dart frogs in the Peruvian Amazon forest found that the auditory discrimination of each species was correlated with how many other frog species made similar sounds. These tiny frogs give repeated *peep* notes from breeding nooks in the leaf litter. After the eggs hatch, the male carries the tadpoles on his back to nearby water. Each species' song has a different rhythm and frequency, although there is extensive overlap among calls. Species whose songs are very similar to one another have much more discriminating ears than those whose *peep* is unique. The same is true for some rain forest crickets. Their auditory nerves are tuned to the precise frequency of the song of their own species. These nerves respond to their own species' song in a rain forest filled with dozens of similar insect sounds. In contrast, the nerves from cricket species from uncrowded meadows in Western Europe have broad sensitivity, firing off in reply to a wide range of frequencies. Acoustic competition, then, seems to have shaped not the calls, but the nerves and behavior of listeners.

Likewise, birds that live in noisy, dense aggregations can extract acoustic details from a hubbub. European starlings can identify the voices of individual flockmates. In the lab, they can pick out their companions' sounds from a confusion of four or more simultaneously singing birds. Penguin chicks have similar abilities, recognizing their parents' calls even when the calls of other adults are much louder, a skill that no doubt ensures the chicks' survival in colonies of thousands. Evolution has performed a double feat here: first, giving each individual an acoustic signature and, second, enabling listeners to extract subtle patterns from a storm of masking and distracting noise. Vocal individuality and auditory discrimination are common in sociable birds and mammals, including, of course, in humans. Infants pick out their parents' voices from a crowd, and adults focus on single conversations in the racket of a cocktail party. Scans of human brains show that listening to voices in noisy environments is demanding. Multiple control and attention centers, brain networks that have only minor roles when we hear speech in quiet surroundings, activate when we listen in noisy places.

Animals use the complex structures of forests to their advantage. Sound delivered from elevated twigs and branches travels farther than it would on the ground. The canopy's crown offers a fine place to broadcast from, especially in the calm of dawn. The forest's structure also allows animals to negotiate social competition among singers with similar songs. By spacing themselves across the forest's complex structure, animals can reduce acoustic masking and competition. Such a process seems at work in the dusk chorus of crickets and katydids in the tropical forests of the Western Ghats in southern India. There, fourteen species call at once, overlapping the annual cycle of their breeding seasons, daily coming into voice as the sun sets. Yet a detailed study of their spacing and hearing abilities showed low sonic overlap between individuals, even for species whose songs had similar frequencies and timing. By singing from perches sufficiently far enough away from others, each individual finds itself an acoustic space. What seems at first a smothering throng of sound contains within it a spatial structure, a microgeography of sound.

While most human music blends sound into a single experience that varies through time in pitch and amplitude, but usually not over space, sound in forests and other habitats lives in rich spatial patterns. Were we to transcribe and notate such sound, we'd need six-dimensional music paper to record variations across frequency, loudness, time, and the three axes of space.

In the cabin at Tiputini, my night vigil tapers into light slumber until my watch alarm stabs me awake an hour before dawn. Time to head out. The trail is a mire of sludgy clay and puddles, winding over ground made uneven by tree roots. The beam of my headlamp lurches as I walk. The wet surfaces of waxy leaves flash then disappear, dozens of forms reeling toward me and slipping away. The humid air is fat with aroma—spicy root and leaf litter, unctuous mud, and the algal scent of lichen-smothered wet leaves. I pass through a knot of calling frogs, *ack ack*, then on into a cloud of insect sound, a dozen pure cricket tones layered onto one another. The crickets encase me in sound, as if I were within a ringing metal bell. A few minutes later, the timbre of insect sounds changes, adding more rough scrapers and whirs to the purer notes.

As my light beam weaves and jogs, furry spiders the size of my fist jump into view on the trail. A bush cricket, its orange abdomen looking oily in the wet air, pounces onto my rubber boot with a click and thud, then leaps away into the dark vegetation. All around, thick ropes of vine and a fine tracery of dangling aerial roots are vivid in the artificial light against the dark surrounds. One of the vines coils and twists: a blunthead tree snake, thinner than my forefinger and nearly a meter long, works through the tangle. Its head is swollen with two large eyes that gleam then slide into shadow. Further on the trail, the dark pools of two more large eyes face me from a low branch. The gecko swallows as it gazes at me, then bobs its head. I pass the buttresses of a giant tree, arching walls that disappear upward into the dark. In the cleft between two buttresses, five whip spiders sit immobile in the glare. Threadlike legs, some topped with pincers, protrude from saucer-sized carapaces. I know they're harmless to me, but my body sends a pulse of adrenaline as the beam of my lamp suddenly brings them in view.

Above, a lone screech startles me. Does a macaw see the first easing of darkness? Over the next half hour, a net of sound weaves across the upper layers of the forest as the predawn gray seeps onto the highest branches. I stand in the gloom below, listening as the light ignites the growl of howler monkeys, the ringing clamor of parrots, the first sawing cicadas, and the incessant piping of flycatchers.

As I walk in the dark, I feel as if I've shrunk to the scale of a mouse struggling through tangled leaf litter. The night forest encloses me in a throng of sound and aroma. Exhilaration and anxiety build in equal measure: delight in the tumultuous sensory diversity, seasoned with little darts of dread as unexpected creatures leap into earshot and view. This is rain forest awe. Admiration and fear, not as detached ideas but as embodied sensory experiences. The forest slaps my body awake. I am immersed not only in the manifestations of life's diversity but in the experience of life's ongoing creativity. The overwhelming press of sound and other senses here is one of evolution's most powerful generative forces.

Sexuality and Beauty

I hear them from a kilometer away, a sound like thousands of small brass bells, mellowed by passage through the wintry deciduous forest. The ringing cuts through the rumble of traffic from the town's bypass road and the sputtering growl of a small airplane. I'm standing in the suburbs of Ithaca, a small town in upstate New York. It is late March and I'm hearing one of the first sounds of spring: a chorus of spring peeper tree frogs.

When I first started visiting these woods, three decades ago, I was a recent immigrant from northern Europe, and the winter seemed to me dispiritingly long. I was used to quickening birdsong and the first sign of garden blooms in January, then a drawn-out intensification through May. Here chill, gray days keep a firm lock on outdoor life until well into March. The season of migrant songbirds and spring wildflowers does not start in earnest until late April. Were it not for the ever-present haze of engine noise, the sound levels in late winter might be among the quietest on Earth.

On windless days, only the gentle palaver of chickadees or the drumming of distant woodpeckers enlivens the air.

Now, after a tepid late-March rain, spring peepers shout their lust to the air in a jubilation of sound. I approach the forest and the merged quality of distant sound clarifies into thousands of individual voices. Each frog gives a sharp *peep*, a pure tone, rising slightly, lasting about a quarter of a second. Mingled among these are longer, raspy calls, *reeep*. I pad along a board-walk through the swampy woods, moving slowly so that I do not startle and silence the singers. Inside the chorus, the sound pressure level is as loud as the blast of a radio turned up high. Visiting amphibian choruses in the springtime has become a ritual that lifts me out of winter's despon-dency. The frogs bathe me in sound. I feel as if every cell of my body is shaken into wakefulness by the force of their voices. I suffuse my body with the energy of a reawakening Earth. We made it. Another winter ended. Thank you.

It is perhaps a measure of how unattuned my senses were to the ecologi-cal rhythms of North America that the frogs sometimes brought me to relieved, grateful tears. Something inside me could not believe that the long gray cold would end, anxiety enhanced by geographic displacement. Now, after thirty spring seasons on this continent, the smiling relief still comes every year. I've also learned to hear more nuance in the amphibian choruses. The rich woodlands of eastern North America are home to more than three dozen species of frogs and toads. These are productive forests, full of frog-ready insect food, fuel for vociferous breeding displays. Every species has its habitat and rhythms. Many seasons are revealed in these sounds, from the chuckle of wood frogs in icy pools to the ear-ringing tumult of gray tree frogs after a summer rain. Frog choruses mark time into finer divisions than the coarse-scaled human chronometer of "spring" and "summer" and offer a portal into how the year is experienced by other species. American toads—endowed with a sweet, whistling trill—start a little later than peepers and sometimes sing all summer. Eastern spadefoot toads give their choruses of explosive *waa*s for a couple of nights only, after summer thunderstorms.

It is not just the experience of time that changes as we listen to the voices of other species. Through the varied sounds of frogs and toads, and those of birds and singing insects, travel becomes an education in the complex geography of life. We humans seem to do our best to impose uniformity on the land, but the tree frogs and song sparrows calling from behind the parking lot or along the edge of the subdivision speak of the complexities that we smother. Every forest or wetland has a distinctive combination of species. Moreover, the voices of individuals within each species often vary from one place to another, revealing some of the fine-grained differences in the character of each place.

Amphibian calls did not, of course, evolve to bring joy or edification to humans. What delights our ears is the expression of the social and sexual dynamics of each species. Sound making mediates breeding, territoriality, and the alliances and tensions of animal social networks. Every species has its own ecology and history, resulting in behaviors and voices particular to each. Much of the sonic diversity of the world, then, is rooted in the divergent social lives of animals.

Standing on the boardwalk, I flip on a small flashlight, holding it inside a translucent red plastic water cup. Frogs have good night vision and can distinguish green and blue in gloomy light that is, to our eyes, a smear of gray. They're less sensitive to red, though, and they keep calling as I pass my dim beam through the tangle of wet vegetation around me. At least ten frogs call within a couple of meters of me, but I see only one. He is perched on a partly submerged stick, his head angled up by extended skinny forelegs. Under his chin, thin, partly transparent skin balloons, a wobbly bulb almost as big as the frog. As I watch and listen, his flanks pulse inward and, a split second later, the sac expands with a *peep*. The frog is about as long as my thumbnail but, at close range, the sound smacks my ears. The spring peeper's call has been measured at ninety-four decibels half a meter away, the loudness of a vigorous bird. Another push from his flanks and the call comes again, repeating once every two seconds.

The peeper calls by jabbing a slug of air from his lungs over vocal folds in the windpipe. The throat sac receives the blast of sound and puff of air. The sac's extended skin broadcasts the call in all directions. The elasticity of the air sac then pushes air back to the lungs, allowing the frog to call again without opening his nostrils to inhale. Amphibians lack ribs and diaphragms, and so they push the air with bands of trunk muscles whose bulk makes up 15 percent of a male frog's body weight.

Why such effort? A single peeper's call is audible over at least fifty meters, an area of about seventy-eight hundred square meters. His body is only two and a half centimeters long, covering an area of just four square *centi*meters. By calling, the peeper has extended his body's presence in the forest by nearly twenty million times, not counting the sound's vertical reach to listeners in trees. By allowing animals to find one another in complex environments, sound helps species to thrive where otherwise they would struggle. The many ecological roles of vocalizing animals—from frogs, insects, and birds on land, to fish, crustaceans, and marine mammals in the seas—are indirectly made possible by the benefits of acoustic communication.

The peeper not only broadcasts his presence and location but also reveals his size, health, and perhaps individual identity. This information mediates social interactions at a distance. Rival male frogs space themselves in the swamp and reduce the dangers of bodily confrontation. Females not only find mates but assess them without coming close and risking either injury or disease transmission. Sound therefore increases the physical range and subtlety of meaning in animal behaviors, substituting for direct combat in territorial situations and allowing more extended and nuanced evaluation of mates than is possible in the tussle of skin-to-skin contact.

When a female spring peeper emerges from her winter hideaway under the leaf litter, thawing a body that was steeped with antifreeze sugars, she listens for the bells that locate the breeding swamp. She likely also remembers the contours and aromas of the land, having lived in the forest, eating spiders and insects, for two or more years before maturing into a breeding adult. In other species, experimenters have shown that frogs have excellent spatial memory and navigation abilities, especially for breeding sites. The same may be true for peepers. Guided perhaps by memory and certainly by sound, the female peeper sets out for the wetland. At this stage in her journey, sound is a guide to the location of potential mates. Finding mates in a vast environment is likely the original function of breeding sounds. For tiny animals in a forest, sound can reduce search time for a mate to minutes, rather than the weeks it would take to wander the forest seeking other individuals by eye. Scent trails also help in this task in some species, leaving hints for keen-nosed suitors to follow, but sound is especially far-reaching and easy to track. Species-specific sounds also increase the precision of mate searching and reduce the risk of predation. To come close enough to mate is to come close enough to be eaten. Sound reveals species identity at a distance, making the search for mates far less dangerous. Exploiters of mating signals underscore the perils of mistaken identity. Predatory katydids in Australia mimic the mating sounds of female cicadas, luring amorous males to their deaths.

The function of sound changes when the spring peeper's trek across the forest floor brings her to the wetland. She now listens for information embedded in individual voices. Males are spaced every ten to one hundred centimeters and so she lops and swims through an array of pealing sound and bulging throat sacs. Most of the calls are *peep*s, but if males get too close to one another, they joust with rough *reeep*s, grappling for territory with sound. The female's inner ear, like those of all frogs, has three separate clusters of sound-sensitive hair cells, unlike the single membrane in our ears. One cluster is tuned to the frequency of the male's sound. A second has a wider range, presumably for detecting the diverse sounds of the forest. The third picks up only low-frequency vibrations. Males, curiously, have ears tuned higher than their calls, perhaps to better tolerate many nights sitting in the cacophony or to listen for the higher rustling sounds of approaching danger. It is also possible that the males' ears are seeking subtle differences in acoustic structure that reveal the identity of neighbors. Bullfrogs recognize familiar calls and respond more vigorously to strangers. Male peepers remember how aggressive their neighbors are, *reeep*ing to those that suddenly ramp up the pushiness. They also call antiphonally with neighbors, synchronizing the timing of their calls so that one frog leads and the other immediately follows—*peep-peep peep-peep*. These synchronous duets occasionally expand into groups of up to five males with closely aligned tempi. We do not yet know whether spring peepers recognize individual voices.

Female spring peepers prefer calls that are loud and rapidly repeated. The sonic vigor and pacing of the *peep* have their evolutionary origins in this preference. Loud males are easy to detect and locate. Evolution has thus cranked the blast of sound about as loud as it can from lungs the size of a pea. At temperatures just above freezing, males give about twenty peeps per minute. On balmy nights, their peeping rate increases to eighty peeps every minute. But regardless of whether the night is cool or warm, some males vocalize at up to twice the rate of others. Females sense these differences and swim or hop over to the faster callers. In doing so, they select the healthiest males in the swamp.

Calling is taxing. Some males deliver more than thirteen thousand peeps in a night, each one powered by strong muscular contractions. Fats stored in these muscles supply 90 percent of the energy needed for calling. Males who cannot supply ample fat to their muscles have little stamina. Compared with their languid neighbors, rapid callers are, on the average, heavier and older, with larger hearts, blood cells better supplied with hemoglobin, and muscles more richly stocked with fat-burning enzymes. They also tend to show up night after night, rather than sporadically through the spring.

Her choice made, the female approaches a male, taps him, then, in a flail of limbs, he clambers onto her back, forelegs clasping her neck. The female oars her way through the water, gluing peppercorn-sized eggs to submerged vegetation, fertilizing each with sperm from the male clinging to her back. Unlike many other frogs that lay eggs in clusters, spring peepers place most of their eggs singly, perhaps to prevent predators from finding and eating them all. Once the eggs are deposited, the parents leave them to their fates. The mother's nourishment in the egg yolk and the parents' DNA is all the inheritance the tadpoles receive. The female's acoustic preferences for extreme calling rates have a practical result for her offspring, uniting her genetic material with that of vigorous males. She may also reap a benefit for herself in the shorter term, avoiding sickly males that might transmit their ailments while they are clamped onto her back.

Over the breeding season, the female spring peeper lays up to one thousand eggs. She endows each one with a supply of yolk, draining her hard-won stores of fat and nutrients. Early spring is a lean time, and so these stores date to the warm, insect-filled days of autumn. The egg yolk supplies energy for the developing embryos and a boost when the tadpoles first hatch. The male's singing is exhausting too, depleting his reserves and exposing him to predators. His investment brings no food or other physiological benefit to the young. Instead, it enforces a kind of honesty in the communication system between males and females. Only healthy males can afford to sing loud, fast, and long. An inexpensive call could be given by any male and the sound would therefore carry no reliable message about body size and condition.

The high cost of calling, then, ensures that the spring peeper's call carries worthwhile information. By using sound to make their mating choices, females select males whose genetic qualities are likely to be helpful for their offspring. The costs of singing have lodged both the females' preferences and the males' songs at the center of the species' breeding behaviors.

This is not how costs usually affect evolution. The spring peeper's body—from toes with adhesive disks for climbing to sticky insect-catching tongues—is built without wasted energy and material. But for the spring peepers' calls, costs are an essential part of the function of the signal. Without them, the communication system would fall apart.

Costs of singing, then, have two opposing effects. For slow, defenseless animal species, making a loud sound likely brings death. This is too high a cost for any sonic signaling system, no matter how much information the sound reveals about the health of the singer. But for species that can leap or wing away from danger, costs of sound making ensure that the sound is meaningful and thus favored by evolution. Evolution will not endow spring peepers with signals so extreme that they all but guarantee death. But it will tax the frogs enough to reveal the vitality of each singer.

Costs play a foundational role in communicative signals across the animal kingdom. The bright colors of bird feathers and lizard throats, and the heavy antlers of deer, reveal the health and vigor of their bearers. The costs of these structures are too severe for feeble animals to bear. Many of these signals are closely tied to the body sizes of animals. The volumes of the lungs and throats of frogs and deer, for example, are revealed by the depth and vigor of their calls. For small individuals, the price of mimicking the call of a large animal is prohibitively large. When gazelles flee predators, they sometimes interrupt their runs with upward leaps. These prances advertise fleetness and tell pursuers that a chase is unlikely to succeed. In plants, large petals saturated with pigments and fruits loaded with colored nutrients faithfully signal the plants' condition to pollinating and fruit-dispersing animals. Even the costly red-colored leaves of autumn may signal the quality of trees. Aphids fare poorly on trees with fiery displays and avoid them when they can.

For vocal signaling, costs take several forms. A calling spring peeper depletes energy reserves, pushes muscles and lungs to the limit, and reveals its location to predators. Singing uses between 10 and 25 percent of a Carolina wren's daily energy budget, and of all the wren's daily activities, only flying requires more energy than singing. A singing wren also pays an opportunity cost because time spent in song is time not spent feeding or preening. Predators such as sharp-shinned hawks may cue in on the wren's song, finding the singer among concealing tangles of vegetation, just as tachinid flies do with katydids. Bird nestlings clamoring at their parents from the nest draw the attention of predators. When a leafhopper blasts its vibratory signals through its legs into plant stems, the insect's energy usage goes up twelve times. A skylark that sings while it flies away from an attacking merlin uses precious breath and time. In each case, listeners receive information about sound makers. Female spring peepers assess the fat stores and muscular condition of potential mates. Wrens infer one another's health. Parent birds understand the vigor and hunger of their nestlings. Leafhoppers communicate body condition. The merlin understands the fleetness of the lark and gives up when it hears song streaming from its quarry.

As we walk our neighborhoods and hear the varied sounds of animals around us, we're participating in a network of flowing information. With a little attention, we can hear some of the meanings in these sounds. In a chorus of insects or frogs, the healthiest animals have the loudest or most persistent voices. Among breeding birds, the individuals with the most diverse repertoire may be those that leave the most surviving offspring. In the song sparrow, a bird species common across North America, for example, males that sing with wide range whistles and trills leave more grand-offspring than those with simpler songs.

Naturalists are taught to recognize animal species by ear, a practice that opens us to the diversity of creatures around us. When I first learned the frog and bird sounds around my home, I felt my sensory boundaries expand. Suddenly I was in contact with the conversations of dozens of species. But, at first, I neglected to go beyond the names of species and attend to the sonic nuances within the sounds of each. Arriving at a name, I stopped. Yet every voice carries meaning. Some individual differences, like the melodic and rhythmic variations of the song sparrow, are easy to pick out after only a few minutes' listening. Others are harder, such as the seemingly infinite complexity of crow and raven sounds or the subtle differences among frog calls. By giving individual animals around our homes the gift of our attention, we can learn much about the meanings of their sounds.

A largely uncharted area for future study is the relation between sound making and the diversity of animal sexuality. Nearly all field studies of song assign sex to individuals with the untested dualistic and heteronormative assumptions that all animals exist in either male or female bodies, and that all pairs are between males and females. Neither assumption is true. Many species have nonbinary individuals, either as third or fourth "sexes" within the species or as unions in one animal body of male and female sex cells, body forms, and behaviors. The frequency of these intersex individuals varies from 1 to 50 percent in most vertebrates. Many "male" frogs, for example, have egg-making cells within their testes. I assumed that the spring peeper whose throat sac I watched was a male, but the animal's hormonal and cellular bodily reality may have been a mix of male and female. A number of frog species also have two types of males: singers and silent "satellites." The silent males are often smaller and sit close to the singer. As I listen to the spring peepers from the boardwalk, about 10 percent of the vocal males likely have a satellite nearby. These males contribute no effort to calling. From a human perspective, such lurking seems perhaps creepy and parasitic. But spring peeper females have their own sexual aesthetic and sometimes choose to mate with these edgy, silent types. In spring peepers, the roles of singer and satellite are flexible, with individuals switching as conditions change. In other species of frog, and in some insects and birds, animals stick with one of multiple within-sex identities for an entire breeding season or lifetime.

Further, in many species, females also sing. Yet the vast majority of scientific studies of song in the breeding season focus on males. The bias against noticing and studying female sounds has both cultural and geographic roots. We project onto "nature" our preconceptions. Victorian naturalists saw quiet domesticity in females and loud, conquering vitality in males. In the Reaganite and Thatcherite years of the 1980s, biologists described song as the result of an economic battle of the sexes. In a free market of competing individuals, silent females assessed which garrulous males might best serve their interests. Now the idea that females are, by nature, mostly silent has been overturned.

The peculiarities of animal behavior in the temperate climate of northern Europe and northeastern North America, where, until recently, most scientists studying animal behavior lived, add to this bias against studies of female vocal displays. There, male birds and frogs dominate the soundscape of gardens and forests. But female birds in the tropics and in warm-temperate regions in the Southern Hemisphere are often just as vociferous as males. The birds of temperate Europe and North America, then, are unusual. A survey of birds around the world shows that females sing in more than 70 percent of songbird families, and a reconstruction of the songbird family tree shows that female song was likely present in common ancestors of all modern species. In the developing embryo, the song centers of the bird brain develop in both sexes. The evolutionary and embryological roots of song are thus present in all adult birds. Among frogs, song is heavily male biased, but females make sounds during social interactions, some of which seem to individually identify each female. In the vibratory world of insects communicating on plants, males and females often duet, passing tremors back and forth along stems or through leaves. Both male and female mice give ultrasonic sounds during breeding interactions, part of a much wider array of sonic communication within their social networks.

In *On the Origin of Species*, Charles Darwin wrote that female birds, "standing by as spectators," might select "the most melodious or beautiful males," causing evolution to elaborate the songs and plumage of males. He was right that evolution sculpts sexual displays, but his cultural context constrained his view of sexual diversity and possibility.

Blinders of our own time no doubt also narrow our views today, underscoring the need to question assumptions about sexual roles. We can expand Darwin's vision and recognize that all sexes—male, female, nonbinary—of vocal species use sound to mediate social interactions. This more expansive view is an invitation and challenge. As we listen to the voices of animals around our homes, can we leave behind our preconceptions and hear the richly varied forms of sexuality in nature? Around me in the glorious pealing thunder of the spring peeper's chorus are not only males and females, enacting a simple story of male advertisement to female spectators. Each individual has its own sexual nature—a blend of "male" and "female" for many—and each its own agency. The sounds that so lift my winter-worn spirits are the information-rich mediators of behavior in this complex sexual web.

There are two great puzzles and wonders about the songs of breeding animals. The first is why any animal would expend energy and advertise itself to predators by making loud and persistent sounds. This seemingly wasteful and dangerous activity allows singers both to reach potential mates over a vast area and, in some cases, to reliably communicate information about health. The second puzzle is the great diversity of sonic forms in breeding displays. A loud and repeated grunt would suffice to advertise the location and vitality of any animal. Yet even among closely related species, sound takes on timbres, tempi, and melodic patterns whose marvelous diversity surpasses what is needed to reveal the location and strength of singers.

The spring peeper's relative, for example, the upland chorus frog, makes a raspy sound, like running a fingernail along the teeth of a plastic comb. Another relative, the northern Pacific tree frog, calls with a rising, two-part *krek-ek!* Sounds of more distant tree frog kin include the rapid tap of flinty stones in the northern cricket frog, stuttering beeps like a Morse code machine gone mad in the European tree frog, and a groaning *waar* from the Mediterranean tree frog. If the sound of these species' calls had been shaped only by a need to show off vigor and fat reserves, the tree frogs would all sound similar, perhaps a *peep* varying only in its pitch in different-sized species. The frogs all call from similar habitats, and so it is unlikely that differing demands of sound transmission would have created such a diverse array of calls.

Consider also the red crossbills of the Rocky Mountains and the animals of the Amazon rain forest. The red crossbill sings with complex, inflected melodies, interspersed with buzzes and flourishes, a song far more elaborate than is needed to merely cut through the masking noise of wind in spruce trees. The nighttime chorus of insects and the dawn salvos of bird and monkey sound in the Amazon are astonishingly varied. These species are adapted to the sound transmission qualities of their forests. Their sounds also reflect the ongoing struggle with predators and with ecological competitors. Yet there is more to the diversity of their sounds than can be explained by adaptation to local vegetative and biological conditions, or the need to communicate the vitality of the lungs, blood, and muscles.

Sexual dynamics among animals are creative forces, diversifying sound. This generative power works in three principal ways, none of them exclusive of the other. The first is the sensory biases of each species. The second is the need to avoid breeding with closely related species. The third, and most creative of all, are aesthetic preferences.

Every listening organ is tuned to particular frequencies of sound. These frequencies are usually those that most reliably signal the presence of danger or food. Sexual displays that match these sweet spots are most likely to be noticed and acted on. The listening organs on the legs of water mites, for example, are tuned to the frequency of the swimming motions of small crustaceans. When the water mites sense this distinctive hum, they grab their prey. Males use the same sound frequency to signal to females, using a preexisting bias in the sensory system for courtship.

Small mammals and insects live in proximity to one another, often in dense vegetation. Their hearing range extends into what humans call the ultrasonic because these high sounds reveal useful information about the close-at-hand environment. Social and breeding signals of these animals are therefore also ultrasonic. To human ears, for example, mice and rats seem almost entirely silent, but these animals have rich vocal repertoires including play sounds, calls from pups to mothers, alarms, and breeding songs. Such high-frequency sounds travel very poorly in air, and so these sounds offer rodents good close-at-hand communication without revealing their locations. For animals that interact on larger scales, like humans and birds, lower frequencies work better for long-distance communication. Their ears—and thus breeding songs and calls—are tuned to lower frequencies. The diversity of sonic expression therefore reflects the varied ecologies of each species.

The evolutionary imperative to avoid interbreeding with other species can also be a potent diversifying force. If two closely related species or populations overlap, then interbreeding will produce hybrid offspring that are sometimes deformed and often poorly adapted to either of their parents' habitats. In this case, evolution will favor breeding displays that clearly differentiate each species, reducing the possibility of ill-advised pairings.

For example, the spring peepers that I listen to in the swamps of upstate New York belong to the eastern population of the species. To the west, in Ohio and Indiana, the peepers are larger and their calls are lower and delivered at a faster clip. Four other populations, one in the Midwest and three along the Gulf Coast, also differ in body size and call style. These six different spring peeper varieties have pedigrees that diverged at least three million years ago, with some hybridization and genetic mixing since. What seemed to human taxonomists one species, labeled with one name, the "spring peeper," is instead a family of six different genetic lineages with subtly different breeding calls. Where the spring peeper clans meet, in areas of overlap, evolution has made the frogs' sounds and preferences especially distinctive, slowing genetic mixing among populations.

The sounds of breeding animals, then, can enforce the boundaries between populations. In doing so, they nudge diverging populations toward more complete breaches. These breakups are one of the foundations of biological diversity: the splitting of one species into two.

These examples should in no way be read as supporting the racist laws and cultural biases against so-called miscegenation in humans. Tree frog lineages have been on different evolutionary paths for at least three million years. Humans show no such deep and wide genetic divides within our species. All contemporary human populations share a common ancestry that dates back only a couple hundred thousand years, at most. Compared with other animals, the genetic geographic differences that exist among our populations are slight. Further, children born of parents from different regions show no propensity to increased genetic illness. Instead, the reverse is often true when inbreeding among closely knit human populations unmasks hidden genetic problems. Last, our commitment to equality and the dignity of all human beings makes any discrimination, even if it were founded on some underlying biological patterns, wrong. The behavior of other species is no guide to human morality.

Avoidance of interbreeding can, in some species, cause breeding songs to diverge. But this process is far from universal. In many species, there is no evidence either that hybrid offspring have poor health or that breeding songs are especially divergent when close species live in the same location. Evolution has another trick up its sleeve, the wondrous elaborations wrought by sexual aesthetics.

In 1915, statistician Ronald Fisher puzzled over the aesthetic tastes of animals in the breeding season. Darwin had proposed that sexual ornaments evolved to satisfy the preferences of mates. But why, Fisher wondered, do animals have such strong desires for "seemingly useless ornament"? His answer starts by noting that the evolutionary success of any animal depends not only on the survival of its offspring, but on how attractive these offspring are when they mature and try to mate.

Fisher reasoned that aesthetic tastes are grounded in the need to distinguish healthy from unhealthy mates. These preferences are shaped by the ecology of each species. Carrion flies, he wrote, love the aroma of rotting flesh, but the same odor on mammalian breath indicates tooth abscesses. Evolution thus favors the development of aesthetic tastes particular to each species, giving animals what he called a rough index of the "general vigour and fitness" of potential mates. Fisher then offered his key insight: Once established, preferences will favor further elaborations in the "splendour and perfection" of the mating display. Attractiveness becomes its own evolutionary force. Animals whose displays meet or exceed the aesthetic standards of their species will leave many offspring because they attract many mates or mates of high quality. Aesthetic preference and exaggeration of breeding displays become linked through evolution, egging one another on, in a process that feeds on itself.

The process of exaggeration continues even if the display "ceases to be any index of vitality whatever." Then the breeding display is favored by evolution only because it is attractive, not because it signals health. Fisher predicted that breeding displays would ratchet up their extravagance until predation or physiological limits put an end to further increases.

In letters to Darwin's grandson, Charles Galton Darwin, Fisher outlined a mathematical demonstration of his idea. He also speculated, without supporting evidence, about how the process might work in humans, viewing sexual choice in our own species through a racist, eugenic lens. He claimed that only the "higher races of mankind" developed standards of beauty that reflected "moral character." Like many early twentieth-century scientists, Fisher took what was a sound insight into evolution, an insight that provides not a shred of support for racist ideologies, then twisted it to fit his white supremacist views. Modern theoreticians have shed and rejected the racism, and confirmed Fisher's mathematical findings about the coevolution of sexual preferences and displays, especially work by Russell Lande and Mark Kirkpatrick in the 1980s, followed by Andrew Pomiankowski and Yoh Iwasa in the 1990s. These biologists concluded that the process of coevolution and elaboration that Fisher outlined has firm mathematical and logical foundations. The evolution of aesthetic preferences and breeding displays can indeed, they concluded, balloon initially modest mating signals into extreme displays. Biologist Richard Prum has even proposed that the theory underlying the process is so "extremely robust" that it should be regarded as the "intellectually appropriate null model" of sexual evolution, the default against which other ideas are tested.

Fisher and many contemporary biologists present this process as one in which female preferences drive male displays. But evolution transcends such restricted views of sexual roles. Any inherited display can coevolve with any inherited preference, regardless of sex. If inheritance happens culturally, when animals learn preferences from older generations, as has been documented in insects and vertebrate animals, Fisher's process of exaggeration can also proceed. In all cases, it is the preference that kicks off and guides the process. Animal acoustic diversity has its roots in the sensory perceptions and preferences of listeners, which are then elaborated through coevolution of preference and display.

A survey by biologist Zofia Prokop and her colleagues of contemporary field studies of animal breeding displays found supporting evidence for Fisher's process. Across ninety studies—with subjects as varied as crickets, moths, cod, voles, toads, swallows, and more—the researchers found that inheritance of attractiveness was more common than inheritance of bodily vitality. If this result holds across the animal kingdom, then the mating preferences of parents can indeed produce attractiveness in offspring, even if such attractiveness serves no other purpose than to increase mating success.

Fisher speculated that his process starts with preferences that indicate the health of breeding animals. But any mating preference can serve as a seed for the process. If the sensory system is tuned to a particular frequency or tempo of sound, perhaps to help find prey, then songs in this range will be particularly attractive. In small populations, accidental changes can also kick off the coevolutionary elaboration of taste and display. For example, when just a few members of a species are isolated from the rest of their kind—colonists on an island or inhabitants of an outpost on the edge of a species' range—they may have mating preferences that are not representative of their species. These small clusters of atypical mating preferences arise through the randomness inherent in picking out a tiny subset of a population. This is exacerbated by genetic drift, the random ups and downs of gene frequencies from one generation to another, fluctuations that are especially pronounced in small populations. Drift also affects behaviors such as the songs of some birds whose forms pass from generation to generation not through genes but by social learning. Any quirk can set off Fisher's process in a direction that depends on the initial particularities of mating preferences.

Drift can, in just a few generations, elevate a rare mating preference to dominance in a small population. For example, after a small group of finches colonized one of the Galápagos Islands, their songs changed rapidly from a simple slight downslur of frequencies to a more pronounced, two-part sweep. Within ten years, colonist songs had almost completely diverged from those of the ancestral population on another island. Likewise, the songs of common birds such as the red-capped robin, western gerygone, and singing honeyeater on Rottnest Island off the west coast of Australia differ markedly from those of the mainland. Despite the fact that many mainland bird populations sing uniform songs across ranges of thousands of kilometers, these island birds sing with their own cadences and rhythms. Island-dwelling robins and honeyeaters have simpler songs than those of the mainland, but the gerygone sings more song types on the island, using rhythms unknown on the mainland. The isolation of these small peripheral populations frees them from the genetic and cultural exchange that enforces uniformity on the mainland. There is a parallel here with cultural change in human societies. Margins are, in the words of essayist and journalist Rebecca Solnit, "where authority wanes and orthodoxies weaken." Islands and other marginal habitats, then, are incubators of novelty and change.

The coevolution of taste and display can be an accelerant for both the diversity of sound and the process of speciation. Small differences are magnified, accounting for the profuse diversity of animal mating displays. But as varied as they are, the differences among breeding displays are not arbitrary; they reflect the particular history and ecology of each species, inflated over time.

There is an improvisational quality to Fisher's process. When musicians improvise, they take ideas, elements of the music, then pass them back and forth, elaborating and exploring as they listen and respond. Evolution works in analogous ways, although it makes its music by shaping the script of DNA and the learned experiences of animals. Each species brings a different set of predispositions and foibles, which are then elaborated through the reciprocal evolution of preference and display.

This view of sonic evolution has a refreshing openness to novelty and unpredictability, contrasting with more rule-based, utilitarian explanations of why sounds are so diverse. Yes, there is order in the sounds of a forest or seashore, revealing the physical and ecological laws of the world. But there is also unpredictable creativity in evolution's work. When I listen to the diversity of birdsong or the varied calls of frogs and insects, I hear exuberant anarchy, evolution drunk on its own aesthetic energies. Other human listeners, though, are more impressed by the order and unity of wild sounds, comparing them with symphonies and orchestras, forms of music whose beauty and creativity emerge through coordinated and hierarchical relationships. Predictable order and capricious whimsy work together to produce the sonic marvels of our world. Human aesthetics, born in our evolutionary path as we developed speech and music, seem to love these tensions between order and tumult, unity and diversity.

The effects of physical laws on animal sounds are easier to measure and document than the unique improvisational history of each species. Fisher's process is wraith-like. Its creative actions left no fossils of sound for us to discover. The ghost left marks of its passage, though, in subtle arrangements of genes and patterns of sound among closely related species.

In Fisher's process, aesthetic tastes and the form of song displays co-evolve. Changes in tastes encourage elaboration of displays, which then stimulate further exaggeration of tastes. This results in a genetic correlation between aesthetic preferences and the form of breeding displays. Animals with genes for extreme displays also have genes for extreme preferences. The limited genetic evidence to date, drawn from studies of fewer than fifty species, shows that, for most species, genes for display and mating preference are indeed correlated. Most of these studies are of insects and fish, animals whose breeding sounds are relatively simple to measure— trills, croaks, and chirps. The genetics of aesthetic preference in more complex sounds—the timbre of a hermit thrush's slow, rich introductory note compared with later rapidly modulated notes, the melodic form of a humpback whale's song, the fine details of the cadence and pacing of a mouse's ultrasonic warbling—are unknown. Uncaged animals live in aesthetic territories whose behavioral genetics are uncharted. For now, we can conclude that in some species, the limited genetic evidence to date is consistent with Fisher's idea.

Fisher's process also leaves evidence more accessible to our everyday senses than statistical correlations among genes. Listen to the animals around us. Spring peepers, chorus frogs, wood frogs, and toads all call from the same American vernal ponds, yet they make a range of sounds that far exceeds the need to tell one from another or transmit sound through vegetation: bell-like peeps, rhythmic rasps, strangled quacks, and sweet trills. The katydids of the Amazon forest tap, chirp, thrum, whir, and whistle, using many tempi, displays whose diversity bears the marks of aesthetic extravagance. The astonishing diversity of birdsong transcends mere utilitarian need to signal vigor.

These everyday experiences can be analyzed more formally using evolutionary trees derived from DNA. Each tree represents the history of origins and splitting of animal species, a family pedigree for the species in question. By mapping the form of songs or other breeding displays onto the trees, we can trace how sounds changed over time. In these trees we read both the predictable marks of physical constraint and the caprices of history. The body size of animals—from the length of bird beaks to the size of chirping insect wings—strongly affects the frequency and speed of song. Larger species, on average, sing at lower pitches, with slower trills and melodies, than their smaller kin. Likewise, the environmental and biological context—density of surrounding vegetation, presence of predators and competitors—shapes the form of songs, molding each species to its surroundings. But alongside these factors there is a sprite-like unpredictability about evolutionary changes in rhythms, melodies, modulations, timbre, loudness, crescendos and decrescendos, and pacing, elements that in a human context we'd call musical form or style.

When songs are mapped onto evolutionary trees, we see that they expand and contract unaccountably through time. Their cadences and timbres shift with seemingly no governing law or direction. A biologist presented with the news that a new species has been discovered might, with the help of an evolutionary tree and information about the animal's body size and habitat, hazard a good guess about the most general qualities of the song of this species, such as frequency and perhaps tempo. But they would be unable to predict other qualities of the song. These evolutionary patterns do not prove that Fisher's process caused the elaboration of sound. But they are consistent with his ideas and, for now, inexplicable by any other known evolutionary process.

In the voices around us, we hear a great meeting of evolutionary forces, like the confluence of lively rivers: Fisher's mercurial processes, the genetic imperative to avoid interbreeding with the wrong species, the benefits of honest signaling of bodily health, the many shapes and sizes of animal bodies, the guiding walls of physical environments, and the diverse ways that animals find their sonic place in complex communities of competitors, cooperators, and predators. The result is a glorious, creative, turbulent flow from headwaters at least three hundred million years old.

H onest signals. Sensory bias. Coevolution of preference and display. What do these workings of evolution mean for living animals?

Every species lives within its own aesthetic. The spring peeper hears the *peep* of neighbors through inner ears tuned to the range of frequencies used in its breeding displays. The sense of hearing is the first gate on the path to aesthetic judgment for the spring peeper, just as it is for all animals that find and select mates by sound. The anatomy and sensitivity of each species' ear frame this portal to aesthetic experience.

The next door is the narrower one, the unique preferences of each animal for the pacing, timbre, amplitude, and melodic structure of the call. A peeper's ear is stimulated by many sounds, including sometimes the sounds of closely related frogs. But only one sound causes her to reach out and tap the bulge-throated singer and initiate mating. Her sonic discernment might serve many ends: picking out a vigorous mate, staying away from transmissible diseases, avoiding interbreeding with other populations, or ensuring that offspring will, when the time comes, have songs that other frogs find attractive. For the frog, though, this long backstory of how preferences came to be resolves into an experience in the moment. Vibrations in air, when they are patterned just right, wake knowledge embedded in the frog's genes, body, and nervous system. She hears and understands.

Aesthetic experience is thus a meeting of the outer world with the knowledge that all animals carry within. The result is subjective, depending on the sensory abilities and preferences of each species and individual within the species. Only a spring peeper truly comprehends the *peep*.

How this experience manifests in froggy subjective experience is unknowable. Even among humans, we cannot project our own experiences onto others. I hear sounds both as aural sensations and sometimes as bodily experiences of light and motion. For others among my family and friends, the same sounds evoke color, and every pitch has its own hue. The senses live in a net of relations, a web whose shape differs subtly among us. Imagining the experience of sound in other humans is therefore difficult. Imagining experience in other species is harder still, best approached in a mode of gentle conjecture. The spring peepers' large mouths and noses are very sensitive to aromas, and so perhaps they experience sounds as odorous vapors or bursts. Or the *peep* may evoke a sense of movement in the chest, echoing its production, in the same way that our body sometimes feels itself in motion when we hear human music. Studies of frog physiology show that sound is transmitted to the inner ear not only through the eardrum but via forelimbs and lungs, making frog hearing perhaps more like the total bodily immersion that fish experience. We live in world of tantalizing otherness. So many experiences coexist, food for imagination and humility.

We humans can reach out to other species with science, empathy, and imagination, but such practices are also subjective, coming as they do from an animal with its own sensory biases and tastes, including our aesthetic preferences for some ideas over others. And so the history of the scientific study of sexual displays cleaves to the values of each age. We hear other animals sing through the filters of our preferences for what is a beautiful or ugly idea.

But subjectivity does not mean that we do not perceive truth. Aesthetic experience can, when it is rooted in deep engagement with the world, allow us to transcend the limits of the self and to understand more fully the "other." Outer and inner worlds meet. Subjectivity gains a measure of objective insight. In an experience of beauty or ugliness is an opportunity to learn and expand.

Biologists seldom discuss aesthetics or beauty. When they do, it is in the context of the evolution of a restricted set of sexual displays, those that we humans find attractive or intriguing: strident songs and bright colors. Quieter sexual beauties are absent from biological theories of aesthetics. We pass over the quiet *chip* notes and camouflaged olive-green plumage of a female bird, even though evolution has likely caused male birds to be highly attentive to these forms of sexual beauty. Further, all animals make sophisticated choices about social relationships, food, habitats, and the rhythms of their activities across space and time. Each of these is mediated by a nervous system that integrates inner knowledge and outer information, resulting in motivation and thus action. Every species has its own neural architecture, but all species share nerve cells and neurotransmitters of the same kinds. Unless evolution has wrought an entirely different product—aesthetic experience—from human nerves than those of all our cousins, aesthetics are at the center of how nonhuman animals understand their worlds and make decisions. To presume otherwise is to suppose that humans and other animals are separated by an experiential wall. There is no neurological or evolutionary evidence for such a divide.

Consider the many manifestations of aesthetic experience in our lives. Almost every important decision and relationship in human lives is mediated by aesthetic judgment.

Where to live? We have profoundly moving responses to habitats, both to houses and their surroundings. In some we find great beauty or ugliness. In others our aesthetic sense yields only a bland *whatever*. These judgments then motivate us to spend a large portion of our resources to locate ourselves in the most beautiful of the choices available to us.

How to judge environmental change? We assess surroundings through our aesthetic responses. This is an especially profound experience if we have years of lived, sensory experience with a place. Sometimes, we are driven into grief by the ugliness of despoiled rivers, forests, and neighborhoods. But we can also feel a sense of gorgeous rightness at the emergence of new life congruent with the biological character of a place. Aesthetics are one of the roots of environmental ethics, powerfully instructive and motivating.

Who does good work? There is beauty in craft, artistry, innovation, diligence, and persistence. We see this in the labor of others; we aspire to it in ourselves; and we have an aesthetic response to both the products and processes of work.

How to behave? We live embedded within webs of relationships and we instantly recognize when actions within that network are beautiful or ugly. We feel this deeply, and our aesthetic response guides both our own behavior and our reaction to others. Moral judgment of human behavior is tightly associated with the aesthetics of relationship.

Are we thriving? We also find beauty in the laughs and smiles of newborn babies, the wise and kindly advice of elders, the astonishing development of skills in children and young adults, and the sense of possibility for the future.

In all these cases, aesthetic judgment emerges from an integration of the senses with our intellect, subconscious, and emotions. A deep experience of beauty draws together genetic inheritance, lived experience, the teachings of our culture, and the bodily experience of the moment. In doing so, an experience of beauty can be a great truth teller and motivator, more powerful than senses, memory, reason, or emotion acting alone. When experiencing beauty, multiple parts of our brains light up, a network of connection among disparate neural centers. The parts of the brain associated with emotion and motivation are activated, as are motor centers. Feeling and action. No wonder that experiences of beauty bond people—as mates, families, cultures—and motivate us to act on behalf of what we've learned through aesthetic experience. Beauty inspires us to connect, care, and act.

Why should this be? In *The Songs of Trees*, I suggest that through profound experiences of beauty, we, in Iris Murdoch's word, "unself." We connect what is within us to the collective experience of others, to both members of our own species and our nonhuman kin. Such opening allows us to partly transcend the narrow walls of the self. Because all life is made from connection and relationship, getting outside of our heads and bodies is necessary to understand the world. Beauty, therefore, is a reward and guide built by evolution to help us pay attention to what matters. The experience of beauty has many forms because there is much in the world that needs attending to, and each context demands its own aesthetic.

The ancestors that bequeathed us their genes found beauty in a safe and fertile landscape, in right relationships among companions, in work done well, in the fruits of creativity, in the body of a lover, and in the giggling smiles of babies. All these experiences of beauty guided our forebears into relationship and action, and thus survival. In giving us an internal glow when we connect to the otherness of people, animals, plants, landscapes, and ideas, beauty feeds and grounds subjective experience via tendrils that run out into the objective world. Aesthetics—the appreciation and consideration of the perceptions of the senses—is a guide and a motivator to find truths beyond the self.

In our unrooted and industrialized world, beauty can also be a deceiver. We often isolate our senses from the consequences of our actions, creating bubbles of pleasing experience built on ugliness elsewhere that might give us pause if we could sense it directly. This is most obvious with international trade. The beautiful objects and foods in our lives sometimes come from places of exploitation. Even soundscapes can be misleading. In the outer suburbs, gentle sounds of insects and birdsong in trees soothe us. Yet this experience is possible only because of the traffic-filled highway that brings us and our goods to sonic oases, and the noise of mines and factories needed to build the extensive infrastructure networks that enable and sustain low-density suburbia. In seeking sensory calm and connection to other species, we can paradoxically increase the sum of human noise in the world. The dislocating power of fossil fuels drives much of this separation between our senses and the consequences of our actions.

One of the perils of our time, then, is that we can find satisfying beauty in experiences that hide fragmentation, destruction, and incoherence. Evolution has built us in thrall to the power of aesthetic experiences. We cannot escape this, our nature. Nor can we easily escape the industrial structures in which our lives are embedded. But we can try to listen, rooting our aesthetic sense in life's community. What a delight it is to feel those roots ramify and learn.

And so I return to the spring peepers' chilly swamp to open my ears. I come here to be renewed by their sound, a spring ritual. I'm motivated by a desire to slake my winter-parched ears with the sounds of the forest. Beyond this immediate pleasure, in ways unknowable in the moment, I also let the lives of other species into my body and psyche. In this opening, there is the possibility of more knowledge and connection. But, mostly, I listen to enjoy. This is evolution's gift to us. The labor of gathering and integrating knowledge, essential work for animal survival and flourishing, is a pleasure. Aesthetic experience rewards us in the moment. In satisfying our hunger for immediate gratification, we also serve evolution's long game. In a world in tumult, might we accept our ancestors' gift and listen?

Vocal Learning and Culture

Midsummer. Bright sun. Yet the air has a snowy bite. Gusting wind and loose rock underfoot make me stumble. I clutch at my breath. In the thin air, my thighs burn and ache with anaerobic effort. In another hour, I'll be trudging four-steps-pause-breathe, four-steps-pause-breathe, a rhythm imposed by the mountain on lowlanders who presume to approach the four-thousand-meter peak, one of the knobby vertebrae along the high spine of the Rocky Mountains.

In the high plains to the east of this Colorado mountain, browned prairie grasses have set seed, and fledgling meadowlarks squall openmouthed as they pursue their parents. For prairie plants and animals alike, the summer season of parental provisioning has come. But here on the mountain, spring just began. Snowfields persist in a few spots. Elsewhere, floral profusion. After nine months of snow and ice, light and water raise from the stony ground a great abundance of blooms, each one a defiant reply to winter's long duress.

No plant grows higher than my knees on this tundra. Alpine sunflower and stemless daisy hold palm-sized gold and lemon flowers on stalks only as long as my finger. Ten paces carry me past hundreds of these glowing disks. Among them, moss campion mounds its narrow dark green leaves, forming spongelike puffs topped with dozens of pink-purple flowers, each bloom the size of a large raindrop. Alpine sandwort displays white flowers of similar size, emerging from a centimeter-high mat of tiny fleshy leaves. Above these creeping forms, mountain buckwheat holds flowers aloft, on slender stems crowned with torch-like clusters of hundreds of minuscule flowers. This buckwheat is a giant among the dozens of wildflowers here, reaching ankle height. Miniature avens, asters, waterleafs, and phlox add varied purple hues. Most plants have stems densely matted with silver hair. This felt protects them from wind and ultraviolet light and, along with the darkness of foliage, traps heat, quickening the plants' inner chemistry in the short growing season. Flowers are heat catchers too, warming their nectar and offering visiting insects sips of sweet alpine hot toddies.

This carpet of miniature flowers is interspersed with shrubby alpine and snowy willows. They grow in tidy knee-high mounds, smooth edged, seeming to flow through swales and to fill small basins, clinging to the lowest, wettest parts of the terrain. Like the wildflowers, the willows' stems and leaves are fuzzy. Every plant is festooned with spiky green baubles that enclose the developing seeds. The willows bloomed before their leaves emerged, when snow first started to melt, welcoming on warmer days the year's first ants, bees, and flies with pollen and nectar.

Subalpine fir, a tree that lances twenty or more meters high at lower elevations, maintains an outpost of crouched, windswept trees here. Every individual is pressed to the ground, growing from trunks turned horizontal. Branches sprout thickly around these recumbent spines, making each tree a flattened, elongate thicket, impenetrable to human limbs. A few of these low trees send up meter-high vertical shoots, testing the air in a bid to escape their creeping life on the ground. All of these sprouts are dead, killed by the ice-blast of wind, and they stand like desolate flag-poles, the prevailing wind's direction recorded in the tattered brown re-mains of twigs pointing leeward.

Thousands of flowers within arm's reach. Tens of thousands within eye-sight. This is botanical hard liquor, alpine amaro, floral wonders distilled into a layer just centimeters high: foliage rosettes, scalloped petal edges, elegant rhythms of stem architecture, and dozens of leaf shapes. My eyes, used to a world on a larger scale, implore me to lie down, to get close and imbibe. Prostration is impossible without crushing the delicate blooms or impaling myself on a jagged stone, and so I hunker down on the worn mountain trail, dizzy from oxygen starvation and the floral marvels of tundra springtime. Many of these diminutive plants are old, some living upward of two centuries, yearly renewing their delicate aboveground greenery from sturdy, deep-buried roots.

We call this place tree line, a boundary, but there is no sharp edge, only a carpeted mosaic of species that thrive where woody vegetation meets its limits. A few plants extend their populations higher, close to the summit, but most dwell in a band where clumps of fir and willow blend with open tundra. Ascending the mountain, it takes an hour, at most, to walk through this world. But this narrow elevational range belies the magnitude of the habitat. Plants here live all along the high Rocky Mountains, then across the vast treeless tundra of the Northern Hemisphere. Moss campion, for example, here confined to a small section of the trail, lives in the mountain ranges of North America, Europe, and Asia, and is common on the open tundra that circles the Arctic.

Wind is the dominant sound here, either its hiss and slap as it scours past my ears or in the roar of fir, spruce, and limber pine that carries up from lower elevations. In the lulls between gusts, animal voices find their way through: the wing whir of bumblebees; the croaks of ravens surfing wind eddies over mountain ridges; the *ewk!* of pikas from adjacent rock screes; and the *pitpitpit* call of American pipits winging across the open tundra, seeking insects to fuel their egg laying and courting. Into these relatively simple sounds, from atop one of the ragged fir flags, comes a more ornate melody. A steady introductory note, a higher buzz, a trill, then three downsweeps, the whole phrase unfolding in just two seconds. The song repeats, then another voice answers from a willow shrub twenty meters away, and a third from a fir thicket downslope. The songs are complex but not jumbled. The purity of tone and finely wrought structure are full of light and delicacy. A figure skater of sound: Two long, sliding strokes, a rise into a spin and twirl, and quick foot sweeps on landing. Control. Speed. Elegance. A striking contrast to the disordered wind.

The singers are white-crowned sparrows setting up territories for a hurried breeding season in the high country. These birds spend most of the year in their wintering grounds at lower elevations in the mountains and, for some, in the open scrubby vegetation south of here in New Mexico and Texas. Their brown-and-black-streaked backs and gray chests blend with the vegetation, but their head pattern pops. Bold black-on-white stripes run back across the entire head, a beacon amid the greens and grays. Even at the edge of my eyes' resolution, gazing across a hundred meters of tundra, I see the banded heads as they bob and fly.

This seems an extreme environment for a songbird, but from their perspective this mountain slope combines many advantages. The brief summer brings a surge of insect food with little competition. The wildflowers and grasses will shortly offer abundant seed, enough to draw forest birds like siskins and juncos from the lower elevations to the summer feast. Moisture is easy to find in the rivulets that run down from melting snow, a rare luxury in this arid continental interior. And although they are conspicuous when they sing from their elevated perches, at the first sign of danger from hunting goshawks they can drop into vegetation as dense as the thickest lowland briar patch, vegetation that also protects nests from the eyes of ravens.

Male and female white-crowned sparrows are indistinguishable to the human eye, and their bold head patterns serve as social and sexual signals for both sexes. The stripes communicate the birds' presence, health, and in subtle variations of raised crown feathers, moods, from spiky-headed agitation, to flat-crowned alarm, to dome-headed relaxation. In the breeding season, most singers are territorial males, and some females also sing to defend their food patches or to drive away rivals.

From my seat on the stony trail, I listen to the birds and am struck by how each song has its own pitch and structure. Individuality is immediately apparent. The first bird, its feet grasping the dead fir shoot, starts high, on a note my sound recorder will later peg at four and a half kilohertz, just above the top note on the piano. This pure, steady introduction flips into a buzzy sound at about the same frequency, then a metallic trill. Three notes at the end dart down from five to three kilohertz. *Eee-bree-tree-tewtewtew*. The singer on the willow starts much lower, three kilohertz, and so is recognizable from the first moment of song. The song's buzz jumps up in frequency, then moves directly to two sweeps, omitting the trill. *Bee-bree-tewtew*. From the downslope fir, the third bird gives another arrangement, starting between the others, three and a half kilohertz, then a higher buzz, a hard chip note, a trill, and five sweeps. *Eee-bree-chip-tree-tewtewtewtewtew*. Over the next several minutes, the birds sing back and forth, sometimes seeming to answer one another, sometimes overlapping their phrases. Each bird sticks with its song, repeating distinctive frequencies and arrangement of parts.

With just a few minutes of attention, I come to know the locals on this patch of tundra.

Mori Point, California, just south of downtown San Francisco. The headland cleaves incoming swells from the Pacific Ocean, presenting a hard edge for waves that have traveled uninterrupted for hundreds of kilometers. The water's energies dissipate in bellows from cliff faces and in the seethe of waves on a pebble beach. Out of the fog comes a row of pelicans, their oaring wingbeats synchronized as they fly north, parallel to the shore.

From one of the many waist-high thickets of coyote brush, a white-crowned sparrow sings. I recognize the introductory pure tone followed by buzzes and sweeps, but the pattern is unlike anything I've heard in the mountains. The introduction is divided into two notes, the trill is gone, and the song concludes with extra notes, tight concluding accents, *eee-eee bree-tewtewtew-chuchuchu*. Another bird answers, again with a two-note introduction. The second part is a little higher, with fewer sweeps and concluding chips, *eee-EEE-bree-tewtew-chuchu*. Like the mountain birds, each bird repeats its song, staying faithful to its variations of pitch and arrangement of phrases. The birds here seem to agree on some stylistic elements, a divided opening and an ornamented ending, but then carve out individual variants.

Later the same day, north of Mori Point, I listen at Crossover Drive in Golden Gate Park, San Francisco. Six lanes of traffic cut through the park. Brake squeals, horn blasts, and an enveloping throb of engines define the soundscape here. In the shrubs next to an encampment of unhoused people adjacent to this traffic artery, a white-crowned sparrow sings. One long note, then seven sweeps. No ornaments or buzzes. *Eee-tewtewtewtewtewtewtew*. I walk west, along the paved promenade, away from the traffic noise. Two more sparrows sing from bushes near patches of unkempt grass. Like the first, they start with a single note, omit any buzzes, and give multiple sweeps, ten or more for both. They also break the string of sweeps into two, the first part a higher and more emphatic *Stee!* than the last, *tew*. One bird gives more repeats of *Stee* and the other more of *tew*.

Back at home, I swing open my laptop and, with the help of thousands of microphone-wielding bird watchers, take an imaginative journey into variations in these sparrows' songs across North America. Two websites are my portals. Both are collections of field recordings uploaded by enthusiasts, assembled into vast databases of sound. Scientists at the Macaulay

Library at the Cornell Lab of Ornithology have been gathering and archiving sound since the 1920s. Their work and that of volunteer contributors now number over 175,000 field recordings. Xeno-canto, a website started by Dutch ornithologists in 2005, gathers recordings from bird watchers and scientists worldwide. Its archive now counts more than 500,000 entries. Within each archive, snapshots of sound are held in the electric charges of billions of microchip capacitors and transistors. In a click, my ears fly across these silicon memories of life's conversations.

The first search result from the Macaulay Library is from the Denali Highway, Alaska. On June 14, 2015, Bob McGuire recorded a sparrow with two introductory notes, the second note with two rapid wavers in the middle of its steady tone, ending in three buzzes that step up then down in frequency. *Eee-eee-diddle-wee-bee-too.* No sweeps, no trills. Compared with Colorado and California birds, this is a reshuffle spiked with the *diddle* innovation. I zoom to Alaska on the Xeno-canto map and click on the colored dots that locate sound recordings in the database. I imagine the recordists standing in the brisk Alaskan summer, breathing willow, fir, and spruce aromas as the birds sing. Each recording is a moment captured and shared as humans reach out to understand and honor other species. The birds in these recordings all sing variations of the song that McGuire recorded, each one distinguished by the frequencies of its introductory note and buzzes, but all sharing the same overall pattern. I scroll west to Nome and east to the Yukon, leaping over mountain ranges with a sweep of my hand, and hear the same overall singing style, with some Nome birds turning the second note into a warble.

Then, south to Oregon. *Eee-diddle-buzz-tew.* Another remix. The buzz comes earlier in the song and a quick sweep is added at the end. Other Oregon birds stick with the scheme but add more sweeps. A little north, near Seattle, they throw in a second buzz and some inflected sweeps, more ornamented than the Oregonians.

The white-crowned sparrow breeds all across the northernmost part of

North America, in shrubby habitat at the edge of the boreal forest and tundra, and also farther south, in the mix of low vegetation and grasses in the western mountains and all along the Pacific coast. This is a vast range, roughly three million square kilometers, home to about eighty million individuals of the species. The diversity of the sparrows' songs reveals some of the complexity of their lives, layers and textures within the multitude.

As I listen to sonic memories from my travels and the electronic gifts left by others from their wanderings, I sense that the diversity of human sound, so rich within our cultures and individual lives, is just one manifestation of sound's creative workings within animal species.

Winter is migratory sparrow season in the southern United States, bringing a taste of the tundra and the boreal forest to fields and gardens in Tennessee. In the brushy edges of fallow cotton and cornfields, white-crowned sparrows glean the remnants of last summer's grass and herb seeds and pluck insects from the soil. These are migrant birds, here only for the darker months before they return to their breeding sites in the North. Their relatives, white-throated sparrows, also winter here, distinguishable by sight with their white bib, yellow daub above the eye, and head stripes that are less crisp than those on the white-crowned sparrow. White-crowns prefer fields, white-throats the denser vegetation of forest edges and rural gardens. These preferences mirror their breeding habitats: open, scrubby land, including treeless areas north of the Arctic Circle for white-crowns, and boreal thickets, swamps, and forest edges for white-throats. Winter in Tennessee is usually a half-hearted business, staying mostly above freezing, and so insect life seldom completely stills. The first flowers of henbit and bitter cress emerge in fields in late February, and edible seeds soon follow. For hardy northern birds, this is easy living.

As the days lengthen, song begins. Light penetrates the birds' skulls, bathing receptors buried in the brain. Aglow, the receptors steep the blood in hormones and signal brain nodes to spark the lungs and syrinx. Birds feel springtime's surging vigor, lift their heads, and sing. In winter, the sparrows communicate through at least nine different short calls, each suited to a different context—*pink* when perched alone or flying, brief trills when meeting other birds, rasps when chasing one another—and only occasionally with a burst of song. In spring, for males especially, song production surges.

The song that emerges from the sparrows' beaks can be an amusing delight. I stop digging in the garden or pause my steps on a rural road and smile. Young sparrows are practicing their songs and, like the charming tumble of human infant voices, their jumbled experiments evoke a sense of newness and play.

A young white-crowned sparrow gives two whistles, like the introductory note of an adult, but each one wavers, seemingly unable to hold steady. Another bird gives a single whistle, also wobbly, then three rough sweeps. *Te-e-rew*, instead of the adults' *tew*. A third has a steady opening whistle, then five sweeps, clear at first, then breaking up into stammers. The birds repeat their phrases, shifting the timing a little, pausing after the whistles or curtailing the concluding sweep notes. The voice of each bird is instantly recognizable as a white-crowned sparrow but, compared with adult songs, these youngsters' sounds are disordered in their arrangement of elements, unsteady in tone, and inconsistent from one repeat to another.

The same muddle and teeter come from young white-throated sparrows. Their adult song is a ringing series of clear tones, two long, then three broken into triplets, *ohhh-sweeeet-canada-canada*. Usually the first note is lower, but some birds start high and work down. The second note varies from steady to slightly stuttered. The species nests in the northern forests of eastern North America and winters all across the southern states. This wide geographic range and the tonal purity of its song make the species one of the best-known avian singers in the region, a sonic mark of the end of winter in the South and the start of summer in the North.

Can-a cana ca. In their first spring, young white-throated sparrows sing shuffled and unsteady versions of the adult song. Their hesitations, innovations, and errors are strongly reminiscent of the babble of human infants. *O-sw-swee. Sweet-cana.* I hear learning, play, experimentation, and maturation. I delight in these sounds because they are the marks of present well-being and future possibility. *Ohhh-swee-ee-eet.*

In Tennessee, we hear only the later stages of the young sparrows' vocal development. While they are still on the northern breeding grounds, shortly after leaving the nest, they give highly fragmented whispers, barely audible even at close range. A trance overcomes the birds as they murmur. Their eyes droop and bodies slump, as if in deep sleep. In this state they may be blissed out on the feel-good hormone oxytocin, a chemical that has been shown to motivate and regulate vocal learning in both birds and mammals. Over months, these first stirrings of song get louder and more organized, and the trance fades from their lives, perhaps briefly reappearing during adult slumber as a sweet memory of childhood.

Like humans, sparrows learn the form of their vocalizations by listening to others. Their songs travel through the generations not as coded strands of DNA but through the attentive ears of young birds listening to the songs of their elders. Sparrows in the Rocky Mountains sound different from those in California primarily not because their genes have evolved different songs, but because the form of the song passed down by learning has diverged.

Social learning of vocalization is rare among animals. In many insects, adults have stopped singing or are long dead before their offspring mature. In other species—fish that spawn in open water or insects that lay their eggs in soil—the young develop away from where adults sing. But even among species whose generations overlap, sound is mostly shaped by genes. Toadfish hatch and spend their first weeks in their fathers' nests, enveloped in his croaking sounds, yet toadfish eggs raised in a lab without a dad grow into adults with normal songs. A cockerel raised with no teachers crows just as well as one raised with adults nearby. Captive fly-catchers exposed by human experimenters only to the songs of other species give perfect renditions of their own, unheard species' song. They do the same even if their captors puncture the birds' eardrums. Deaf squirrel monkeys also vocalize normally. Periodic cicadas sing without tutelage, seventeen years after their parents last filled the air with song, an extreme example of the genetic inheritance that characterizes all known sound making in insects. Some species can learn to distinguish the songs of others—frogs, for example, identify rivals by ear, and primates are experts at learning the meanings of sounds—but few learn their own sounds by listening and imitation.

All the exceptions known to date are birds and mammals. Humming-birds, parrots, and some songbirds learn their songs. These branches of the bird family tree are separated from one another by tens of millions of years, and so represent three separate inventions of vocal learning. Among the majority of mammal species, social learning is focused on predator avoidance, foraging, mediating social dynamics, and mate choice. Sound making in these species is mostly innate, although many species learn to modify their use of these inborn calls in different social contexts. The exceptions include bats, elephants, some seals, whales, and one great ape: humans. Our close relatives, chimpanzees, bonobos, and gorillas, have sophisticated cultures, but these are not founded on vocal communication. These groups of vocal learning mammals are not closely related, and each group likely independently evolved vocal learning. Because birds are easier to study in the field and manipulate in the laboratory than whales, elephants, and seals, it is from species like the white-crowned sparrow that we have learned the most about vocal learning in nonhuman animals.

It is a puzzle why many birds and a few mammals are champion vocal learners, but their close kin and most other animal species communicate with mostly innate, unlearned sounds, even among species that have extensive abilities to learn other behaviors. It is possible that learning is only favored when the information imparted by vocalization varies considerably from one generation to the next, and that this is only true in a few species with complex social networks. In such situations—sounds revealing the identity of individuals and the ever-shifting nature of clans and other social groups—learned vocalizations might allow animals to more effectively navigate social dynamics. Among other species, where sounds' meanings are relatively fixed—territorial songs, calls to signal the discovery of food, alarm at the sight of a predator—vocal learning offers no advantage and may instead merely impose a costly delay as youngsters learn.

The kinship I feel with the youthful palaver of birds is therefore a bond of analogy, and not of direct shared ancestry. Differences in the details of how learning works in birds and humans underscore the divergent evolutionary paths that we took to arrive at vocal learning. Yet there are also some surprising parallels, a unity of process despite our divergent histories.

The birds I hear in the gardens and fields are experimenting with sounds that they first heard last summer, more than six months ago. As nestlings and recent fledglings, they listened to the songs of parents and neighbors. They sang only in their whisper trances, although they also made squalling begging sounds and diverse forms of *cheep* notes and trills. The memories of the adult songs they heard last summer now serve as a standard against which to judge their own attempts. Over weeks, the birds try different combinations, converging on a final version, the song that they will henceforth use as their own. This is a feat of memory alien to how humans learn sound. We hear and vocalize in the moment, a back-and-forth that refines sound as it is made, although infants also comprehend many sounds before they are able to repeat them.

Listen to human parents and their toddlers: The child makes a cute attempt. Parents smile and repeat in adult form. The child insists. The parent repeats. A duet that, over months and years, eases the sounds of infant speech into adult form. For sparrows, hearing and vocal production are largely separated in time and space. A song heard in June in northern

Quebec lives through the winter in a young sparrow's brain, meeting the bird's hesitant attempts to vocalize later that year in Tennessee. Months-long memory is the sparrow's primary teacher as it matures its song.

Among white-crowned sparrows, there are exceptions to this wide separation of hearing and singing. On the California coast, the sparrows do not migrate but instead live in dense, stable communities where they defend territories year-round. In these populations, a young sparrow setting up its first territory will learn the songs of neighbors, aligning its song to those of this new home rather than to those it heard where it hatched. This extension of learning into early adulthood is common among songbirds that live in permanent settlements and allows young birds to more closely fit into the acoustic milieu of the neighborhood. Neighbors often negotiate territorial disputes by matching phrase to phrase in back-and-forth singing contests that seem to boast of each bird's local knowledge of song variants. If you sound different, you cannot compete.

In all vocal learners, genes guide the process indirectly, producing brains eager to and able to learn, and predisposing each species to the sounds of its own kind. These predispositions are activated by social connection. Sparrows isolated in the lab and fed sound via speakers learn their songs, but they can do so only for a few weeks after hatching. Sparrows embedded in the rich social life of a flock keep listening and learning for months.

Testosterone shuts down the learning process. In the sparrows' first springtime, the percolation of the hormone through their blood causes the ebullient experimentation of youth to solidify into a final adult song. Artificially removing testosterone, through either physical castration or chemical neutralization, extends the learning period. Testosterone-fueled territorial display is a yoke whose weight seems to expunge creativity.

Individual white-crowned and white-throated sparrows sing only one song variant, repeated tens of thousands of times throughout their lives. These repeats differ somewhat in the emphasis placed on different parts of the song and are embedded in a repertoire of call notes that also vary by context, but the basic form of the song is, to human ways of classifying sounds, singular. This consistency helps with communication. Every bird knows all their neighbors' sounds. If everyone is singing from their patches, then all is well. If an unknown song pipes up or if a familiar song comes from a different territory, then the birds rise in an aggressive fury.

Other bird species learn not just one but multiple song variants. Song sparrows, common suburban and rural birds across North America, sing eight to ten different variants of their jaunty song of accented notes and trills, repeating each variant several times before switching to another. Every bird has its own repertoire. By listening carefully, we humans can build a sparrow-sound map of the neighborhood, drawn in air by the ephemeral ink of birdsong. Their repertoires are just rich enough to challenge human memory. From one spot in a garden in Tennessee I can hear five males and about forty song variants, a delight as I try to notice and keep each songster's collection in mind.

The brown thrasher, though, defeats human ears. Each singer has up to two thousand phrases in its quiver. These it shoots out for hours in volleyed pairs. The thrasher also croons a softer, whispered song when close to its mate or newly fledged young. Some of the sonic variants are mimicked versions of other species, suggesting that learning continues throughout life, but most are the birds' creations. Lifelong vocal learning is well known among parrots and starlings. Such flexibility presumably helps to mediate complex social lives among long-lived species, but despite centuries of imposing human words on caged birds, we know little about the many nuances and meanings of these birds' learned sounds in the wild.

Social learning—animals listening and observing others, then using this knowledge to shape their own behavior—is the gateway to culture. Genetic inheritance passes from parents to offspring at a tempo set by the length of generations in each species, the time it takes for an embryo to develop into a reproductive adult. Cultural inheritance can flow in directions unconstrained by kinship, and its speed is limited only by the time it takes animals to notice and copy one another. Animal species that learn their sounds, then, have opened new creative possibilities for sonic elaboration, refinement, and diversification, free from the rigidity and sluggishness of genetic inheritance.

When a white-crowned sparrow settles on its final adult song, it does not produce an exact replica of what it heard from its elders. Rather, the bird finds a song that fits within the norms of the neighborhood but also bears its own mark, perhaps a distinctive inflection or opening frequency. This balance between individuality and conformity is essential to the function of the sparrow's songs. A song that diverges from local custom is unattractive to potential mates and a weak counter to territorial rivals. But complete mimicry of another bird invites confusion within the social order.

Changes, even minor ones, to any inheritance open the door to evolution. In genetic evolution, changes are introduced through mutations and the reshuffling of DNA in the coordinated dances of sexual cell division and union. These genetic variants then rise or fall in abundance within a population, either by chance or by Darwinian selection. When white-crowned sparrows listen, remember, then re-create songs that are not-quite-faithful replicas of what they heard, the birds fuel cultural evolution.

The speed of any cultural change depends on how conformist or innovative the learning is. White-crowned sparrows can be traditionalists. In some of the year-round resident populations in coastal California, the birds have been singing the same kinds of songs for at least sixty years. If current rates of cultural change hold into the future, some of the song variants of North American swamp sparrows will last for hundreds of years. In contrast, the songs of indigo buntings, birds of forest edges in the eastern United States, are more labile, always on the move. Indigo buntings sing by cycling through a repertoire of about six different song types.

Young males establishing territories pick up song types from established local males, then sing them back to these older birds. But the newcomers also add new embellishments to their songs. Year by year, these novelties add up as the old guard dies off and new cohorts arrive. In just ten years, the song types in any given place are entirely replaced. The speed of cultural change is faster yet among yellow-rumped caciques in Panama. These garrulous, black-and-sulfur birds with daggerlike ivory beaks nest in colonies, dozens of nests in a single tree. Within a colony, birds sing from a repertoire of five to eight songs, copying from one another but also inventing novel variants. Three-quarters of the song varieties that were popular at the start of the breeding season are gone within a year. Yellow-rumped caciques invent their own sounds—whistles, clanks, and toots— and also mimic the sounds of frogs, insects, and other bird species. Rapid cultural evolution in this species is driven by minds that listen carefully to social context, copying songs of colony members and the many voices of the surroundings. Careful listening is then fuel for sonic innovation.

Vocal learning not only allows sound to change over time, independent of changes in DNA, it also creates geographic diversity. Every colony of yellow-rumped caciques has its own collection of sound, actively curated by the tastes of its members. Colony members mediate their alliances and disputes through noisy, squabbling interchanges of a shared repertoire. When a bird disperses from its natal colony and joins a neighboring group, it tosses its old lingo and quickly adopts the sounds of its new home. For these birds, every nesting tree is a distinctive cultural unit, its boundaries created by the imperative for all colony members to learn and use the same sounds.

In the white-crowned sparrow, the extent of geographic variation in song depends on the migratory behavior of the birds. On the California coast, where the birds live year-round on stable territories, the songs are structured into small neighborhoods, sometimes of no more than a few territories. All the birds within a neighborhood share a similar pattern of whistles, buzzes, and sweeps, although each male adds a signature of his own. This fine-grained sectarian geography of sound is, like the tree-centered cultural world of yellow-rumped caciques, a product of the behavior of incomers, itself a result of the sexual preferences of females and the rules of territorial engagement with males. When a young male establishes his first territory, he must match the style of his song to fit with social norms, a strong pressure to conform. Each neighborhood was likely founded when vegetation grew back after fire cleared the land and drove out resident sparrows. Sparrow colonists to the rejuvenated habitat brought their own quirks of song, which were then passed down as cultural variants particular to each patch. The small sizes of cultural units on the coast, then, are a product of small-scale disturbances that create a patchwork of song types. In the years without fires or other calamities, song differences within each patch are maintained by vocal learning within a conformist sparrow society.

White-crowned sparrows from the mountains or the edge of the boreal forest migrate south every winter and don't live in highly stable communities. Their songs vary from place to place but on the scale of hundreds of kilometers, not tens of meters. This is a pattern typical among mobile bird species with wide distributions. One of the joys of traveling is to hear the regional variants of each bird species. Songs that are familiar at home take on new inflections or add peculiar elements when we step outside our familiar paths. This geography of sound varies in its scale and texture by species, depending on the particular balance of creativity and conformity of each. It is often the homebodies, those that seldom move and whose young settle nearby, that have the most compact and parochial geographic distributions. A morning walk in the San Francisco Bay Area takes us through several neighborhoods of white-crowned sparrows. To hear the same for the song sparrow, we'd need to drive several hundred kilometers. The white-crowned sparrow has no distinct regional dialects, although a novel song variant that turns *ohhh-sweet-canada-canada* into *ohhh-sweet-cana-cana* has been sweeping across the continent in the last two decades, the rapidity of its spread helped by the wide-ranging migratory behavior of the birds.

These geographic variations of birdsong, regardless of scale, are usually referred to as dialects. But this word is perhaps loaded with too much human meaning to help us hear the many layers of cultural variation among vocal learning birds. The caciques innovate and change on a week-by-week basis, and every colony tree has its own changing flow of popular sounds, more like Top 40 music playlists than a dialect. California white-crowned sparrows cram more patois into a small area than even the most highly structured human languages. White-throated sparrows sing with such consistency across their range that the new variant is perhaps analogous to the spread of a single idea or catchphrase.

Culture, then, can diversify sound, in forms unique to each species. In doing so, culture combines its powers with those of genetic evolution. In white-crowned sparrows, for example, the rate of trilling is partly a product of culture and partly the result of the genetic evolution of beak size. Birds trill where this sonic ornament is popular and stick to whistles and sweeps where it is not, a learned behavior. Birds with large beaks—a result of genetic adaptation to local foods—cannot trill very fast and so their song is partly a reflection of their beak size, a character largely shaped by genes.

Vocal culture can also fold back into genetic evolution. White-crowned sparrows in coastal California settle into stable neighborhoods as early as their first autumn. They therefore need to rapidly slot into the sounds of the home where they will likely live out their lives. Sparrows in the mountains, though, migrate away from their parents' nesting site, then return in spring not to their place of hatching but to a breeding site whose location the young birds cannot predict. They'll settle, depending on the vagaries of opportunity and chance, in one of a wide range of locales spread across the vast breeding range of the species. The brains of each population of white-crowned sparrow have evolved learning mechanisms to match the demands of their life histories. Coastal birds start learning songs relatively late, extending their learning into the autumnal period, when they match their songs to the new territory. Their learning is focused and accurate, picking out the single best option for their social context. Birds from the mountains learn song earlier, picking up a variety of potential songs during the few weeks available to them between hatching and when they migrate. They remember this wide variety and, when the time comes, practice multiple variants, settling on their adult song when they arrive in their breeding territory. These differences in the timing and breadth of learning persist among captive birds in the lab, indicating that evolution has shaped the nervous system of each to match the social context in which its song is delivered. Sparrows from different populations also have a genetic predisposition to pay attention to and learn songs from their own region, a preference that presumably evolved to help them focus on the most relevant and useful sounds. Genes make culture possible, by providing a blueprint for animal bodies that are able and eager to learn. Culture, once developed, then shapes genes, favoring the blueprints most suited to their cultural milieu.

The most dramatic way that culture might affect genetic evolution is by causing species to split. Songs during the breeding season serve both to connect animals with similar preferences and songs and to exclude those with different tastes and vocal displays. As with genetic evolution, if animals with similar preferences and songs stick together, these sexual dynamics can cleave a population, producing two or more gene pools. Over time, these differences can create new species. It makes no difference whether the inheritance of the preferences and songs is genetic or cultural; what matters is whether a link develops between the forms of the songs and the sexual preferences for them. If they do, populations can be broken into cliques that breed among themselves but not with others.

For more than half a century, scientists have studied the question of whether song learning can cause speciation. Their work reveals that cultural differences in birdsong types are widespread, but these are only occasionally associated with genetic differences among populations. White-crowned sparrows provide one of the clearest examples. In northern California and southern Oregon, the resident population of the California coast meets the migratory population of the Pacific Northwest. Each has its own song "dialect," with the northerly birds singing with longer whistles and shorter sweeps and trills. Playback experiments show that birds respond more vigorously to songs of their own dialect, suggesting that shared songs unite each population and keep them separate from others. But in border areas where the two populations mingle, these behavioral differences were slighter. This suggests that although cultural differences in song do seem to keep populations apart, this force may weaken in areas of extensive contact.

Vocal learning also affords a degree of flexibility that connects divergent populations, delaying speciation. Female birds sometimes prefer the song types of their home region, but this preference is not universal and can be erased by exposure to other song varieties. In a neighborhood near San Francisco where song types are uniform, females will therefore likely enforce conformity, preferring familiar songs. Farther north, on the Oregon border, females hear many song types and will have more catholic and flexible tastes, potentially selecting males from other areas. For males, too, culture can smooth over geographic differences. By molding his song into the style of the neighborhood, a young male setting up his first territory can partly break free from his parental inheritance. He's stuck with his genes but can find a new vocal identity through learning.

In addition to its role in promoting or slowing the evolutionary splitting of populations, vocal culture can make endangered species more vulnerable to extinction. If population densities drop too low, it is harder for animals to find one another and young birds fail to learn the species' full songs. In the Blue Mountains of Australia, for example, the population of the regent honeyeater, a black-and-gold nectar-drinking bird, is down to just a few hundred birds. In recent years, many of these birds have started singing atypical songs, including the songs of other species. Compared with recordings from previous decades, contemporary birds also sing simpler songs. Lacking suitable tutors, young birds pick up snatches of sound from other bird species or invent their own songs. Males with these malformed and often stunted songs are less attractive to females. At the edge of extinction, then, social learning of song can become a liability. As Hawaiian honeycreepers on the island of Kaua'i declined in numbers, the diversity of their songs plummeted, likely due to the loss of social connections that formerly sustained the cultural richness of song learning. In endangered whales too, it seems that cultural diversity is lost when populations shrink. Such losses have been heard in endangered and declining sperm and orca whales. But we have no record of the vocal diversity of pre-twentieth-century whales, and so the full extent of the loss is unknown. The decline of vocal diversity may have been severe among those species that were scythed down to 10 percent or less of their former abundance.

Among all nonhuman animals with vocal learning and cultural evolution, the white-crowned sparrow is one of the best understood. The species' geographic variation in sound is obvious even to human ears unaccustomed to parsing the details of birdsong. The species offers us an imaginative window into the possibilities of culture in all vocal learning species, most of them unstudied by science. Wherever vocal learning happens, cultural evolution can unfold, driven either by the creative impulses in animal minds or by the simple accumulation of copying errors as each generation learns from its elders. These cultural changes cause sound to change through time and to fan out in a richly textured geography.

Birds offer the most well-studied examples, but geographic variation is common among other vocal learners such as marine mammals. New song variants of humpback whales, for example, spread in just months across entire ocean basins, often originating with whales that live in an innovation zone—an incubator for whale sonic creativity—off the coast of Australia, then spreading worldwide. Why one patch of ocean should be the origin of so much new whale song is unknown, as are the causes for the sudden spread among whale singers of one song variant and not others. Cultural variation in the sounds of toothed whales, like sperm whales, orcas, and dolphins, reveals subtle hierarchies of affiliation within each species, from parent to offspring, to clans, to large regions. Sperm whales, for example, live in matrilineal groups that range over thousands of kilometers. These matrilines remain stable over decades, likely held together by shared patterns of vocalization, learned by youngsters from elders within each group. Sperm whales communicate with short bursts of loud clicks. When the whales are close together, these pulses of clicks are like the excited chatter of human friends gathering at the weekend, overlapping one another in a frenzy. Individual whales seem to have distinctive voices or accents—unique ways of using click groupings. This individuality is embedded within a larger spatial and social structure. Matrilines have their own distinctive clicking styles and are themselves part of regional "dialects." In the Pacific, these dialect groups overlap in range, but whales within each do not associate with one another, seeming to disdain the company of whales with the "wrong" way of clicking. In the Atlantic, whales in each dialect group stick to their own nonoverlapping subregions of the ocean. When a sperm whale clicks, other whales presumably can immediately identify its region, family, and individual identity, just as we humans can infer the identity and biography of those we hear speak.

Sometimes cultural evolution crosses species boundaries. Parrots, lyrebirds, mockingbirds, and many other birds take snatches of the sounds of other species and weave them into their own sonic creations. In the case of Australian lyrebirds, these sounds are then passed down through the generations culturally. When lyrebirds were introduced by humans to Tasmania in 1934, they remembered and repeated the song of the whipbird as part of their mimicking display, even though whipbirds did not live in their new home. Thirty generations later, descendants of the colonist lyrebirds still sang the whipbird song, passed down to them by previous generations of lyrebird.

Nonhuman animal sounds also jump the fence into the culture of our own species. When recordings of humpback whales inspired a generation of ecological activists, when musicians from Sibelius to Pink Floyd weave bird sound into their creations, when our onomatopoeic verbs croak, twitter, and bellow, or when police sirens evoke the howl of wolves, fragments of animal sound from other species lodge into human imagination and spread through our own webs of listening, remembering, and responding.

Cultural evolution connects animals—both within and across species—in networks of learning more expansive than the parent-to-child inheritance. This weblike flow of information reawakens flexibility of evolution that the DNA of vertebrate animals has lost. Billions of years ago, our bacterial ancestors exchanged genes promiscuously through their watery surrounds, a back-and-forth that carried the DNA of one cell into another and back again. These movements were unconstrained by the rituals of sexual cell division and parental inheritance that would later control the genetics of complex animals. Cultural evolution, in breaking free from the rules of genetic inheritance, regains this lost rapidity and fluidity of evolution, allowing behaviors to jump from one animal to another through the process of learning. There are limits, of course. Genes and anatomical constraints set bounds on what animals attend to and copy. Sparrows will not learn the calls of ravens, and whales do not mimic toadfish. Within these boundaries, cultural evolution samples, remixes, and connects, reclaiming a little of the evolutionary nimbleness of our bacterial ancestors.

Songbirds and humans last shared a common ancestor more than 250 million years ago. The brains of birds and mammals have each taken their own paths since this split, resulting in parallel worlds of sensation and experience. Birds cram a higher density of nerves into their skulls than mammals, giving their small brains as many cells as much larger primates. The folds and layers of the forebrain have different geometries, hierarchically layered in mammals and clustered into nodes in birds. But despite the long separation of our lineages, our vocal learning converged on some similar processes. Social learning has some universal qualities.

The first of these parallels is evident when we hear human infants and young birds babble. My parents tell me that half a century ago I could not manage the sophisticated tongue and lip movements needed for *cat* and *chocolate*, and so in my infant voice felines were *vuff* and treats were *clockluck*. The trills of white-crowned sparrows are likewise beyond the capabilities of youngsters, and so the young birds squeak and waver, gradually building proficiency. But motor control is not the only aspect of maturation. The order, pacing, and form of sound in young birds and humans are more diverse than those of adults, coming in streams unconstrained by the rules that allow meaning to be conveyed. Maturation prunes this wide-ranging juvenile sound into precise adult forms. Vocal learning also gets harder as birds and humans age. An older white-crowned sparrow will not pick up new songs. Human adults struggle to grasp the rudiments of new languages, even though as infants we readily master any language we are immersed in.

The winnowing and clipping that birds and mammals experience when learning to vocalize also shape forms of growth and maturation in other beings and at other time scales. Twigs on a tree ramify in tens of thousands of directions. Only a few mature into stout branches, and the rest are dropped, food for worms. Animal bodies develop partly through expansive early growth later trimmed by the programmed death of cells. Evolution by natural selection first increases genetic variability through sex and mutation, then narrows this range of possibility as the physical and social environment picks out winners. The words on this page, too, are the few left behind after countless others were culled, along with hundreds of permutations of narrative and analogy. Arthur Quiller-Couch's oft-quoted advice to writers, "Murder your darlings," was inadvertently an insight into many of life's creative processes.

In both birds and humans, vocal perception and memory are controlled by different parts of the brain than vocal production. Listening, memory, and action are each sequestered into their own spaces, and their activities are similar in humans and birds. The perception centers of the brain are tuned by unknown means to the sounds most relevant to each species. These centers feed sonic information to the parts of the brain that control muscles and nerves. Underlying these feedback loops in the brain are the genes that build the brain. The *FOXP2* gene that is so important for human speech is also important in the early development of vocal learning pathways in the songbird brain.

When sparrows and human infants babble, we hear a deep-buried unity. The same genes build parts of the nerve network needed for vocal learning in humans and songbirds, despite the very different forms of the mature brains of humans and songbirds. The patterns and processes of learning, too, are similar. To smile when we hear the tumbling, inchoate songs of birds is not mere sentimentality, then. The pleasure rising within us is a reminder of kinship across difference.

Kinship, but also particularity. We're a peculiar species. Among our close kin, the primates, there are no other species nearly as adept at vocal learning as we are. The complex behaviors and cultures of these other primates are based on visual and tactile observation, not vocal learning. These nonhuman primates also seem to have different brain functions. Those brain regions that are essential for vocal learning in humans play only a minor role in vocal production in other primate species. There is uniqueness here, one seized upon by those seeking to carve out a special place for humans in the natural order. But the cultural evolution of sound in birds, whales, and other vocal learners suggests that human vocal learning is not so much unique as parallel. There are multiple paths to vocal learning and culture within the animal kingdom.

Like the evolution of wings in bats, birds, and insects, evolution has produced vocal learning through bodies of different design. In any convergent evolution of this kind, we expect each independent invention to have its own features. Ranking one as superior to the others seems absurd. Yet humans like to reserve "language" for ourselves. Other animals make sound, but only we have language—as if bats fly, but birds and insects only flap, soar, and flutter. On what basis do we make this distinction? Humans are not unique in learning, intentionality, possession of vocal culture, evolution of culture over time, encoding meanings in sound, or representing in our speech external object or internal states. Every species has a logic, a grammar, to its sound making. It is not clear why only one of these grammars should qualify as language, nor is it obvious which dimension of grammatical refinement should be used as a yardstick. Birds, for example, are superior to humans at discriminating the subtle nuances within individual sounds, seeming more attuned to the rules and syntax contained in syllables than the arrangements among strings of syllables. If this capacity were the measure of language, we'd be ranked below sparrows. Experiments with rhesus macaques and European starlings show that even the purported uniqueness of recursion in human syntax—our ability to create and understand a wide, perhaps limitless, range of expressions from a finite set of elements—is not limited to our species.

We have only a rudimentary knowledge of sound making and vocal learning in other species, a hazy and imperfect gaze into the complex vocal lives of others. Yet even amid this fog of ignorance, we clearly see our own species as just one of a multitude of speaking, cultured beings. Perhaps our species' special quality is not the achievement of a state unattained by others—language or culture—but a convergence of abilities. Many animals learn their sounds, and these sounds help them to prosper within their social worlds: finding mates, resolving tensions, and communicating identity, belonging, and need. Many animals also learn the practical physical and ecological skills needed to thrive. This knowledge usually passes from one generation to another through close observation, not through elaborate sounds. Young vertebrate animals often spend years studying their elders in order to learn how to find and process food, where to migrate, how to build shelter, what to do when a predator arrives, and how to navigate the cooperative and competitive social world. Without this knowledge, they are lost. These two aspects of culture—vocal communication and learned practical skills—are mostly separate in nonhuman animals. In humans, the cultural evolution of sound and that of other forms of knowledge unite. For us, learned sound is an aesthetic experience, a mediator of social relations, and a source of detailed information about how to navigate and manipulate the world. Other species use culture in all these ways, but we knit them together in a union so far unknown among others.

In the last five and a half thousand years, we've taken another step. By carving into clay, inking onto pages, and thumb tapping against screens, we have captured and frozen what was ephemeral, giving speech long-lasting material substance. The invention of the written word broke the constraints that confined all previous vocal communication. When I read an ancient poem, the minds of the dead resurrect within me and speak. When I immerse myself in a book penned on a different continent, I travel across space and time and hear the author's voice. The possibilities for accretion and interconnection of knowledge are vastly increased over the powers of the spoken word alone. Written notation did the same for human music. The score on my music stand carries a melody across centuries.

Text is a crystallization of sound, a diamond compared with the gaseous carbon in breath. A beautiful gem. But a hard one, too, in the powers that it gives us. In the face of some of the productions of the written word—machines, changed atmosphere, human appetites empowered to take and control—other animal cultures are in decline. The population of the white-crowned sparrow, for example, has shrunk overall by about one-third since the 1960s. This population change is uneven, with pronounced declines in California and Colorado, most likely because of the fragmentation and degradation of its preferred scrubby habitat, but with increases in the northern Rocky Mountains and Newfoundland, for unknown reasons.

Among other cultured species, the loss is yet more catastrophic due to habitat loss, pollution, and hunting. Half of all parrot species are in decline worldwide. The last half century has seen bird abundance decline by one-third, about three billion individual singing birds gone from North America, a decline also found on other continents, especially in agricultural areas. About one-third of all whale and dolphin species are threatened with extinction. Wherever human activities preempt land—agriculture, forest clearing, mining—songbirds are in steep decline, and forest fires and desertification claim yet more.

Birds have likely been learning their songs for at least fifty-five million years, when songbirds and parrots shared a common ancestor. Mammals perhaps for about the same time, dating to the origins of bats and whales. Over this long time, vocal learning and cultural evolution were both soil and fertilizer for the growth and blossoming of sonic diversity. In humans, though, these processes turned and started to erode life's diversity, an abrupt change from the expansion that learning and culture previously encouraged. Perhaps part of the cause of this switch from flourishing to destruction lies in our inattention. We humans, distracted by our newfound powers, turned inward and largely forgot how to learn from the voices of other animal species. If this is true, then by reawakening the practice of attending to the voices of others, we will dim the destructive impulse and renew the creative powers of listening and learning.

The Imprints of Deep Time

When introducing students to the practice of attentive listening, I ask them to sit quietly and focus their attention on minute changes in the sounds around them, sending their ears "out" into the world to forage for acoustic experience. Part of what we learn is how hard it is for our frazzled modern minds to keep our attention on any sensory experience without inner distraction. But repeated practice opens a space where the clamor of the mind quiets and the sonic richness of the world blooms. In just fifteen minutes, we each hear dozens, sometimes hundreds, of different sounds in places where usually we would notice, at most, a handful. By listening in the same location over months, we find that these short exercises excavate not only an impressive count of different sounds but also patterns and relationships among them, fragments of earthly music with many layers and tempi.

This subtle complexity underscores how inadequate a few words are to summarize the soundscape of any place. A single hour could fill a book if each timbral, rhythmic, and spatial variation were adequately described. But even a sketch, however incomplete, can perhaps glimpse ways in which sound lives in the moment and has been shaped by history.

Contrasts among soundscapes are most obvious to us when their divergent sounds are the product of markedly different physical energies or human noises. We find it self-evident that wave-pummeled shores sound different from forested valleys, and that suburban streets have a different acoustic character than airports. Not so superficially obvious are differences among the sounds of living species. For ears not attuned to the voices of insects, birds, and other vocalizing creatures, variations are easy to miss.

Just as sounds from ocean waves or mechanical engines reliably disclose their sources, so, too, do animal sounds. The most obvious differences among the many calls and songs of living beings reflect broad taxonomies. Corrugated tymbals of cicadas rasp and whine, rubbing wings of crickets chirp, and membranes in bird chests whistle and trill. Within each one of these categories of sound, with some help from DNA and fossils, we can also discern the evolutionary history of each group of species: where they came from and which other species are their kin. In the soundscape of any one place, we hear the sounds of many species and thus many biographies. This is the biological equivalent of wandering through a busy city, hearing a multiplicity of languages and accents. In these sounds, patterns of human indigeneity and migration are revealed, some recent and others dating back tens of thousands of years. For nonhuman species, we hear even deeper into the past, sometimes hundreds of millions of years.

When we sit and listen to our animal cousins, we open ourselves to the experience not only of the moment but to the marks of plate tectonics, the history of animal movements, and the echoes of evolutionary revolutions.

Three forest edges, on three different continents. Each one is just under 32 degrees of latitude from the equator. In the divergent textures, cadences, and rhythms of their soundscapes, we hear the imprints of deep time.

Mount Scopus, just outside the Old City of Jerusalem, fifty kilometers east of the Mediterranean coast. I wander through the Botanical Garden of Hebrew University. Walkways, dusty with limestone, thread through plantings organized by habitat, representing twenty-two of the many different ecological zones found in the region. It is July and so the early summer rains have ceased, yet the vegetation remains green, helped by both moderate temperatures on this limestone ridge and trickling sustenance from irrigation pipes. Trees and shrubs seem to grow directly from the crumbling, ivory-colored stone. Boulders and small rocks lie all around the walkways. A cluster of two-thousand-year-old tombs is carved out of the rock face, further exposing the mountain. Without the care of horticulturalists, most of the plants here would wither on the thin soils. All around are buildings, roads, and, at the university, irrigated lawns, a startling sight in so dry a land. The gardens are an island refuge in a growing urban sea. Birds and insects find welcome on the diverse collections of well-tended indigenous plants.

A squeaking sound, like that of a cork being twisted into the neck of a wine bottle, pulses from the saw-toothed leaves of a Syrian ash tree. I cannot see the singer, but the tight rubbing sound likely comes from the wings of a marbled bush cricket. On the ground, in the jumbles of stones around the base of cypress, pine, and redbud trees, Mediterranean crickets chirp with sweet, vigorous notes, pumping out calls two or three times per second. Both of these insects are mostly night singers, but in midsummer, the height of their breeding season, their songs linger into the morning. From the branches of olive and oak trees, the day's first cicadas wake, rasping at a lower pitch than the other insects, like a ratchet or windup clock being cranked once per second. Theirs is the sound of dusty air and unrelenting sun. Under the heaviness of afternoon heat, they are often the only animals making sound. Now, as the morning warms, the insects give three-dimensional form to the soundscape: a sparkling cloud of cricket chirps hovers over the ground, bush crickets mark out higher spaces, distinct spheres around the trees from which they sing. Cicadas stitch the treetops into a crackling canopy.

Birds thread their voices into this matrix of insect sound. A greenfinch, its gold-edged wings shining in the dark recesses of the twisted branches of a pine, gives a high trill, switches immediately to a series of rapid whistles, back to the trill, then a string of chirps and sweeping whistles. Like that of its relative, the canary, the tone alternates between sweet slurs and sharp fibrillations, delivered at a caffeinated pace both within each phrase and in the quick flips from one phrase to another.

In the same pine tree, tearing at cones with its stout beak, a house sparrow gives a string of monosyllabic *cheep* notes, answered by kin on the ground. Sparrow bones from archaeological sites show that this species has lived in the region alongside humans for millennia. Following the rise of agriculture in the Middle East, the house sparrow colonized the first cities, feeding on waste grain and nesting in cracks in buildings, and has since followed humans to urban areas worldwide. The *cheep* we hear on city streets across the globe is a continuation of a relationship that began here in the stone walls of the Middle East, just like those in this garden.

A Eurasian blackbird's mellow warbling provides a melodic and tonal counterpoint to the sparrow's incessant staccato, an undulation of clear, sometimes sliding notes, edged with a melancholy burr, like a wistful folk tune. The sound is characteristic of the thrush family, a group whose flute-like songs are common in wooded areas across Eurasia, Africa, and the Americas. I'm used to hearing Eurasian blackbirds in the gardens and cities of northern Europe, but here the bird throws open its yellow-orange beak from among the branches of an olive tree. In late autumn, the blackbird will turn its attentions to the tree's oily fruits. Blackbirds and other thrushes are the fruit-dispersing partners of plants wherever the birds live, a collaboration that sustains the birds and ensures the vitality of plant communities. In the Mediterranean, Eurasian blackbirds and other thrushes are the wild olive trees' original dispersers, a role usurped in the last eight millennia by humans who have bred plumper fruit, convenient for us but a challenge for bird gullets.

Four white-spectacled bulbuls, working their way through the trees in a tight flock, have a sharper timbre and sing in short phrases interspersed with chattering, a convivial exchange unlike the blackbird's more solemn solo recital. In their sounds, I hear a lively society, each bird continually checking in with its flockmates, a roving web held together by bright threads of sound.

A spotted flycatcher sallies from an oak twig, snatches a small dragon-fly, and loops back to the same perch. The bird strips the victim of its wings and swallows the body, then returns to its vigil, standing erect, darting its head from one side to another as it scans for more flying insects. As the flycatcher watches, it gives soft *zeep* sounds, slightly raspy, a sound like that of the bush cricket. This brisk, sweeping sound is characteristic of muscicapid flycatchers, a family of insect hunters found all across Europe, Asia, and Africa.

A hooded crow mumbles as it pokes the edges of the garden's paths. Although crows and their raven and jay kin, the corvids, are known worldwide for their raucous, boisterous cries and caws, they also have a rich repertoire of soft whistles, squeaks, chuckles, and murmurs. Some-times these mediate social interactions among pairs or within family groups, but just as this hooded crow is now doing, the birds are also vocal when, to human eyes, they seem to be alone. For corvids, sound seems to be ruminative as well as communicative.

The avian soundscape is given a percussive element by a Syrian wood-pecker on a dried oak branch. Slamming its beak back and forth, the bird pounds a drumroll from the resonant wood, a tremulation that is vigorous and clear at first, then sags. Woodpeckers are found across Africa, Asia, Europe, and the Americas. They have good ears for the acoustic properties of wood and other solid materials in their territories. Unlike other birds that sing using their unaided bodies, woodpeckers use hollow trees, siding planks on houses, drainpipes, and chimney caps to amplify and broadcast their territorial drumming signals. They are selective in their choice of tympanic aid, sampling the qualities of neighborhood materials, then using those that give the most resonant response. On Mount Scopus, the choice of dead wood is limited by horticultural management, but the vegetation is just unruly enough to supply some choice dead tree limbs.

My visits to Mount Scopus in spring had a similar aural character to summer, although the insects had not yet begun to sing. As new leaves unfurled on trees, cadences of blackcaps wove into rattles and trills from Palestine sunbirds, cheery notes of great tits, and the soothing bamboo fluting of the laughing dove. This is a gentle soundscape, or so it seems to human ears: tapping, warbling, and trilling birds brightened by sweet and squeaky crickets. Cicadas roughen the edges, especially late in summer when they frazzle the air, as do the squabbles of rose-ringed parakeets or European jays. In my half dozen visits, I have never heard amphibians here. In wetlands away from the city, green toads trill and tree frogs grunt, although seldom in large choruses.

Saint Catherines Island, on the coast of Georgia in the southeastern United States, lies 10,300 kilometers west of Jerusalem, and a mere 16 kilometers south. I stand in the early morning on the dock where I previously lowered a hydrophone to immerse myself in the shimmer and bleat of snapping shrimp and toadfish. It is midsummer and already the back of my neck trickles with sweat. Air humidity is close to 100 percent, and by midafternoon, we will reach an oppressive 38 degrees Celsius.

Plants exalt in the hothouse. In the abundant moisture, they throw open the breathing pores on their leaves and, their chemistry stoked by the bath of sultry air, feast on sunlight and carbon dioxide. Their growth rate is four to ten times higher than that of unirrigated plants in the Middle East and southern Europe. Annually, the coast here receives two to three times more rain than Mount Scopus, moisture that comes all year, rather than concentrated in winter as in much of the Mediterranean. As I stand on the dock, I look through the island's fringe of sabal palms into a forest of live oak trees festooned in Spanish moss, mixed with towering loblolly and longleaf pines. Despite the sandy soil, which is leaner than the rich soils farther inland, these trees are lush. A young pine, unshaded by competitors, can lance upward at a meter or more every year.

Animal thunder is one fruit of this productivity. To ears used to less fecund lands, the vigor of insects, frogs, and birds here is astonishing. In wetlands and temporary puddles, fed by the generous sky, thirty-one different species of frogs and toads sing in Georgia.

Each frog species has its preferred season and habitat, creating distinctive ensembles for every month of the year and location. At the edge of a forest of live oak, in a shrubby swale, I hear the signature of July in a coastal wetland, a diverse assemblage of tempo and tone: irregular, growly grunts from pig frogs, whining bleats from eastern narrowmouth toads, pulses of tinkling from cricket frogs, and *enk enk* honking from green tree frogs. The tree frogs crescendo until they obliterate all other sound, then snap into silence when they see me move. I wait, hunched against a barrage of probing mosquitoes, and the tree frogs ramp up again. As with the crickets on Mount Scopus, these frogs are usually nocturnal, but the warm day has drawn their chorus into the morning hours.

Last night, katydids throbbed as loud as a waterfall, their sounds dominated by the unified *cha-cha-cha* of the common katydid, spiced by the lisping rasps of the angle-winged katydids, and the swirling high clicks and trills of the aptly named virtuoso katydid. Now, as the sun reaches the tree canopy, cicadas erect a wall of a hissing, crackling sound. Unlike the katydids, cicadas are patchily distributed in the forest. As I walk, I find quieter pools of grassy forests where the high trills and chirps of crickets replace the cicadas, a sound more soothing to human ears. The insects here have similar timbres and rhythms to those of Mount Scopus, although species diversity and abundance of individuals are higher in this lush American forest.

At the sulfurous, muddy junction of land and marsh water, boat-tailed grackles, birds whose plumage shimmers with purple-black iridescence, clamor from palms and oaks. Their sounds, which keep the flock united and convey news of predators and new food sources, are like electric buzzes overlain on a jangle of metal flywheels. Red-winged blackbirds, resting in the reedy water edges, puff out their crimson epaulets as they blast their territorial signal, *conk-a-ree*, a fist of a song accented by a sweet trill at the end. The elaborate combination of high jingles and guttural chatters of grackles and blackbirds is characteristic of the icterids, a family that comprises the American blackbirds, caciques, grackles, and cowbirds. The more than one hundred other species in this family make songs of great complexity, usually ornate juxtapositions of slides, whistles, and harsh cries.

From the low spreading branch of a live oak tree, its nest perhaps hidden in drooping veils of Spanish moss, a northern parula works its buzzy song up the frequency scale, then ends with a quick downslur. The bird belongs to a sibling family of the icterids, the parulids, or American warblers. This is perhaps the least fitting of all bird family names. The more than one hundred species of parulids give tight, energetic lisps and buzzes, often arranged in short, repeated phrases, but do not warble. More than thirty parulid species nest, winter, or pass through this island on migration. Their changing sounds are one of the primary acoustic markers here of the passing seasons: territorial songs in spring, followed by the gentle *chip* notes as they feed in migration.

A brown thrasher perched atop a young pine delivers a boisterous stream of inventions and mimicked snippets of the local soundscape, showing off to rivals and potential mates alike. Like its close kin, the mockingbird, the thrasher is a listener and an innovator, assembling a rapidly delivered collage. The taxonomic name for these birds, mimids, belies the sophistication of their craft—they do not mimic but instead sample, remix, and add novelties, a process more creative than simple repetition. A boisterous *wheep* from near an old woodpecker hole in a longleaf pine tree comes from a great-crested flycatcher, joined by the sneezy *pit-ZA!* of an Acadian flycatcher sitting in the low branches of the pine, both members of the tyrannid flycatcher family. Their simple, emphatic songs are characteristic of this diverse American bird family.

From a neighboring oak tree, an American robin warbles his singsong phrases, groups of four or five whistled notes. Two fish crows wing overhead, cawing at each other. Barn swallows chitter as they arrow and twist in their pursuit of flying insects. These sounds signal where productive food patches are located. A Carolina wren, skulking in the saw palmetto that grows knee high, sings a rolling *tea-keetle-tea-kettle*, answered by a scolding call from its mate, *tssk-tssk*. Unlike many other songbirds here, the wrens duet, presumably to maintain the pair bond, and sing all year, a bright tumble of notes.

This acoustic melee is distinctive of the humid forests of eastern North America. Many of these sounds give us a northern taste of the American tropics. Especially in forests away from the poison clouds of airplane-sprayed agricultural fields and the herbicided quiet of industrial tree plantations, the loudness is like that of the South or Central American rain forests. No temperate forest can rival the outrageous number of species in the tropics, but the summertime exultation of sound is just as forceful. The timbres and rhythms here make up sounds also found in Eurasia—cicadas, toads, tree frogs, thrushes, and wrens—but also include, especially for birds, voices unique to the continent. The short, tight songs of American flycatchers and warblers are avian minimalists, their energies and meanings compressed into repeated exclamations and phrases. The icterids are like experimental electronic musicians, pushing bird sounds into modulations of whizzes, buzzes, and clangs, instantly recognizable to naturalists as an acoustic signature of the Americas. To human ears, these sounds combine the wild jumps among frequencies and timbres of electronic music—Milton Babbitt's *Composition for Synthesizer* comes to mind, as do the repetitions and leaps of electronic dance music. The brown-headed cowbird, for example, sweeps up ten kilohertz in less than a second (about twice the range of a piano keyboard), a feat that takes the birds two years to learn. Other icterids, such as oropendolas, caciques, and grackles, make similar sweeps, mixed with harsh chatters or bell-like notes. Once mastered, the birds repeat these sounds tens of thousands of times over their lifetimes.

Crowdy Bay, New South Wales, Australia, is located 10,300 kilometers east of Mount Scopus and about the same distance west from Saint Catherines Island. The latitude is the same as that of Jerusalem and Saint Catherines but transposed to the south. I walk just after dawn through a mix of tall eucalypt forest and open heath, just inland from beaches on the Pacific Ocean. Although it is August, wintertime, I'm in shorts. The seasons here cycle between warm and hot. Rain, on average, falls year-round, with a peak in late summer, but droughts and deluges often interrupt this rhythm. The vegetation is evergreen, and most plants have leathery leaves, suited to summer heat, nutrient-poor soils, and unpredictable dry spells.

A family of four pied butcherbirds gathers in the limbs of a blackbutt eucalyptus. The contrast between their black hoods and wings and their white backs and bellies makes a strong visual mark against the tree's dark green leaves. One bird looses three slow notes of extraordinary richness, flowing gold, lit from within by warm light. The bird repeats, downslurring the highest note, then adds another pure, steady tone at the end. A companion takes up the fluting, answering with higher notes, also languid and clear. The two then sing in call and response, and a third joins, overlapping the pair with its repeated five-note, undulating melody. They continue for several minutes, the calls serving to keep the birds in constant sonic contact and, presumably, to communicate danger, the location of foods, and ever-shifting social dynamics within the group. Then the fourth gives a harsh call, like a human blowing over a blade of thick grass clamped between thumbs, and the group wings into the adjacent heath, disappearing into the shrubbery.

The rich tones of the pied butcherbird's song are gorgeous, and the song's tempo is mellow enough for human ears to catch every note and inflection. There is an open-ended quality to the melody as the birds seem to pass it around and respond to one another with twists and elaborations of their themes. My brain's aesthetic processes are aglow, maxed out by tonal quality, melodic creativity, and sound that speaks of a lively and intelligent web of relationships among the birds. For them, these sounds no doubt mediate their family lives and communicate to neighbors, as vocalizations do in other bird species worldwide. For my ears, the stunning sound is also a signature of this continent, its timbres and dynamics unlike anything I have experienced in the Americas, the Middle East, or Europe.

I walk on a sandy dirt road, away from the blackbutt eucalyptus and into the dense leathery-leafed *Banksia* shrubs of the heath. Here the sound of birds becomes less tonal but no less striking. A pair of little wattlebirds, colored like a chocolate cake decorated in streaks of white piping, creak like old hinges on a swinging gate. They intersperse goose-like honking amid these grating sounds, a raucous medley. A white-cheeked honeyeater flies into the shrub and the wattlebirds clatter their bills, perhaps as a threat. The honeyeater hops to an adjacent branch, at the crown of the shrub, and rips a series of *tew tew* sounds like blasts from a child's toy laser gun. Black and gold wings flash as it leaves.

A noisy friarbird lances in from behind me and lands with a fluster of wings in the same shrub. Its red eye blazes from a bare-skinned black head. The bird seems more interested in thrusting its dagger of a beak through the foliage than singing, but it chatters as it works, a stream of sounds jumping from shrieks to harsh grunts to resonant *ak* sounds. Four yellow-tailed black cockatoos fly over. They giggle as they pump their wings, then whine *wee-ar wee-ar*. On the path in front of me, a dainty willie wagtail prances after insects, flicking its tail sideways, singing with urgent low-to-high repeats, like rubbing a finger on clean, wet glass. The bird throws rattles into this high squeaking, like a string of camera shutter snap sounds.

My experience at Crowdy Bay is typical of the shrublands and forests of temperate eastern Australia. Leave your windows open here and you wake to the ethereal caroling of Australian magpies, followed as the sun hits the trees by the scolding bickering of some of the dozens of species of honey-eaters. Lorikeets and parrots turn the air into a thicket of grating, thorny sound, loud enough to drown human conversation. Figbirds, gorging on fruiting trees in flocks of dozens, yelp at one another then burst into rich whistles. On higher ground, in temperate rain forests, whipbirds duet, one bird holding a single tone absolutely steady for two seconds then ending in an ear-splitting slash from high to low frequency, instantly answered by its mate's sweet *chew chew*. Green catbirds sing with a strangled nasal waver, a sound like a mightily distressed cat or human baby.

Singing perhaps the world's most complex and richly timbred birdsong, the lyrebird both mimics other species and adds its own flutes, whistles, crackles, and trills, a performance that lasts sometimes for hours, delivered so loud that the voice carries for up to three kilometers. Olivier Messiaen, the French composer who spent decades listening and responding to the music in birdsong, wrote that the *nouveauté*, novelty or strangeness, of the lyrebirds' rhythms and timbre was *absolument stupéfiante*, absolutely astounding. Nothing he'd heard in Europe prepared him for this. The lyrebirds' sounds, along with honeyeaters' and butcherbirds', inspired passages in his last orchestral work, *Éclairs sur l'au-delà*, premiered by the New York Philharmonic in 1992, six months after the composer's death. The lyrebirds' song was striking enough to carry it, via France, to the stage of Lincoln Center.

In my walk at Crowdy Bay, I hear no frog sounds and only one species of crickets gently pulsing from deep under the shrubs. In summer, though, cicadas rival the loudest birds, joined by katydids and more crickets. From wetter parts of the forest, when rain gathers in depressions and ditches, eastern dwarf tree frogs and striped marsh frogs call. Compared with Mount Scopus and Saint Catherines Island, Crowdy Bay's insects have similar timbres and rhythms, instantly recognizable as the chirping of crickets or the harsh whine of cicadas. The frogs here, too, pluck, fibrillate, and pop with sounds reminiscent of those on other continents, but without the ear-splitting vigor of American choruses.

The energy and textures of the soundscape here are dominated by birds. A few species—the silvereyes and superb fairy wrens—give soft warbles and gentle trills, but these flow into an otherwise clamorous, muscular stream. Butcherbirds, magpies, honeyeaters, and others produce a sonic confluence of virtuosic leaps through rich harmonies and jarring, atonal pulses and eruptions. Angels are playing woodwinds alongside performances of *musique concrète* and industrial found sounds. *Absolument stupéfiante.*

The vigor and tonal diversity of Australia's birds struck many nineteenth-century colonists. Naturalist William Henry Harvey wrote in 1854 of "several chirpers, a few Whistlers, many screamers, Screechers, & yelpers, but no songsters." To human ears used to European sounds, Australian birds are "exotic," "disruptive," or "ugly" according to surveys of recent émigrés by anthropologist Andrew Whitehouse. Some people are impelled to return to Europe, unable to bear the cacophony of birds that "crash into your consciousness." These reactions are partly founded in our affinity for the sounds of our youth. Psychologist Eleanor Ratcliffe and her colleagues found that familiarity of timbre and melody are predictors of how restorative we find birdsong. Andrew Whitehouse's surveys found that Australians living in the United Kingdom hanker for the sounds of their former home, sometimes playing recordings to awaken aural memories. The power of bird sounds to forcefully evoke in us feelings of alienation or belonging is partly a reflection of how divergent the sounds of different continents can be. These feelings are also reminders that the sounds of other species are lodged deep within us, carried within our subconscious as aural compasses, orienting us toward home.

I t is perhaps an absurd overgeneralization to characterize and compare the sounds of entire regions or continents. Summaries belie inner complexity. After all, every habitat has many sonic variations and textures. Walk a kilometer or two through any forest and your ears will encounter variegations of tone and rhythm from the combined voices of sometimes hundreds of species. Yet alongside this fine-grained local texture, the voices of Earth also differ on a continental scale.

Some of this sonic diversity emerges from the varied physicality of the world. Earth has many forms of wind, mountain, rain, wave, beach, and river. Raindrops are larger in the Amazon than in North American skies. Northerly coastlines retain the mark of scouring glaciers, and their rocky headlands have more assertive voices than the sands and muds of unglaciated subtropical shores. Rivers meandering through continental interiors are slurred and languid compared with water coursing down mountain slopes. The geologic history of the world has created varied surfaces and flows for unvarying physical laws to play against.

Evolution adds two more creative forces to this global diversity of sound. The happenstances of history have populated different regions with varied branches of the tree of life. Each branch has its own stories of origin, migration, species diversification, and extinction. Combined, these stories yield diverse geographies of sounds. Overlain on this, every species experiences its own path of aesthetic innovation and sonic adaptation to place. Because these evolutionary paths are guided by forces that are often fickle and improvisational, the sounds of each species diverge in unpredictable ways. Over millions of years, divergences scale up to give whole regions different sonic characteristics. These processes contrast with those that shape the sounds of water, stone, and wind: A raindrop of a given size makes the same sound whether it lands on rock in America, Israel, or Australia. The songs of animals in these places, even species of very similar sizes and ecologies, cannot be deduced from physical law. History and the quirks of animal communication add delicious layers of contingency and caprice to life's voices.

On any place on Earth, we hear the voices of both indigenous and colonist animals. Some of this mix is recent—European starlings singing alongside American crows across much of North America, for example—but most stories of animal biogeography have deeper roots. When we look back tens or hundreds of millions of years, we find that the modern distribution of every group of animals results from some species cleaving to home and others striking out for new land. A few of each type then split into new species, producing a rich tangle of geography and taxonomy.

The oldest singing animals, the crickets and their now-extinct kin, evolved on the supercontinent Pangaea. It is not surprising, then, that the sounds of crickets today are so similar among continents. Each place inherited crickets from a singular landmass that then split. But crickets are hardy too, and can withstand ocean journeys on floating vegetation. Some of the unity we hear is the result of more recent dispersal. The familiar chirpers of fields, gardens, and parks—the gryllina "subtribe" of crickets—are found on every continent except Antarctica and have colonized many oceanic islands.

A similar pattern of ancient unity and more recent colonization accounts for the distribution of other singing insects. Katydids or bush crickets likely originated on the southern supercontinent Gondwana, one of the landmasses formed when Pangaea broke apart. They then repeatedly jumped among landmasses, producing a family tree with close cousins on different continents. The marbled bush cricket that I heard in Jerusalem belongs to a clan that invaded the temperate regions of Europe and then North America from Australia. The common katydid that pounds the night air on Saint Catherines Island belongs to a different branch of the family tree, one that colonized the Americas from Africa. Cicadas also have a global distribution, their present form dating back at least to the time when Pangaea broke up. Since then, they have repeatedly jumped among continents, with close kin on widely separated landmasses. The periodical cicadas of North America, for example, are taxonomic cousins to some Australian cicadas.

The ancestry of most living frog species is also rooted in Gondwana. There, two main branches formed. One, on land that would become Africa after Gondwana split apart, led to the pond frogs, Australasian tree frogs, and narrowmouth toads. The other, South America, gave rise to all American and European tree frogs, toads, and Australian ground frogs. To this day, South America and Africa are home to the majority of frog taxonomic families. Away from these centers of origin, we mostly hear the few families that managed to cross oceans and colonize new lands. How these frogs crossed the ancient oceans is not known, but the small number of families that made it across—about 10 percent of South American and African taxonomic diversity—suggests that rafting across salty water was a rare event.

The original homeland of the songbirds is the Australo-Pacific, an area now divided into Australia, New Guinea, New Zealand, and eastern Indonesian islands. An ancestral group of birds split into two in this region about fifty-five million years ago. One descendant lineage led to the modern parrots and the other to modern songbirds. Both groups are highly vocal and comprise species with well-developed vocal learning and culture. Combined, these two branches of the bird family tree comprise more than half of the nearly ten thousand living bird species. In many soundscapes, they are the dominant singers alongside the insects.

The extraordinary sounds that I heard at Crowdy Bay, then, are rooted in the evolutionary homeland of songbirds. Cockatoos and parrots, common birds all over Australia, have lived here since their ancestors split from the songbirds. Butcherbirds, Australian magpies, and willie wagtails all also belong to deep branches of the Australo-Pacific songbird family tree, kin to ancestors that left the region and evolved into modern crows. The lyrebird stem of the family tree dates back nearly thirty million years, and its complex song is evidence that ancestral songbirds were accomplished singers. Wattlebirds, friarbirds, and honeyeaters belong to another deep branch, one whose descendants live only in the Australo-Pacific and are now among the noisiest and most diverse birds in the region.

In genealogical terms, the songbirds elsewhere in the world are a subset of this diverse array of Australo-Pacific birds. The sounds we hear outside of this region are elaborations of the legacies of small groups of emigrants, dispersing birds whose descendants produced marvelously diverse soundscapes across the world. But to my ears, no continent's songbirds have quite the range of timbre, rhythmic pattern, and vigor as those of the Australo-Pacific.

Songbird emigration from the Australo-Pacific happened repeatedly, but two waves stand out in their long-lasting effect on the distribution of birds around the world. The first wave populated Asia and then the Americas but left no living descendants in the Middle East and Europe. On Saint Catherines Island, the great-crested flycatcher and Acadian flycatchers are among this number. The second wave founded a lineage that now comprises more than half of all living songbird species. Most of the familiar voices of songbirds across Asia, Africa, Europe, and the Middle East belong to this group of emigrants: thrushes, larks, swallows, finches, weavers, Eurasian and African sparrows, starlings, and "Old World" warblers and flycatchers. A few of these bird families also came to the Americas. But American soundscapes owe much of their character to the flourishing of just one offshoot of this second wave. The American blackbirds, warblers, tanagers, sparrows, and cardinal-grosbeaks are all descendants of this clan.

This view of Australia as the crucible and exporter to the world of songbird diversity, grounded in the latest analyses of bird DNA, upends some traditional views about evolution. Biologists long assumed that Australian animals and plants originally came from Asia, side branches of a story that they believed was firmly rooted in the Eurasian landmass. As Australian biologist and writer Tim Low pithily states in his groundbreaking exploration of Australia's birds, *Where Song Began*, nineteenth- and twentieth-century

biologists, including luminaries such as Charles Darwin and Ernst Mayr, believed in "a version of terra nullius, an empty land filling with good things
from the north." This colonial view of biogeography is still lodged in taxonomic language: "Old World," "New World," "Oriental," and "Antipo-
dean," as if geologic time and the tree of life were rooted in northern Europe.

Songbirds are not the only voices in the avian soundscape. The combative
chirping and frenzied wing whirs of hummingbirds are unique to the
Americas. But thirty million years ago, hummingbirds were European, as
attested by fossils from Germany. This ancestral stock then colonized
South America. The European hummingbirds went extinct, but in South
America the birds found a congenial home and, in partnership with flowering plants, rapidly diversified.

A taste for sugar may have stimulated the evolutionary flourishing of both
hummingbirds and songbirds. Genetic changes to taste receptors happened
early in the evolution of both groups, repurposing an umami receptor to
taste sugar. This new taste for sweetness allowed the birds to seek out and
benefit from flower nectar and the sugary exudations of sap-feeding insects.
Just as the origin of flowering plants forever changed the sounds of Earth by
giving a boost to the diversity of many singing insects and other animals,
the diversity of birdsong is partly founded on a link between birds and
Australo-Pacific plant sugars. Songbirds, hummingbirds, and parrots are
all highly vocal, and many of them learn their songs and have vocal cultures. In the rich voices of birds, we hear the sweet gifts of flowers and sap.

In every one of the ancient dispersal events from the Australo-Pacific, the
arrival of a small band of ancestral birds seeded a later flourishing. As we
listen, we hear the legacy of chance events from millions of years ago. Had a
different group of birds been blown to Asia from the north coast of New
Guinea or wandered across the Bering land bridge to the Americas, bird
soundscapes would have a very different geographic structure. Overlain on
these quirks and accidents of history are millions of years of speciation and
adaptation in each of the descendant populations. Each species experienced

its own story of sexual elaboration and environmental adaptation. Combined, these are the stories of evolution's creative manufactory of sonic diversity.

The stories of dispersal and kinship come from analyses of the DNA of modern species, supplemented by information from fossils. These studies also reveal something about human senses and affinities. We have approximately one hundred times more genetic information available from birds than from insects, and so reconstructions of the avian past have wider and sturdier foundations than those of insects. Insects do not lack DNA. What is missing is research funding and scientific attention.

Birds are popular subjects of scientific study partly because they catch our eye. Bird colors entrance us, and their bodies are large enough for the sight of them to evoke the human imagination. Icarus flew on wings of feathers, not of insect exoskeleton. The Christian Holy Spirit descends as a dove, not a cicada. Birdsong more closely approximates the frequencies, timbres, and tempi of human speech and music, further linking them into our senses and thus our aesthetic affinities. Were insects as mellifluous and colorful as birds, we'd devote more attention to their study.

Just as the breeding displays of animals often plug into the preexisting sensory bias of their mates, our fondness for birds reveals our sensory biases, born in the ecology of our primate lineage—a fondness for red to see ripe fruit and healthy flushed skin, a love of elegant motion to judge another's vitality, and ears eager to hear the information carried in human sound. Birds' prominent roles as poetic, religious, and national symbols are a product of these particular tunings of human eyes and ears. If we communicated by ultrasound, as rats do, or by scent, as many salamanders do, we'd have rodents and newts on our coinage and in our sacred texts. Our sensory proclivities also spell doom for many bird species. One in five vertebrate animal species are captured and traded worldwide. Species with feathers and songs that please the human eye are especially popular. A few insect species are captured and kept, especially crickets in parts of Asia, but wildlife trade is an insignificant threat to most, unlike bird species whose evolutionary path led them to the unhappy end of being attractive to humans. Yet alongside peril is the power to provoke change. Human aesthetic responses prompt moral concern. *A Robin Red breast in a Cage / Puts all Heaven in a Rage*. Our senses lead to desires for both consumptive possession and protective care. Perhaps by appreciating the origins and fragility of the marvels that delight us we might tip our desires and actions toward the conservation of wild beauty?

The sounds of Mount Scopus, Saint Catherines, and Crowdy Bay seem so ephemeral and light, dissipating as soon as they are made. Despite being fleeting, they are also layered records of history. Every voice carries the imprint of its clan's origin and dispersal. A soundscape is therefore an accretion built over hundreds of millions of years. As I listen, I am often caught up in the moment-by-moment melodies and tonal layers of soundscapes: cadences of whistling birds and textures of insect sound, the varied pulses and timbres of species playing against one other, or the antiphonal calling of rivals or mated pairs. Alongside these delights of the instant is an invitation to hear the stories of evolution's past. These legacies of animal movement and plate tectonics are often older than the ground under my feet. Saint Catherines Island is made of Pleistocene sand and more recent dune deposits, none older than fifty thousand years. The sandy soils of Crowdy Bay are as young as those on Saint Catherines Island, underlaid by two-hundred-million-year-old lava. The limestone of Mount Scopus is an uplifted seafloor, the remains of salty ooze sixty-five million years old. The sounds atop these soils and stones are often tens or hundreds of millions of years older.

Sound, made of breath and gone in an instant, can be older than stone.

Listening to the animal voices around us, we hear the legacy of a sonic geology made of vibrations in air, diversified by plate tectonics and the ancient movements of animals across continents. Unlike stone, no durable physical substance carries sound's many shapes through time. Instead, the form of animal sound has traveled in fragile strands of DNA, remade every generation and, in species that learn their songs, through an unbroken chain of connections between youngsters and their elders.

PART IV

Human Music
and Belonging

Bone, Ivory, Breath

Forty thousand years ago, in ice age caves in what is now southern Germany, a new kind of sound was born. This sound was simple, just a string of whistled notes, seemingly unremarkable compared with the complexity and range of birds and insects that sang outside the caves. Yet the sound was revolutionary. In the moment of its creation, Earth's generative powers leaped forward, powered by cultural evolution.

Listen: primate lips blow into shaped bird bones and mammoth tusks. A chimera emerges. Hunter's breath animates the skeletons of prey. The air vibrates with melodies and timbres from a source previously unknown anywhere on Earth: musical instruments.

Time has honeyed the whiteness of bone and ivory. Millennia spent buried in dust and rubble have imbued a stain the color of pinewood. In a dark room, resting on black cloth in glass cases, the objects glow under gentle spotlights. I'm in the Blaubeuren Museum of Prehistory in southern Germany, gazing at flutes crafted nearly forty thousand years ago from bird wing bones and mammoth tusks.

The flutes' seeming fragility astonishes me. In preparation for this visit I've pored over technical papers and studied photographs. On paper, the objects look substantial, like sturdy bones familiar from a dinner plate or zoology lab. In their presence, though, I'm confounded by how old and delicate they appear. Their timeworn hues, papery-thin walls, and tiny fracture lines teach my senses the meaning of great antiquity. My body and emotions finally understand what my mind has tried to grasp.

I'm in the presence of our species' deep cultural roots. These objects are the first known physical evidence of human instrumental music. They are three times older than human agriculture. Two hundred and forty times older than the age of oil wells and gasoline. No other species makes musical instruments, although a few come close. Some tree crickets cut holes in leaves to amplify the trill of their wings, and mole crickets shape burrows to act as trumpets. The insects are amplifying existing voices, not creating new ones. Orangutans press leaves to their mouths to make kissing sounds. Palm cockatoos use trimmed seedpods or sticks to drum on hollow tree limbs, a singular example of nonhuman tool use.

A griffon vulture wing bone: at one end, V-notches cut into the bone, like modern end-blown bamboo or wood flutes. Along the convex side of the bone's gentle curve are four holes. Part of a fifth hole is visible in the broken, unnotched end. The holes are spaced so that the fingers of two human hands would easily rest against their openings. Each hole is beveled, and the precise knife marks left by a stone tool are still visible in each depression. The beveling creates a dimple exactly the size of a human fingertip. Every cut speaks of intent. This is a bone sculpted to fit the human hand and mouth.

The maker used the bird's radius bone, the slimmer of the two forearm bones of the vulture, and so the flute is thin as a twig, only eight millimeters across. But it is nearly as long as the top of my forearm. Griffon vultures spend their days soaring in search of carrion and have huge wingspans, wider than eagles', making their wing bones an excellent source of long tubes for Paleolithic flute makers.

Fine fracture lines divide the bone's smooth surface into a dozen pieces. These fragments were recovered from cave deposits, then reassembled and interpreted by University of Tübingen archaeologists Nicholas Conard, Maria Malina, Susanne Münzel, and their colleagues. A gash on the right side of the flute, just above a finger hole, speaks of the ephemerality of this thin-walled bone and how improbable was the flute's journey from the Paleolithic to our modern world.

This is one of four bird-bone flutes from the caves of this region, all recovered from deposits dating to the early Aurignacian, a time period immediately after the first arrival of anatomically modern humans in what we now call Western Europe. Two of the other flutes are known only from small finger-hole-bearing pieces. The third is made from a swan radius, incomplete but with three clear finger holes, reconstructed from twenty-three fragments.

Here in the Blaubeuren Museum, adjacent to the griffon vulture flute, sits a flute of a stouter design. It has three beveled finger holes on the concave side of its curvature. One end seems deliberately notched into a deep U. A splinter extends down from the third hole, suggesting that the flute was originally longer. Unlike on the bird bone, two seams run down the length of this flute. Each seam is crossed by repeated short lines, like suture marks on a long incision.

This flute is made from mammoth ivory, a material unfamiliar to my modern eyes. The griffon vulture radius is easily recognizable as a bird bone, a giant version of chicken and turkey bones. Mammoth ivory, though, has no everyday contemporary analog. Its surface has the patina of well-worn leather, an illusion emphasized by thin walls of the flute that look like tanned animal hide. The finger holes and ends, though, appear cut into solid bone. The object seems exotic to me, but for Paleolithic people the mammoth was a staple of both diet and craft. Their caves are littered with mammoth ivory and bone: tools, ornaments, cooked bones, and partly worked tusk pieces. Mammoth ivory was multifunctional, and judging from the remains left in caves, often discarded or abandoned. The plastic of the Paleolithic, perhaps, but locally sourced from free-range animals.

Bird bones are hollow and fit readily in a human hand, good matches for flute making. But a mammoth tusk is solid and hard to carve. Whoever made the mammoth-ivory flute spent days at the task.

Close study of cut marks on the flute and experiments by modern archaeologists and reconstruction experts suggest the manufacturing sequence used by the ice age craftspeople. First, they used sharp stone cutters to excise a portion of a large tusk, making a stave or blank. Thousands of tool remnants in the caves show that they also used this technique to carve blanks for hunting projectiles from reindeer antlers. Ivory is not easily turned into a tube, and the artisans lacked drills. So they pared the stave into a cylinder, split it lengthwise, then scooped out each half before reassembling the whole, now as a tube. To do this, they exploited the growth form of the ivory. Mammoth tusks have an outer layer, cementum, around a thicker inner core, dentine. By carefully carving the stave from the junction between these layers, the makers crafted a stave that was half cementum and half dentine. The junction was a weak spot that could be eased apart with blades and small wedges, bisecting the cylinder along its long axis. The hollowing of two halves took commitment and, judging from the result, great skill in creating two thin-walled half tubes from a solid column.

Before splitting the ivory, they cut regular deep grooves down the two sides, perpendicular to the axis of the column. These marks guided the flute's reassembly once the halves had been hollowed out. Tree resin and animal sinew likely held the pieces together. The result was an airtight fit, ready for the addition of beveled finger holes and a notched end for human breath.

Even after breakage and burial for forty millennia, the flute is an impressively precise construction, its halves fitting snugly, notches aligned. Its thin walls give the illusion of coming from a natural tube, like a bird bone, belying the labor that went into its production. The flute on display here is the most complete of the four mammoth-ivory flutes so far unearthed from the region. Tool marks on the fragmentary remains of the others indicate similar methods of construction.

Life was undoubtedly hard for the makers of these first-known instruments. They lived just north of the glacier-smothered Alps and south of the ice that covered the north of Europe. Animal remains from that time are creatures of the tundra, cold steppes, and mountains: woolly rhinoceros, wild horse, ibex, marmot, arctic fox, arctic hare, and lemmings. Pollen and remains of wood in caves show that vegetation was mostly grasses, sagebrush, and a few boreal shrubs and trees. Every bite of food, stick of fuel, and piece of clothing had to be wrested from a landscape often snowbound and always cold. Yet these people devoted the highest forms of their technologies to making music. The flutes, the mammoth flute in particular, emerged from the application of the most sophisticated craft possible at the time. Their work evinced deep understanding of material properties and skillful use of tools. Soundless, solid animal tusks were transformed by human hands and imagination into hollow, multipitched wind instruments. Precisely wielded stone tools carved voids, spaces where human breath could enter and reanimate the dead.

Musical instruments, then, did not originate as ornaments for well-off aesthetes whose material needs had been met. Instead, people living arduous and undoubtedly insecure lives gave the world the first known instrumental music. When our modern schools cut music programs, polemicists from the left and right argue that art is decadent or an excess to be trimmed, and academics dismiss music as fundamentally unnecessary to human culture, they might look back to finely crafted flutes from ice age caves and reconsider.

I sit with the flutes in the museum room for a few hours. Twenty people pass through. Three look at the flutes. The others hurry straight to the wall of buttons, each one provoking from a loudspeaker a brief melody from a reconstructed flute. To my consternation, the objects themselves elicit little visible wonder or interest.

To be fair, the flutes have competition. The museum is also home to exquisite carved figurines. Wild horses with nostrils flaring, birds diving with wings folded, lion people standing erect, dozens more, all evoked by hands that knew how to imbue thumb-sized pieces of tooth or bone with the living presence of animals. Instrumental music was not the only human art preserved in these caves. The patient brushes and delicate probes of archaeologists have excavated dozens of carved animals and human-lion hybrid forms. The cave sediments contain bodily ornaments too: ivory and antler pendants and beads. The inhabitants of these caves were creative people, transforming everyday bone and ivory into what we now call art.

The most famous of the sculptures is just down the museum hallway from the flute displays. It has its own room, a dark space with one illuminated object at its center. Every visitor here has likely seen its photograph in news reports or museum videos, posters, and websites. No wonder visitors tend to hurry past the flutes. We're in a museum whose narrative builds toward a hallowed object.

On a plinth stands an impressively plump female human figure. Instead of a head, the ivory carving has a small ring, delicately worked. This ring presumably received a cord, and the palm-sized figurine, six centimeters tall, served as a pendant or amulet. Polish from such a cord is still visible within the eye of the ring. The figure's limbs are short and part of the left arm is missing. Breasts, buttocks, and vulva are swollen and slightly lopsided. The waist is pinched and the belly flat. The hands are finely rendered, resting above her hips. Incised lines run across the figure, perhaps suggesting a wrap or other covering, although sculptures from this era of nonhuman animals are often also decorated with similar surface markings.

The object is named in the museum and the technical literature as a Venus, like the figurines from other caves, such as the famous Venus of Willendorf unearthed in 1908. These other Venus Paleolithic female figures are at least five thousand years younger, and so the connection to the one here in this museum is distant at best. To modern eyes, the figure appears to emphasize sexuality. But the meaning for Paleolithic peoples is unknown. Religion, protest, porn, humor, selfie, game piece, toy, portrait, artisanal training exercise, supplication, or gift? We have insufficient context with which to judge. Projecting the name of a two-thousand-year-old Roman god, Venus, back nearly forty millennia reveals more about our culture than it does about the intentions of the ancients.

People gather in the darkness around the illuminated figurine. This carved mammoth ivory is the oldest known figurative sculpture in the world. Until the discovery in 2019 of a nearly forty-four-thousand-year-old cave painting in Sulawesi, an Indonesian island east of Borneo, the figurine was the oldest known figurative art of any kind.

In the cave, the figurine was buried three meters below the present-day surface. It lay an easy arm's reach from the griffon vulture flute, in the same layer of cave sediment. In archaeology, layers of sediment are records of passing time, each passing century adding its film of dust and detritus. The laminations of dust tell us that the flute and the figurine were contemporaries.

How old are the flutes? Carbon dating suggests at least thirty-five thousand years old for the griffon vulture flute and the more fragmented mammoth flutes. The more complete mammoth flute and the swan radius may be as old as thirty-nine thousand. The lowest layer containing the debris of human settlement is just over forty-two thousand years old. These dates are confirmed by both the radioactive decay of carbon and time-sensitive changes in crystals trapped inside buried animal teeth. New techniques may, in future, further refine the dates. Likely these German caves were not the only places in which instrumental music loosed its early notes to the world. Any instruments made of wood or reed decayed into forgotten oblivion long ago. Or they wait buried in places yet to be excavated. For now, though, these German caves yield the earliest physical evidence.

Human music is older than any instrument. Our voices surely played with melody, harmony, and rhythms long before we carved any tusk or bone. People in all contemporary human societies sing, play music, and dance. This universality suggests that our ancestors, too, were musical beings, long before some of them invented musical instruments. Today, across known human cultures, music emerges in similar contexts: love, lullaby, healing, and dance. For humans, then, social behavior is often mediated by music.

Fossil evidence also shows that ancestors five hundred thousand years ago possessed hyoid bones that would enable modern-sounding speech and song. Human throats thus had the capacity to utter spoken and sung words hundreds of thousands of years before we manufactured musical instruments.

Whether speech or music came first is, at present, unknowable. The neurological prerequisites for the perception of both speech and music are present in other species, suggesting that our linguistic and musical capacities were elaborations of preexisting qualities. Like humans listening to the spoken word, other mammals process the sounds of their own species mostly in the left hemisphere of the brain. Other sounds go to the right, the primary locus of human musical processing, or are shared between the two hemispheres. The left brain uses subtle differences in the timing of sounds to understand semantics and syntax. The right brain uses differences in frequency spectra to grasp melodic and timbral content. But this division is not absolute, suggesting that no firm line separates speech and music. The intonations and prosody of language activate the right, but the semantic content of sung music lights up the left. Sung music and poetical language, then, braid the operations of our two hemispheres. We hear this in the form of music across human cultures: all incorporate words into song, and the meanings of all spoken languages emerge partly from their musical qualities. As babies, we recognize our mother by the pacing and pitch contours of her voice. As adults, we communicate emotion and meaning through changes in pitch, timing, vigor, timbre, and tone. As cultures, we pass down our most precious knowledge through a union of music and language: Australian song lines; Middle Eastern and European cantillations, hymns, and psalms; the San's "calling narratives" during trance dancing; and the many manifestations of chant in societies worldwide.

Instrumental music, then, has a special quality that separates it from both song and spoken language. It is a form of music entirely free of language. The first flute makers, perhaps, discovered how to make music that transcends the particularity of words. In this they were possibly finding kinship with nonhuman animal species—for insects, birds, frogs, and others, sonic expression, of course, exists outside the framework of human language, although each species may have its own forms of grammar and syntax. If instrumental music does allow us to experience sound in ways analogous to the experiences of nonhuman animals, it is a paradoxical experience. Through tool use—manufacturing musical instruments, a recent and uniquely human activity—we experience sound as animal kin may still do, and prehuman ancestors surely also did, as a sonic experience full of meaning and nuance beyond and before human words. Instrumental music perhaps returns our senses to an experience that predates tools and language.

Percussive forms of music, too, may be older than speech or song. Given that drumming often uses fragile and rapidly decaying everyday objects such as pieces of skin or wood, the archaeological evidence is scant. The earliest known drums are only six thousand years old, from China, but it

seems likely that human drumming is much older. In Africa, wild chimpanzees, bonobos, and gorillas all use drumming as a social signal. These ape
cousins use hands, feet, and stones to beat against other body parts, the
ground, or tree buttresses. This suggests that our ancestors, too, may have
been drummers, perhaps communicating identity and territoriality, as well
as drawing social groups into cooperative, rhythmic unison. Compared
with other great apes, human drumming has a more regular and precise
beat. Intriguingly, for many chimpanzee populations, beating rocks against
trees has a ritual component. The chimpanzees focus their efforts on particular trees, resulting in an accumulation of stones at each site. Chimpanzees don't merely deposit the stones; they toss or hurl them to produce a
boom or clatter from the tree. Often, as they thump the stone against the
tree, they also give the loud pant hoot vocalization and bang on the tree
trunk with their hands and feet. Both chimpanzees and humans, then, unite
percussive sounds, vocalization, social display, and ritual. This suggests
that these elements of human music existed before the origin of our species.

The exact timing of the growth of the deepest roots of human music is,
for now, a mystery. But the connection between instrumental music and
other forms of art is clearer. The world's oldest known musical instruments
are entombed right next to the oldest known figurative sculpture. Both come
from almost the lowest layer of human deposits in the caves. Under them lie
layers of sediment devoid of human presence, then, deeper, Neanderthal
tools. In this part of the world, instrumental music and figurative art emerged
together, as anatomically modern humans first arrived in the icy landscape
of Europe.

Musical instruments share with figurative sculpture the idea that three-
dimensional modification of materials can yield mobile objects that stimulate
our senses, minds, and emotions, what we now call experiences of art. The
juxtaposition of the flute and the figurine suggests that, in the Aurignacian,
human creativity was not channeled into one activity or function. Artisanal
skill, musical innovation, and representational art were connected.

Evidence of such linked forms of creativity also comes from the very earliest human art. The first known drawing is abstract, not figurative. It comes from layers seventy-three thousand years old in Blombos Cave, South Africa. There, someone used an ochre crayon to draw a cross-hatched pattern onto crumbly stone. This drawing comes from a level that also contains other evidence of creative work: shell beads, bone awls and spear points, and engraved pieces of ochre.

So far, though, the record shows that the craft of three-dimensional art objects in southern Germany may have developed at a different pace than figurative art that uses pigment. The flutes and figurines show no evidence of being specially colored. The caves in which they were found are unadorned by wall paintings. In this region it is only much later, in the Magdalenian, twenty thousand years after these flutes, that there is strong evidence of stone decoration with ochre pigments. Another European Aurignacian site, the Cave of El Castillo in northern Spain, shows a different trajectory. A wall painting of a disk dates to more than forty thousand years ago and, on the same wall, a hand stencil is more than thirty-seven thousand years old. But as yet, there are no three-dimensional artworks known of this age from this region. Likewise, the figurative paintings on cave walls in Sulawesi are not associated with any known sculptures. These differences may tell us more about the incompleteness of the archaeological record than they do about the story of human art. But for now, three-dimensional artworks—sculptures and flutes—seem to have first developed in different places and times than paintings.

This deep history reframes our experience of more recent art. Gazing at the Paleolithic flutes and figurines, I think of the crowds at the British Museum, the Metropolitan Museum of Art, and the Louvre. We stand in line sometimes for hours to glimpse important moments in human art and culture. But here in a small museum in rural Germany we experience art's deeper roots.

I stretch out my arms. If this span were the extent of known human musical and figurative art, the ice age flutes and carvings would sit on my left fingertips, joined by the Sulawesi cave paintings. Most of the canonical pieces of art in major museums sit on the extended fingers of my right hand, products of the last millennium. In no way does this diminish the importance of the artworks of the last few centuries. Instead, the field sites and museums that record early human artistic flourishing complement these more recent works and root the story of human creativity. Art was born in relationship with the animals and physical spaces of each region, elevated by Paleolithic human technological prowess and imagination.

I take two vulture bones in hand. I intend to make flutes patterned on the proportions of the ancient griffon vulture flute. My bones' original owner was a North American turkey vulture, killed on a road. Its salvaged body became part of the zoology collections at the University of the South in Sewanee, Tennessee. For Aurignacian artisans, the griffon vulture bone was likely easy to find. These birds scavenge hunters' kills and nest near caves. Their bones are commonly found in cave deposits. But not so for swans. Their bones were specially procured, perhaps from wetlands far from the caves.

In the lab, I pluck two forearm bones, the radius and ulna, from the turkey vulture's cardboard ossuary. They're shorter by a third than the bones of the wide-winged griffon vulture, but have about the same shape and proportions, twice as long as my thumb and thinner than a pencil.

After an overnight soak in warm water—the bone had been stored in a dry room for a decade—I grasp the radius and bear down with a crude flint knife, sawing in an attempt to cut the bone's head from its shaft. I made my small stone tool by bashing a hard cobblestone onto a nodule of chert, breaking away a flake. The result is very sharp, but this edge is of little use in my unskilled hands. My efforts yield little more than blurred scratch marks on the bone's surface. Bird bone is surprisingly hard and its surface is slick. My blade slips around, even when I hold it steady with my thumbnail.

I feel embarrassed to be a descendant, as we all are, of masterful stone-workers, yet unable to complete the simple task of lopping the end from a bird bone. My clumsiness with an unfamiliar tool is one cause. The other is the unrefined nature of my toolmaking. The cave deposits in which the flutes were found contain hundreds of stone, antler, and bone tools: daggers, scrapers, awls, scalpel-like bladelets, chisels, knives, borers, and burins. These tools were made with precision and, judging from the artwork they created, wielded with great skill. An hour or two fumbling with my primitive flake teaches me how sophisticated was their craft and rude are my attempts.

I give up and resort to a more familiar tool, the blade of a modern coping saw. With steel teeth born of mines and smelters, I cut into the bone. First one end then the other, slicing off the bulbous ends that connected elbow to shoulder. The bone is surprisingly tough. I have to press down hard on the saw blade to make the incision. Shorn of its bulky heads, the bone immediately feels different in my hand. It's lighter and pleasantly balanced. No longer dominated by the heavy, knobby ends, its weight rests evenly throughout, easy to turn, inviting my hands to explore.

The bone absorbs heat from my fingers and takes on a mild, welcoming glow. I feel paradoxical animacy in this eagerness to absorb and emit warmth in the remains of a dead vulture. The surface is smooth but variably so. There is a slight roughness on one face, like a sprinkling of dusty sand. Some fine ridges run lengthwise. One of these diverges into two, creating a facet. The bone speaks readily to my hands, quickly revealing details that my eye passes over. The most delightful feature is the curvature, a suggestion of an *S*, more bowed at the elbow end than the wrist. The two ends differ in cross section. An irregular pentagon at the elbow end, a clean *D* at the wrist.

My hands twirl and stroke. They interlace the bone between fingers and squeeze gently, then harder. A springy yield but no hint of brittleness. I rest the bone on my palm and bob up and down, feeling its slightness as a surprising absence. Hands beckon my mind into the vulture's flight. We're both creatures of bone and muscle, possessing bodily understanding of what it means to move, to exert force on the ground and air. This kinship is the common language that my hands understand. But what they learn is shockingly alien. The impossibly light bone startles my earthbound mammalian body. This is what it takes to fly, my hands exclaim, this awesome weightless strength. Reliving in memory and recounting the experience later, I recoil, not trusting ecstatic claims to knowledge coming from mere hands. The seat of the mind is up here, in the cranium, I insist. But I cross the room and open the vulture box. There the bones lie, and, yes, I exult in holding them once again. My hands are given another taste of how the air lovers fly.

No exultation, though, when I lift the bone to my lips.

At first, all I get is the coarse whoosh of a stream of air hitting an obstacle, like blowing on the end of a pencil. I play with the angle of the cut bone end against my pursed lips, seeking that sweet spot where flowing air finds the flute edge and resolves into a clear sound. The turkey-vulture bone is frustratingly thin, skinnier than a drinking straw, and my lips feel like clumsy pillows against its narrow end. All I get is breathy noise. Not exactly a moving evocation of the dawn of instrumental music.

The next day I try again and hit the spot. A wheezy, high-pitched whistle. A sharp sound, focused and insistent.

I've also prepared a second flute, this one made from the ulna of the turkey vulture. It is the same length but twice as wide, almost as fat as my index finger. Ten bone nodules run along one side, the attachments for some of the vulture's wing feathers. This bone feels better on my lip, and I quickly find a tone. With a strong puff from my mouth, a loud, single pitch flows out. It's high and as I play around, I discover another, slightly lower one that pops out with a gentler breath, but this is a slippery note, hard to catch and hold. These two sounds are pitched like the higher octaves of a modern flute. There are no low mellow sounds.

We should expect as much. Flutes work by enclosing within themselves a seemingly paradoxical phenomenon, a stationary wave. This air pressure wave inside a flute is like an ocean wave frozen in time, one that transmits to the rest of the ocean the form of its crests and troughs. In the flute, the crests and troughs are air molecules that oscillate at the flute's ends, but they are unmoving in the center of the flute's bore, a still point where pressure flowing from each end is exactly balanced. As long as the player keeps blowing, the wave holds steady. The pulsing air molecules at the end of the tube push onto those outside, sending sound into the world. The length and thus frequency of the enclosed sound wave are determined by the length of the flute. Stubby flutes like my turkey-vulture bones create short waves that we hear as high-pitched notes.

Each flute is therefore a vessel that captures and holds what is normally fleeting, the human breath and sound waves in air. Breath is understood in many cultures as the foundation of life. The first discovery of the flute's properties must have been stunning: Spirit briefly held, shaped, and sent into the world. In this age before machines, likely the cave-enclosed flute was also one of the loudest sounds the Aurignacians heard, awesome in its power.

My turkey-vulture bone flutes are about the length of a short pen, just thirteen centimeters. A Western concert flute is five times longer, a piccolo more than twice as long. When I plugged these dimensions in the relevant equations, the lowest sound coming from my flutes should be about 1,200 hertz. The lowest note on the Western concert flute is 262, middle C. The turkey-vulture flute has a shrill voice.

Wind instruments, though, do not conform to simple predictive equations, especially not equations that treat them as mere tubes. The swirling, pulsing flow of air is shaped by the details of the instrument's form and how it is played. The angle and sharpness of the edge that meets our breath alter the crispness and pitch of the sound. Flare at either end of the flute, curvature within the bore, or interior imperfections can choke, squeeze, or expand the sound waves within. The keenness of finger-hole edges and the placement of the holes themselves rework the sound. The player brings the shape and skill of their body into relationship with the instrument. End- and side-blown flutes have no fipple to direct the air flow from mouth to instrument, as do penny whistles and recorders. Instead, the player uses lips, tongue, facial muscles, and teeth not only to precisely direct a fine stream of air to the flute's edge but also to sculpt the sound with subtle oral changes. This embouchure interacts with the rhythms and forcefulness of the player's lungs and diaphragm to create music. If flutes were the simple tubes described in elementary physics textbooks, musicians would not need to spend years working on their craft.

I'm no flutist. I bring unschooled embouchure and breath to the bone edge of the flutes that I've made. What would a professional make of the Paleozoic instruments?

Writing about what drew her to work with replicas of ancient flutes, Anna Friederike Potengowski says she felt a bit lost with her work in contemporary music. She sought an experience of roots, of beginnings. With bone and ivory replicas made by Friedrich Seeberger and Wulf Hein, experts in Paleolithic reconstruction, she set out to explore the sonic possibilities of Paleozoic bone and ivory. Seeberger's and Hein's artisanal and research efforts informed much of what we know about how the flutes were made. Potengowski took this experimentation into the sonic realm.

I slide headphones over my ears and enter a space of sonic imagination. We cannot know for sure how the ancient flutes sounded, but these recordings open our senses to possibility. Sound works its power, carrying ideas and emotions from one consciousness to another. Potengowski's playing is not time travel but rather offers experimental connections across the divide that separates us from ancient people. All of the dozens of her sound samples and compositions are modern imaginings, but a few surely capture the edges of musical innovations from long ago.

The artifacts do not disclose to the eye how they were played. But experienced mouths, facial muscles, and lungs can teach us what the eyes cannot discover. Two methods of playing seemed possible to Potengowski. In the first, she blew a tight stream of air from closely pursed lips across the top of the cut bone, almost whistling across its end. To direct the air without lips getting in the way, she held the body of the flute at an oblique angle, somewhat like a Middle Eastern ney flute. The second method worked only on notched flutes. Holding the flute vertically, with the unnotched end against her lower lip, she blew across the top of the flute, hitting the notch with a stream of breath from lips slightly parted in a horizontal smile. This is like the embouchure used for notched wooden and bamboo flutes such as the Andean quena.

Given that notches are widespread in modern flutes, she expected the second method to be more successful. The notch creates a sharp edge that slices the narrow stream of air, causing the stream to fibrillate, rapidly alternating its flow on either side of the edge. This air-against-edge is also the principle used in pipe organs, recorders, and many whistles. But Potengowski found that playing the notches on the Paleolithic flutes gave sounds that were, at best, indistinct. The notch on the mammoth-ivory flute gave warm but blurry sounds. Despite much effort, the notch on the griffon vulture flute would not evoke a clear sound, only wheezy puffs. The notches on these flutes, then, may be artifacts of breakage. Or their fragmentary state may distort our idea of the original shapes.

The oblique method of playing, though, worked for all the flutes. The first time Potengowski put the swan radius to her lips using this method, her breath woke two simultaneous notes from the instrument. Two equally strong waves coexisted within the flute, one a harmonic of the other. The effect is a fulsome sound, one that offers a taste of tonal harmony rather than a single pitch. This is unusual for a flute, an instrument that normally plays one predominant pitch at a time. Potengowski thought that the sound must indicate a "mistake" in her approach. She quickly changed her opinion and came to appreciate the double tones as "wonderful and a tool for musical expression." Multitoned sounds were perhaps one of the foundations of Paleolithic music.

Single tones, too, have curious properties in these instruments. From the swan radius came a crisp whistle. Potengowski then slid the whistle up a full octave, then back down, a smooth incline of pitch changes. The sound is a little like a modern piston whistle, swooping up and down. But there's no slider in these flutes changing the pitch. She used nothing but the shape of her tongue, facial muscles, and lips, a technique she termed the *oral glissando*. This glissando works only with the oblique playing method, with the flute's end held against pursed lips. Potengowski found that the glissando was better at changing pitch than were the finger holes cut into the flutes.

The mammoth-ivory flute, when played on its notch, is obnoxious, a shrill squeak. I find it hard to listen to the whole thirty-second track without lunging to turn the volume down. When she plays the instrument with the oblique method, though, the tone is gorgeous. The low sounds are like a distant train whistle, the higher ones like a sweet piping note from a bird.

Like all wind instruments, the flutes can be overblown to find higher registers by increasing the force of the breath. Potengowski found that she could readily make all three flutes leap like this, giving each a range of about two and a half octaves. The highest notes, pitched close to the highest possible on the piano keyboard, were the hardest for her to create and their unpleasantly piercing sound wavers as her breath pushes the instrument to its highest limit.

Potengowski's work shows that our explorations must leave behind modern preconceptions. The bird-bone and mammoth-ivory flutes may look to us like close kin of contemporary wooden and tin flutes, but this visual similarity is deceiving. Pitch changes from these modern analogs come mostly from changes in fingering. The breath energizes and shapes this sound, but it is not the main source of the melody. For the Paleolithic replicas, Potengowski found the reverse. Fingering had only a modest effect on pitch, but by changing her mouth and breath, she could evoke any tone within the instrument's range and thus play in any scale.

What might be learned from further experimentation with replicas of Paleolithic flutes? After reading of their work and listening to sound samples, I contacted Hein and Potengowski. We agreed that a new experimental reconstruction of a mammoth-ivory flute would be an interesting avenue for research. The replica that Hein had crafted and Potengowski played was a copy of the ancient flute from the cave. But the Paleolithic flute appears to be broken at one end, suggesting that it originally was longer. An uncarved stave that looks like a blank for a flute has been recovered from the same cave deposits as the flutes. This stave is longer than the ancient flute—thirty centimeters for the stave and only nineteen for the flute—again suggesting that the artifact from the cave is a broken part of a longer original. Hein works on archaeological reconstruction projects for museums all across Europe, and had on hand a piece of mammoth ivory from a previous project. He agreed to construct a new mammoth-ivory flute matching the Paleolithic stave's length.

Hein's videos of the manufacturing process reveal the material properties of mammoth ivory. To human hands, the ivory is hard and impossible to scratch, let alone cut into. But the edge of a flint tool slices readily, scoring the surface or gouging shavings like a metal plane skimming over soft wood. Watching his hands at work, I realized that stone tools not only made the work of Paleolithic people faster and more precise, but it allowed them to craft substances that are otherwise entirely beyond our capacity. The technological distance between our bare-handed ancestors and those who invented stone tools seems far wider than the gap between Paleolithic and modern metal tools.

Hein built the instrument with seven finger holes, their spacing matching those on the longer bird-bone flutes. This is not so much a replica as a hypothesis about the form of a longer mammoth-ivory flute. Once the flute was complete, Hein sent it to Potengowski for further sonic explorations. As with the other ivory flutes, the oblique method of playing worked best, directing a narrow stream of air to the top edge of the instrument. The timbre and frequency range were similar to those of the other flutes but extended a little further into the lower regions. What struck me most were her descriptions of how difficult the instrument is to play. Any bodily or mental tension interfered with the sound. Cooler, wetter days were harder. Some days the instrument burst into sound; on others, sound had to be coaxed. Later, when I tried the flute, I could draw from it only occasional whistles. My ineptitude is not surprising, but Potengowski has spent most of her life playing flutes.

Perhaps musicianship was highly advanced in the Aurignacian. Long winters in ice age caves provided ample time for practice. Or maybe embouchure was different and easier in those days. Hunter-gatherers have a strong edge-to-edge bite in their front teeth, unlike the overbite of soft-mouthed agrarians. Perhaps this gave Paleolithic players better control of facial muscles and the flow of breath? It is also possible that the ivory we recovered from the caves is only part of the instrument. Strips of grass or bark may have served as reeds. If so, the instrument was not a flute but a clarinet or oboe. Scraps of plant material are unlikely to survive for tens of thousands of years, so the record of artifacts in caves cannot answer whether or not they were used. Reeds evoke sounds from pipes even in unskilled hands, offering a less arduous path than fickle flutes into tonal music. When I held a sliver of modern oboe reed against the instrument's beveled top end, I was immediately rewarded with a loud whistle. If Paleolithic children were as enthusiastic as modern youngsters about blowing onto grass stems to make squeaky sounds, it would take only a small leap of the imagination to attach these vibrating scraps of plant material to hollow tubes.

These experiments, and Hein and Potengowski's earlier work, show us that ancient music must be understood through bodily engagement. The instruments' challenging embouchure, multitonal qualities, the oral glissando, and the effects of overblowing are all discoverable only by participation. These experiments open our imagination to the music of the past.

Curiously, discoveries from the Paleolithic have not greatly influenced the creative work of contemporary music. This contrasts with the discovery of Paleolithic visual art, which inspired artists and art curators in the early twentieth century. In 1937, the Museum of Modern Art in New York City put on an exhibit titled *Prehistoric Rock Pictures in Europe and Africa*, with photographs and watercolor copies of rock paintings alongside work by contemporary artists such as Paul Klee, Hans Arp, and Joan Miró. The Institute of Contemporary Arts in London followed in 1948 with *40,000 Years of Modern Art*. It was understood that Paleolithic art had something to contribute to the creativity of the present moment, that it lived in vital relationship with contemporary work. These connections were on vivid display in the 2019 exhibit *Préhistoire, une énigme moderne* at the Centre Pompidou in Paris. There, in the works of Paul Cézanne, Pablo Picasso, Max Ernst, and dozens more, were exhibited the fruitful influences of Paleolithic artifacts on modern art. When I visited, I was taken aback by the physical juxtaposition of ancient ivory carvings and sculptural work by Henry Moore, Joan Miró, and Henri Matisse. The similarities of form were astonishing.

So, too, was the absence of Paleolithic sound. Visual art from the distant past is in lively dialogue with the present. But in our leading cultural institutions: mostly silence from the deep past.

Partly, this is a result of the recency of the discoveries. The Paleolithic flutes of southern Germany were found more than a century after the first figurines and cave paintings. But flute pieces were unearthed in the 1920s from the Paleolithic layers from the Isturitz cave in southwestern France. Perhaps the fragmentary nature of these finds accounts for their failure to spark interest among contemporary composers or musicians?

Music also has difficulty traveling across deep time. Millennia later, we can see that ivory carved into a figurine is visual art. On viewing Paleolithic carvings, sculptors can immediately relate what they see to contemporary work. Twentieth-century modernists, in particular, saw parallels between Paleolithic art and cubism, minimalism, and lyrical abstraction. Although the cultural context of the original artist is lost, the objects still speak directly to us. But an ivory flute unearthed from a cave is silent. Instrumental music requires a musician to bring the art to life. Music is always ephemeral and relational, animated by the connection between instrument and player. Its essence and form cannot be captured and displayed in a collection of artifacts. Written music notation, itself an imperfect means to communicate the subtleties of sound, is a relatively recent invention, with the earliest known example from Ugaritic clay tablets of the fourteenth century BCE. The advent of electronic forms of sound making in the twentieth century also likely contributed to the disinterest among composers and players in Paleolithic instruments. Electronics gave musicians vast new powers. Compared with this, the discovery of bone flutes superficially similar to other flutes worldwide was a modest spur to imagination at best.

But Paleolithic instruments offer marvelous possibilities for living connections across time. Music's ephemerality places the living artist at the center of discovery. Music requires the presence of an artist in active, bodily conversation with the materials and ideas left by long-dead predecessors. Experiments in Paleolithic music making will always be imperfect replicas in their form—we'll never know the exact tones and melodies of ancient musics—but they quite literally reawaken creative processes that have slept in the rubble of caves for millennia.

Resonant Spaces

Springtime has come to southern Germany and I am sunning myself on a partly wooded slope, my back to the mouth of a cave in a limestone escarpment. In front of me is a steep incline, suffused with the aromas of reawakening wildflowers, maple and beech leaves, and grasses. The canopy cover is sparse, admitting the gentle afternoon light. From where I sit, the slope drops to a small river weaving around fields, wooded copses, and scattered buildings on a level valley floor.

The cave sits at the foot of the limestone wall, a pocket the size of a large, high-ceilinged room. Archaeologists recovered three flutes from the sediments of this cave: two swan-bone flutes and the most well preserved of the mammoth-ivory flutes. The pit from which they were removed is now backfilled with coarse stones, and its coordinates are marked with vertical strings hung from the ceiling, preserved and mapped ready for future explorations. A latticed steel fence keeps out visitors.

As I sit on the chalky soil in front of the cave entrance, a Eurasian blackcap gives me a lesson in acoustics. The small bird wings to a low branch a few meters away and looses a melody, a string of ten fast, clear notes, each one inflected up or down. After a pause, he gives a variation of the original, this one with a couple extra sweeping notes. For the next five minutes, he unspools these phrases and rests, switching among variations. The song has a rich timbre, a rapid flow of fluty notes, a performance lauded in bird field guides as one of the finest in Europe. But most striking to me today is how the sound comes alive in this space.

The blackcap chose a perch at the edge of a natural bowl, a partial enclosure for sound. Limestone buttresses extend on either side of the cave mouth, ribs of stone that have resisted erosion. The cliff overhangs here too, forming a high partial roof. The cave itself is a modest indentation in the limestone wall. Its forecourt is a high-walled limestone yard. This enclosed shape likely gave the cave its modern name, Geißenklösterle, the "goat-chapel," where herders could pen their livestock. The view to the valley is through a gap in the buttresses. No doubt this natural enclosure afforded protection for ice age inhabitants from wind and unwelcome visitors. It also created a space in which sound blooms. The space cups each note of the blackcap's song, causing them to linger and ripen.

The blackcap's notes reflect back to me from the limestone walls, the reflections arriving about fifteen milliseconds after the sound that flowed directly from beak to ear. Because the reflections arrive so soon, my brain perceives them as part of the original sound, not as separate echoes. The reflections give a feeling of great clarity and richness. The architects and acoustic engineers who design modern recital halls pay special attention to what they term these "early reflections." Large baffles above and to the side of the stage shoot early reflections straight to the audience, giving a feeling of intimacy and verve even in larger spaces. A few natural spaces do the same, notably Red Rocks Amphitheater at the foot of the Rocky Mountains near Denver. There, sedimentary rocks from the Paleozoic form a bowl and high side walls that combine to produce a spectacular performance space, a larger version of this cave entrance in Germany. The walls of "shoebox" recital halls produce a similar effect, bouncing sound from the players seated at one end of the narrow box all the way down the length of the room. Geißenklösterle cave and its buttresses act as reflectors for the blackcap's song and, perhaps, long ago, the notes of a swan or mammoth flute.

Enclosures also add reverberation and thus a sense of depth and richness to sound, as every bathroom singer knows. The polished hard ceramic tiles of bathroom walls are excellent reflectors of sound, and so each sung note ricochets over and over. These reflections meld into a reverberation that prolongs the life of each note. The effect at the cave mouth is subtler than a bathroom, maybe half a second of slight reverberation. But this is enough to add a touch of tonal gold to the bird's voice.

Half an hour's brisk walk south of Geißenklösterle is another cave, Hohle Fels, "the hollow rock." The cave entrance is a dark maw at the base of the slope, wide and high enough to admit a small truck. In the past, farmers stored hay inside, and during the Second World War, military vehicles were stashed here. Now the entrance is protected by a metal gate hung with signs naming visiting hours. In front, the narrow river meanders across a meadow glowing with thousands of dandelion blooms. The cave entrance is at the bottom of a smooth-faced limestone cliff, a wall about six stories tall.

Inside the cave mouth, beyond the cabinets of maps and artifacts that line the entrance, a passageway heads straight back into the hillside. As I walk, the walls and ceiling close in. The smell of damp limestone dust and algae displace the aromas of trees and meadows. After a minute's walk, the cave floor drops precipitously, and a metal walkway carries me on. Below my feet is a pit about four meters deep, illuminated by scattered spotlights, its walls bermed with sandbags. This is the site of an archaeological dig underway since the 1970s. The bags protect the unexcavated layers below, ready for work to recommence later in the year.

I look down from my perch on the metal walkway. Propped on the sandbags are laminated paper signs naming and dating the cultures associated with the sediment "horizons" or layers. The deepest, "Neanderthal, 55,000-65,000 vor heute [before the present]"; then, rising along the side wall of the cave, "Aurignacian, 32,000-42,500"; "Gravettian, 28,000-32,000"; and "Magdalenian, 13,000." The slow accretion of sediment has captured and preserved artifacts from sixty-five thousand years of domestic life. First Neanderthals, then the changing culture of anatomically modern humans in the ice age. Fragments of memory, layered into the earth. In one of the the deepest, oldest layers of human presence, the Aurignacian, lay the female figurine and the griffon vulture flute, now on view at the Blaubeuren Museum ten minutes' drive from here.

My feet on steel mesh, I hover over the excavation, gazing on this record of human life. I surprisingly experience not awe or temporal disorientation—feelings that accompany much of my reading about the Paleolithic or other ancient times—but a sense of calm. In this taste of humanity's long prehistory, some deep-buried anxiety unknots. My life is almost entirely embedded in the tempo of modernity, living by the minute, focused on hours and sometimes years, living in houses that will likely fall apart this century, using electronic tools that will not see out the decade. Our culture is on track to remake itself and much of the Earth by century's end. Almost nothing draws our senses, imaginations, and aspirations further than a few years. And when we do think on the scale of thousands of years, it is hard to imagine any continuity of the human story between now and this distant future. The past, too, is alien, out of reach of the senses and thus bodily understanding. The physical presence of tens of millennia of humanity tells my body: there is another, longer narrative.

The vast majority of our species' time on Earth was experienced by people with bodies and brains just like ours, living and sometimes thriving through their relationships with one another and the land. The form of these relationships differed on different continents, but whether in Africa, Eurasia, Australia, or later, the Americas, the record speaks of persistence across spans of time incommensurate with my experience of the everyday. This long life as hunters, gatherers, and agriculturalists is part of our identity and inheritance, now almost entirely obscured by the technologies and preoccupations of the moment. For a few moments, it feels good to breathe the scents of old Earth, and I feel at home. This is not nostalgia. I don't hanker for a return to an illusory Eden. Instead, the pit recalibrates within me the sense of what it means to be human. In these long, almost forgotten millennia lies much of our history. A fragment of truth about identity is revealed here. I'd known this, of course, but our species' past seemed abstract, a disembodied set of ideas. This pit, this exhumation of time, spoke not only to ideas but to the lived, embodied experience of our species.

I linger, savoring the sight of so much human life condensed into one place, then move deeper into the cave. From the walls of this tunnel, the clang of my feet on the metal gratings echoes, a harsh and confined sound. But there's a softening up ahead, a spaciousness that intrigues my ears. I duck at the passageway's end, pass through a constriction, and, treading on dust and gravel, enter the cave floor beyond the excavation.

I lift my head and gasp. I've stepped into a vast cavern. A few spotlights directed at the walls suggest its size, but it is the sound of water drops that drives home the point. They fall from the high ceiling onto puddles and wet stone. Each *tok* of their landings fills the space, quiet snaps that reverberate for more than a second. Even the scuffs and crunches of my feet on the cave floor are magnified. The cave sounds like a Romanesque church or a large unadorned rotunda.

There's no singing bird here to demonstrate how whistled notes behave, and so I use my voice and hands to explore sound. I clap and the impulse comes back to me as a stretched decay, loud at first, then tapering over a second or two. Later, when I deliver the same clap outside, it is a lash of sound, gone in an instant. In the cave, I whistle and each note remains strong for a second or two after my breath stills. The effect is sonic animacy, as if the cave imparts afterlife to sound.

This drawn-out reverberation is the acoustic signature of capacious, hard-walled spaces like cathedrals, empty factories, or huge cisterns. The walls reflect sound, sustaining reverberation as sound bounces from one side of the enclosed space to another. But even a good sound reflector like stone drains some energy from sound waves. In a voluminous space, sound has long airborne intervals where it flows with little attenuation between its draining collisions with walls. A large volume thus creates a sound that lingers in the air as waves travel from one distant wall to another, sometimes for many seconds, especially if the space lacks sound-absorbing material like heavy curtains. Hohle Fels cave has a volume of six thousand cubic meters, like a big church.

This cave's reverberation is much more drawn out than that of Geißenklösterle. As a consequence, very rapid and nuanced sounds are quickly blurred. If I'm just a few meters away from other visitors, their speech turns to a velvety smear. This would be a terrible place to give a lecture. Likewise, a complex violin piece would sound disastrous here, the swiftly changing notes would melt into one another. But simpler melodies sound gorgeous. I've never heard my whistling lips sound so good. Outside the cave, in the meadow, my hand claps and whistles are like thin, dry bread. Inside, they fatten and expand into luscious slabs of cake. Flute music would be gorgeous here.

In parts of the cave, reverberation of my voice hits sweet spots and resonates, amplifying frequencies of sound whose wavelength matches the size of the space. Especially in the smaller side chambers, the lowest frequencies of my voice balloon. This resonance is a general property of sound in enclosed spaces; from wineglasses, to bathrooms, to halls, the dimensions of each space boost particular sound frequencies. In the cave, this resonance combines with echoes to create an expansive feeling, an acoustic luminosity.

Paleolithic people surely chose the Hohle Fels and Geißenklösterle caves for protection from the elements, not for their sonic qualities. But alongside their utility as living quarters, both spaces have rich acoustics. In the afternoon that I spent in Hohle Fels, I watched dozens of visitors come and go from the large inner cavern. On entering, every adult immediately hushed their voice to a whisper. Whoops and whistles came from the children, playful salvos of sound. These are places that immediately assert their sonic exceptionality.

The first known musical instruments were made in places well matched to their sounds. Or so it seems to modern ears. Today many live performances and recordings of flutes use electronics to add reverberation, placing the sound within a simulated cave or chamber. Did the reverberant qualities of caves somehow catalyze the invention of the first flutes? I imagine a child sucking marrow from a bird bone and delighting in how the sound bloomed within the cave. Skillful parents might then have taken up familiar tools and experimented. Bird-bone flutes perhaps then planted the idea for the sophisticated tool work needed to create a mammoth-ivory instrument.

These are speculations. All we know for sure is that rich acoustics of space and the first evidence of instrumental music co-occur in the same cave. This coincidence resolves into something more like a pattern when we also consider evidence from other Paleozoic caves in southern Europe.

In the 1980s in France, musicologists and archaeologists Iégor Reznikoff and Michel Dauvois used their voices to explore caves with notable Paleolithic wall paintings. By singing simple notes and whistling, they mapped their perceptions of the caves' acoustics. They found that paintings were often located in places that were particularly resonant. Animal paintings were common in resonant chambers and in places along the walls that produced strong reverberation. As they crawled through narrow tunnels, they discovered painted red dots exactly located in the most resonant places. The entrances to these tunnels were also marked with paintings. Resonant recesses in walls were especially heavily ornamented.

In a 2017 study, a dozen acousticians, archaeologists, and musicians measured the sonic qualities of cave interiors in northern Spain. The team, led by acoustic scientist Bruno Fazenda, used speakers, computers, and microphone arrays to measure the behavior of precisely calibrated tones within the cave. The caves they studied contain wall art spanning much of the Paleolithic, dating from about forty thousand years to fifteen thousand years ago. The art includes handprints, abstract points and lines, and a bestiary of Paleolithic animals including birds, fish, horses, bovids, reindeer, bear, ibex, cetaceans, and humanlike figures. From hundreds of standardized measurements, the team found that painted red dots and lines, the oldest wall markings, are associated with parts of the cave where low frequencies resonate and sonic clarity is high due to modest reverberation. These would have been excellent places for speech and more complex forms of music, not muddied by excessive reverberation. Animal paintings and handprints were also likely to be in places where clarity is high and overall reverberation is low but with a good low-frequency response. These are the qualities that we seek now in modern performance spaces.

The convergence of cave visual art and sonic qualities suggests that people were noticing and responding to caves as acoustic spaces, not only as shelters and painting canvases. If so, then like other animals whose sounds are molded to the acoustic shapes of their homes—treehoppers whose sounds match their host plant, birds singing in mountain winds, whales calling through the deep ocean channel—the form of human music is partly a product of its sonic context.

The first musical instruments were well suited to their homes. Whether by design or happy coincidence, the bone and ivory flutes fit the acoustics of limestone caves in which they were crafted.

The flutes fit the cave, not the reverse. There is no evidence that Paleolithic people changed the shape of caves to adjust their sonic qualities. Like almost all other species, human sound making found its home within the constraints and opportunities offered by preexisting spaces. But this one-way relationship would change. We are one of the few known species that deliberately sculpt sound-making spaces. Prairie mole crickets are our companions in this innovation. In this threatened species of the North American prairies, every courting male builds a bulbous underground chamber leading to a funnel that opens to the aboveground world. The males sit in the chamber and make repeated croaks by rubbing their wings together. They face away from the funnel, directing sound into the resonant chamber and out through the funnel to the world. Males gather in clusters on the prairie and blast their combined sound to the sky, an arthropod fanfare sung through trumpets made of sculpted prairie soil. Males are flightless, but winged females home in on the sound. In remnant patches of prairie habitat where this species lives, the chorus is sometimes loud enough to be heard four hundred meters away.

Humans are mole crickets on a magnificent scale. We build not small burrows but concert halls, worship spaces, lecture rooms, and headphones, each tuned to the particular needs of the sound they contain. This ability to adjust the spaces in which we make sound has kindled a creative triangle: human musical composition, the form of musical instruments, and the space in which we make and hear music. Within this triad—composition, sound making, space—no one member is dominant. Instead, which one leads or follows has shifted over time. The story starts in the Paleolithic but is alive and accelerating in our modern concert halls, earbuds, and streaming online music services.

Flames and swirls of color by muralist Eli Sudbrack dance across the building's brick facade. Down the street, light reflected from the East River gleams from the glass and metal of new condominium towers. Most other buildings in the neighborhood are under scaffolding or have already been upgraded to expensive offices and retail. This building, though, is one of the survivors of Brooklyn's raze-and-build boom, an architectural hold-out from the neighborhood's industrial past. White block print runs above the new mural's bright colors: National Sawdust Co. In the 1930s, wood was pulverized and bagged here, sent off to sop blood in butcher's shops, soak up barroom spills, and pack stored blocks of ice. The sawyers' blades and blowers long gone, National Sawdust is now a performance venue and, through its residencies and programs, a catalyst for new music. I've come here to hear how the ancient relationship between acoustic space and music is taking novel forms.

It's September 2019, opening night of National Sawdust's fifth season. There are a dozen performances on the program, crossing genres from chamber to experimental electronic music, solo voice to large choruses, and classical piano to contemporary instrumentalists. But it is not just the diverse program that gives the evening its power. The room, too, takes on a different sonic form for each performer, transforming from spacious, to warmly intimate, to tight and loud. We are experiencing the launch of a new way of shaping sound within the space.

Above us hangs an array of sixteen microphones. On walls and ceiling, 102 speakers wrap the room, some visible, others out of sight. This system—installed weeks before by an audio company, Meyer Sound, sculpts the sound of the venue, taking the ancient creative triad of musicians, acoustic space, and instruments into its next iteration.

This sound system is not merely amplifying sound, although for music created on laptops or for very quiet instruments, that is part of its role. The system allows performers and sound designers to decide how sound will behave inside the venue, opening new possibilities for composition and performance. Following the touch of buttons on an electronic tablet, the performance space can now sound like a cave, recital hall, or a space so far unimagined. Walls move in and out. Sound shifts its points of origin in the room. Reverberation expands, then contracts.

As I listen to the concert, I'm carried from one place to another. The air glows as soprano Naomi Louisa O'Connell's voice lingers above us. We're in a sun-warmed atrium, looking out on an expansive vista. When the Young People's Chorus of New York City surround us, lining the walls, each voice is clear and distinct, yet they also merge and swell. Walls seem to shiver with their rising, hopeful energies. Rafiq Bhatia and Ian Chang are on stage, but somehow we're inside the sound of guitar, percussion, and electronic samples, immersed in the knotted, turbulent flow of their stories. The melodies of flutist Elena Pinderhughes live partly on her lips and in her flute, then they fly across the room, a bird's motion briefly come to life as sound. Music from the National Sawdust ensemble comes directly from their instruments, but lingers in the air for a fraction of a second, as it does in a classical recital hall. Then a short announcement and the room has the clarity of a university lecture theater.

These transportations are achieved by playing back into the room what's happening on stage, with subtle alterations to the sound: adding and changing the duration of reverberation, brightening or darkening the tone, and shifting the spatial origin of sound. The system works like the reflectors, baffles, and curtains of concert halls, but the reflection has passed through microphones and speakers, not bounced from wood, stone, or cloth.

The idea of electronically shaping a venue's sound is at least seventy years old. In 1951, the reverberation and bass response of the newly built Royal Festival Hall in London were too weak. The music felt anemic, clear but lacking rich tones. Rather than gut the interior to remedy excessive sound absorption, the hall was equipped with microphones and speakers, allowing engineers to boost reverberation and low frequencies without giving an obvious sense of amplification. This "assisted resonance" system was remedial and not intended as a tool for elaborate sound design. In the late twentieth century, similar sound reinforcement systems were installed in concert halls worldwide, complementing the acoustics of rooms and doubling as amplification systems for speech or plugged-in instruments. Now, better microphones and speakers, combined with software that allows us to model and manipulate sound, make the system at National Sawdust a creative instrument in its own right.

Is such electronic shaping of sound a defiling artifice for "acoustic" instruments like cellos or flutes? Are we sullying the purity of the musical experience by adding a touch of electrical power to the sounds in a room? *The New York Times* music critic Anthony Tommasini writes that "natural sound has always been the glory of classical music." He was "dismayed" by the 1999 addition of an electronic control system to the New York State Theater, then home to both the New York City Opera and New York City Ballet, writing that "a line has been crossed, and I fear the worst." Conductor Marin Alsop, commenting in 1991 on an early version of the electronically enhanced concert space in the Silva Concert Hall in Eugene, Oregon, said that "to rely on a sound technician for your balance is completely antithetical to the role of a conductor."

Yet all music is a product of its context. The sound of the human voice or violin that we hear in a recital hall is not an unmediated experience of vocal folds or bow on strings. Rather, the sound is partly constructed by centuries of analysis and experimentation by "technicians" with the acoustics of interior spaces. If we're listening in a large modern concert hall, our experience is the product of hundreds of thousands of dollars of architectural artifice to bring us the sound we hear. The New York Philharmonic, for example, plays in a hall at Lincoln Center that was built in 1962, then renovated to improve acoustics half a dozen times over the next twenty-five years. A major redesign of the hall is now underway that will, in part, once again overhaul its acoustics, at a cost of more than half a billion dollars. "Natural sound" in these spaces is an expensive contrivance.

The Meyer system, and those of other companies with similar products, builds on a long-standing tradition of engineering the relationship between music and acoustic space. To be fair to skeptical late twentieth-century commentators such as Tommasini and Alsop, early versions were crude compared with what can be achieved today. In 2015, Alex Ross, the *New*

Yorker's music critic, wrote admiringly of the possibilities of these electronic systems and concluded that "although no amount of digital magic can match the golden thunder of a great hall vibrating in sympathy with Beethoven's or Mahler's orchestra, the Meyers may have come closer than anyone in audio history to an approximation of the real thing." Whether or not electronically enhanced sound is any more "real" than other sound in concert halls, these new systems upend how the relationship between music and space can evolve, adding rapidly adaptable electronics to the protracted architectural work of changing the physical form of buildings. Meyer has now installed its system in concert halls from Vienna, to Shanghai, to San Francisco, mostly for subtle adjustment of reverberation. The grumblings of the 1990s have quieted down as active electronic enhancement has been accepted as another form of architectural modification in concert halls.

The most obvious and immediate benefit of these electronic systems is to vastly increase the versatility of the space, thus serving many needs in a community and increasing the financial stability of a venue. The "natural sound" of specialized opera houses or other single-use halls is a luxury enjoyed only where the wealthy congregate, mostly in large cities. Electronic adjustments to the acoustics of performance halls potentially bring sonic art to a wider audience, allowing spaces that were formerly limited by their poor and inflexible acoustics to now become diverse hubs in local cultural networks.

In a single week, National Sawdust hosts opera singers, jazz, a movie and lecture, a classical ensemble, solo piano, and electronic rock. Each has its own acoustic requirements, some of which are incompatible in a single space. For opera, we need a balance of reverberation and clarity. Classical ensembles require a little more liveliness from the walls. Medieval church music was written for long, cave-like reverberations. For cinema, absolute deadness is ideal, letting the soundtrack flow into the room with minimal sonic reflection. Rock music needs amplification, only slight reverberation from the room, and no odd frequency spikes or feedback as sound bounces back from the room to the microphones on stage. A lecture benefits from a hint of reverb to enrich the voice but not so much as to blur intelligibility. Electronic adjustment allows one space to meet all these needs. Other parts of the sensory experience of music venues—the grand vistas offered by an opera house, the aromas of old stone and incense in a cathedral, the pleasant tension in your legs as you climb the tiers of an amphitheater, the stickiness of spilled beer underfoot in a club—cannot of course be molded by microphones and speakers. But carefully designed electronics can open and diversify the sonic qualities of space.

A few months after the opening concert, I visit National Sawdust during the day to better understand how its new sound system fits with the organization's mission. I sit at a small table in the center of the empty performance space with Paola Prestini, cofounder and artistic director; Garth MacAleavey, technical director and chief audio engineer; and Holly Hunter, director of projects and artist residencies.

As we talk, Garth touches the screen of a small electronic tablet. Tap. We're talking in a recital hall, our words clear and rich. Tap. A cathedral with soaring resonance. Tap. A reverberation that goes on for five or more seconds, like standing inside a vast empty oil tanker. Tap. Dead. The warmth of our voices shrinks. We're suddenly pushing harder to be heard. System reverb is off. Curtains hidden behind the paneling that forms the shell of the room absorb sound waves and eat our voices. Tap. A lecture hall, suddenly our words are clear and lively. We laugh nervously. The sudden flip is disconcerting. We feel completely natural, yet a click of a button transforms how it feels to hear one another and speak. A lesson: our voice comes from the larynx, but its sound and feel are born in relationship to the surroundings. Tap. A brook runs down one side of the room and four singing birds perch across the ceiling above us. Tap and slide. The brook moves to the center. Tap. We're back in dead space. More astonished laughs.

For millennia, music has evolved with space. This close relationship is now mostly hidden because we hear music in spaces engineered for a good match. Opera in the opera house. Film score in the cinema. Rock in a club or through earbuds. Gregorian chant in the stone-walled church. Switch any of these pairings and the music is garbled, muddied, or deadened.

These close relationships reveal some of the reciprocity between space and innovation in the history of music. Instruments discovered in later Paleolithic caves—flutes, rasps, bull-roarers—are well suited to gatherings of a few dozen people. Louder instruments appeared when human societies grew and sound needed to travel farther. Drums and horns called people to war, the hunt, and religious gatherings. The first documented drums are from the millet- and rice-farming Dawenkou culture of eastern China, from about 4000 BCE. The first known trumpets are from the powerful eighteenth dynasty in Egypt in about 1500 BCE. When societies became large and hierarchical enough for political and religious rulers to build large spaces, the coordinated playing of many instruments filled these buildings with sound. In the third millennium BCE, harps and lyres appeared in royal tombs in Mesopotamia. The royal tombs of ancient Egypt were often stocked with instruments numerous enough to create ensembles. Wall paintings from these tombs and from temples show groups of dozens of musicians playing wind and stringed instruments. The grave of Marquis Yi of Zeng, from the fifth century BCE in China, contained an especially grand instrument, a three-tiered, chromatic-scale set of sixty-five large ornate bronze bells, sonic markers of prodigious wealth. The great philosopher of that age, Mozi, complained of the drain imposed on society's time and resources by the "great bells and rolling drums, zithers and pipes" of the ruling classes. The first pipe organs were invented in Greece in the third century BCE and soon spread to the homes of the wealthy and public performance spaces of ancient Greece, Rome, and Alexandria.

Humanity's creative exploration of sound through instruments was inspired by the tones and timbres of new materials and technologies—ceramics, strings, brass, bellows, valves—and each culture used its most sophisticated craft to build new instruments, just as Paleolithic ivory carvers had done. Increasing potential for loud sounds was one consequence of these technologies.

The present-day diversity of musical instruments reflects the importance of acoustic space in guiding culture and technology. This is most clearly seen when spaces change, opening new possibilities and needs for instruments. In Europe, the advent of large public concert halls in the nineteenth century demanded louder sounds than the small recital halls of the aristocracy. Instruments evolved in response. Compared with the first pianos of the sixteenth century, modern pianos are thunderous. The vigor of their sound increased as the sizes of concert halls increased and new discoveries in metallurgy allowed for stronger wires. The tension in the wires of a modern piano is ten times that of early instruments, an increase made possible by the nineteenth-century addition of solid metal internal piano frames. Tighter-wound metal wires also made violins louder, starting in the late seventeenth century. By the nineteenth century, the tension in violin strings was such that the bass bar, bridge, and fingerboard of older instruments had to be adjusted. The violin bow, too, was refashioned, making it longer and giving it a concave arch, the better to tighten horsehair and give players control. The concert flute was extensively modified in the nineteenth century, mostly through the work of one man, Theobald Boehm. He engineered larger tone holes, better keys, and a reshaped head and embouchure. Although Richard Wagner complained that the vigor of the new flutes made them "blunderbusses," Boehm's work established the flute's place in the modern orchestra. Improvements in valves and keys also loudened and stabilized the sound of other woodwind and brass instruments. The great size of symphony halls became embodied in the forms of the instruments on stage. Orchestras expanded too, from Baroque orchestras of a few dozen to the more than one hundred players put on stage by Wagner and Mahler in the late nineteenth century.

Electric amplification also changes the relationship between instruments and space. The guitar, formerly an instrument suited to parlors, campfires, and other small gatherings, can now, with a mere brush of the hand, fill a stadium with sound. The guitar moved from a rarity in large public venues to near ubiquity in Western popular music. The nature of human song, too, was changed by electric amplification. Now a whisper or throaty croon into the microphone suffices, no projection or push from the diaphragm needed, a radical break from millennia of performance that required unaided lungs to fill places of worship, palaces, and concert halls. Just as the modern piano's sound was partly born from the vastness of symphony halls, the breathy notes and throaty growls of contemporary popular music have as their parents the furnaces of electric power stations.

We create sonic space every time we press "play" on our smartphones and CD players at home. Because we have an abundant choice of music, albums and tracks are set into competition with one another for our attention. The loudest ones usually win, even if we think we have no preference for loudness. Our brains consistently judge louder music as "better." More, our brains also prefer music that has had its quiet passages cranked louder. This psychological quirk sparked the "loudness wars," starting with CDs in the 1990s and continuing to the present day. Producers increase the amplitude of every part of the music, turning the variable loudness of a piece of music into what they call a brick wall, a final product in which every part of the track is boosted to the highest level possible. The resulting sound file on a computer screen shows a tall and unvarying wall of intensity instead of the ups and downs of the volume of most live music. The overall impression is of louder, more present music. But the process eliminates the pop of percussive effects like snare drums, creates a sense of boxed-in tightness, and, in extreme cases, fuzzes the music with white noise.

Producers often disdain the process of "brick-walling" their albums but are pressured by musicians and marketers to push loudness upward. Two infamous examples are the albums *Californication* by the rock band the Red Hot Chili Peppers and *Death Magnetic* by the heavy metal band Metallica.

Both were subject to petitions from fans demanding remastering to undo extreme brick-walling. Digital streaming services—another new sonic space—are now relieving some of the pressure. These platforms automatically adjust volumes to avoid jarring changes in loudness between tracks. This removes some of the incentive to push up amplitude on recordings. Many albums now are produced in two ways, one for digital streaming and one for CD. The digital version is often produced "as if for vinyl," hearkening back to a world where the sounds of recorded music came from the physical motion of industrial diamond on rotating plastic. The cutting equipment for vinyl disks cannot cope with brick-walled sound, and so requires a subtler touch from the producer.

Earbuds and lightweight headphones, too, make new forms of sonic space. Like physical space and acoustic instruments, earbuds and portable music systems have coevolved. The evidence is here in my desk drawer. A thin metal headband connecting two foam-covered minispeakers connects to a 1980s-era pocket cassette player. White-wired earbuds dangle from the plug of a matchbox-sized MP3 player from 2005. Black over-the-ear headphones tangle their wires with a red-and-black set of plastic earbuds, listening devices intended for the three generations of smartphones that have passed them by. Each system is portable and convenient, encasing me in private experiences of music and voice over the decades. Each one has poor sound quality, delivering the outlines of music but not its subtleties. Low and high frequencies are mostly absent. Ambient noise penetrates the thin foam or plastic and washes out quieter sounds. And so on my flimsy 1980s headphones, music that arrived on cassettes from friends after multiple rounds of copying sounded pretty much as good as the original cassette. Later, with MP3 players and smartphones, there was little noticeable difference through inexpensive earbuds between CD-quality sound and highly compressed digital sound files.

The bootleg culture of cassette tape copying and, later, the early popularity of highly compressed digital audio files, many also pirated, were made possible in part by the low quality of earbuds and small headsets. The devices we poked onto or into our ears created a new sonic space and, as it always does, music changed according to the particular demands and possibilities of the space. Technology mediates this relationship, as it does in the analog world. Now, as noise-canceling headphones and better earbuds improve the "personal" listening space, richer music flows into our ears, aided by cheaper and faster data transmission.

The intimacy of headphones changes the relationship between music and listener too. Singers whisper directly through our earbuds and headphones. Compare the Grammy Song of the Year awarded in 2020 with that from 1970: Billie Eilish's "Bad Guy" is a conspiratorial murmur. She's right there, her lips to our ears. Joe South's "Games People Play" is reverberant, distant. He's on a stage with his band, the sound seeming to flow into an audience. The snap and shimmer of the instruments behind Eilish's voice sound great on my coin-sized laptop speakers. The same speakers lop off the depth and blur the inflections of the violins, organ, and drums on South's track. Music from 2020 sounds great on portable cheap speakers, but recordings from 1970 only sound good on more sophisticated audio equipment. The plastic capsules in our ear canals have changed the form of music.

Electronic sculpting of sound in performance venues, like that at National Sawdust, carries the digital revolution into three-dimensional spaces where people gather together to hear music. The technology weakens the link between form and sonic qualities of space, for the first time in the long history of musical evolution.

One effect will be a closer relationship between audience, musicians, and composers. When a performer works in a space mismatched to their music, they're fighting against the room's acoustics, as if trying to get their sounds, and thus their feelings and ideas, through a headwind. Tuning a room to the particular needs of music therefore activates connections among artists and audiences.

Acoustic flexibility of space turns what was fixed—the sound of a venue—into another part of the instrumentation that composers use to sculpt sound. This is an extension of stereo, quadraphonic, or "5.1" sound—systems that use two, four, or six speakers to build immersive listening experiences—into realms where sound has a fine-grained spatial structure, its positions and movements controllable on-the-fly from an electronic tablet. The flight that I heard in Elena Pinderhughes's flute music was one example. She played her flute from the stage, but the music drifted and swooped through the performance space in service of narrative and emotion.

Composer and electronic music pioneer Suzanne Ciani, interviewed after using a Meyer system at Moogfest, put these possibilities into context. She said that the first uses of quadraphonics in the 1970s "didn't have the content, there was no real viable reason to do it." But today "we have a new generation of kids who are playing electronic music that just wants to fly all over the room and be sculpted and moved." She emphasized the emotional heft of spatial design in music: "Powerful . . . until you feel it, you don't really know what it is."

Spatial audio technology has a natural affinity with dance, which, by its nature, moves through all three dimensions of space. Wherever the dance is participatory, rather than watched by a seated audience, these new audio systems will allow music to move along with human bodies. From ballrooms to clubs, composers and performers can now make music dance, literally. Combined with strap-on haptic devices that pump low-frequency sounds into our skin and body tissues, the line between body motion and music is blurred. This builds on the link established hundreds of millions of years ago when our fish ancestors first evolved inner ears that detect both motion and sound, a design that we and all other vertebrates have now inherited.

The application of these methods to electronic dance music (EDM) is clear. Movement by listeners is part of the EDM experience, and new technologies are readily embraced by performers and participants alike. But spatialized sound technologies also offer an opportunity to understand traditional instruments in new ways. When we hear a violin, guitar, or oboe, we receive an integrated sound that flows from the instrument's entire surface and volume. This is the intent: to animate the air with a coherent tone and texture. But when your ear is close to the instrument, you realize that its sound has a topography. Might we now, as part of the narrative of an instrumental piece, travel across the varied terrains of a violin's belly, the bore of a flute, or the surfaces of a piano? Instruments would then be experienced as three-dimensional objects full of tensions and harmonies, just as a musical score is. Form of instrument and form of music can now converse not only through time, a single dimension, but within the three dimensions of space.

Our ears could also be given what live musicians have, a position on the stage. Sit with the violas. Fly to the brass at just the right moment. Pause a moment between the bass and the banjo in a bluegrass concert, then, as the music demands it, sweep to the fiddle, then pan out to the whole.

Such compositions would bring to the concertgoing experience some of the same spatial dynamics of walking in a forest or through a sound installation in an art gallery. Moving through an ecological community is an experience in which sound has form and texture within space. The same is true when sound is used as a sculptural form in gallery spaces or outdoors. In the Museum of Modern Art in New York City, for example, the electronic sounds of David Tudor's *Rainforest V* come from everyday objects—a wooden box, an oil drum, plumbing parts—suspended in a large room. As we move within the space, sound takes on different rhythms and colors. Unlike the living species in a rain forest, though, Tudor's objects lack long evolutionary histories of sonic coevolutionary haggling and dealmaking among themselves. Instead, the physical forms of manufactured objects in the installation are animated by electricity, an effect augmented by the objects' responses when sensors within them detect and reply to sounds made by visitors. Spatially nuanced work such as this can now, with the help of electronics, enter the concert hall.

Most human music is experienced as a temporal flow from one point within a field of sound. We take a seat in a concert hall or slide headphones over our ears. Even when we are walking with earbuds, the sound does not track our movement, but instead arrives in a way previously unknown to any living being: a seemingly stationary sound source reaching a body in motion. Composers can now bring more spatial dynamics into their creations, integrating sound and motion. This work is an extension of more traditional forms of composition and performance. Processionary and marching music creates spatial narratives, as do instruments and voices ranged around halls and worship spaces.

Music is relationship. It connects people, but it also engages us in the physicality of the spaces that we occupy. Every instrument and form of music is thus made partly from its acoustic context. In this, human music does not differ from the communicative sounds of other species. Each species has, through evolution and animal learning, found its sonic place in the world.

Humans, though, actively shape our acoustic spaces in ways impossible for almost all other species. Singing birds cannot modify the reverberations of the forest. Snapping shrimp do not turn a knob to brighten their crackling choruses. A katydid in a rain forest is incapable of adjusting the amplitude or frequencies of the dozen other insects singing around it. Even the mole cricket does not rework its burrow to fit its song. But human music making allows creative reciprocity among our compositions, instruments, and the acoustics of space. Electronics in our ears and concert venues have now opened new possibilities for these fruitful relationships, the continuation of a process that began in the sonorous caves of the Paleolithic.

Music, Forest, Body

The plaza at Lincoln Center in New York City has been stripped of all signs of nonhuman life. Contrasting black-and-tan paving slabs mark out a geometric design centered around a 317-jet, illuminated fountain. The architectural narrative aims to honor and elevate high art but also to exclude, forcefully declaring that human power and ingenuity are entirely in control here. The rest of life's community has been erased, save for thirty London plane trees, planted away from the main plaza, arrayed in soldierly rows in a gravel-topped concrete rectangle. Memory of the thriving human community whose neighborhood was leveled in the 1950s to build this place—seven thousand Black and Latino families who received no assistance for relocation—is also obliterated. This is a place seemingly for those who believe themselves to be *maestros*, or "masters," from the Latin *magister*, "he who is greater." Much beauty, artistry, and meaningful connection happen here, but this is also a place of fracture and erasure.

We walk into the concert hall, home of the New York Philharmonic, the oldest symphony orchestra in the United States. Here, too, the space conveys the message of dominance by a single human architectural plan, as do almost all places where humans gather to be fed by the fruits of culture: performance venues, lecture halls, museums, cinemas, and places of worship. Upholstery. Metal railings. Wood panels so smooth and glossy they seem made of plastic. The doors to the concert hall close, sealing out sounds from the rest of the world. On stage, the musicians' bodies are veiled by uniformly black shirts, trousers, and dresses. The aesthetic is formal and signals wealth.

Every part of the journey to this concert impresses on the listener that they are engaged in a shedding of the messiness and particularity of the city, the community of life, and even of human flesh. The audience sits apart from musicians in a darkened space, muscles and nerves resisting any urge to become entrained in or contribute to the music. The experience of sound here will, it seems, transcend this time and place, focusing our attention on a sonic experience of creativity, artistry, and beauty unshackled from Earth. This release promises an experience of God, in sacred music, or into the realms of human ideas and emotions.

But this escape is an illusion. We can pave over the living soil, displace human and nonhuman life, occlude views of the human body, and close the doors in a sound-proofed vault, only to arrive back in human flesh and the diversity of the living world. The concert hall delivers a powerful experience of embodied life, a union of the human and more-than-human world almost unmatched in its bodily intimacy and richness of ecological relationship. There are few other places in our culture where the boundary between "human" and "nonhuman" is so thoroughly erased, even if we do not usually celebrate this merger in our external representations. Perhaps the sensual power of interbeing experienced here is why we must use pavement, sealed-in chambers, and shrouded bodies? These trappings of concertgoing mediate the entry of music's earthly power into our bodies and psyches, easing a union that might otherwise be discomforting in its raw openness, vulnerability, and animality.

Lights dim. Paper programs rustle, like a strong breeze over dry oak leaves. Conversations hush as heads and torsos orient to the stage. Tonight's concertmaster, Sheryl Staples, steps onto the stage with an eighteenth-century Guarneri violin in her hand. From a position below the conductor's podium, she signals to Sherry Sylar, principal oboist for this concert, who lifts her cocobolo wood instrument and sounds an A. The note sails out into the hall from the oboe's bell, drawing in its wake a flotilla of notes from other instruments. Then silence: the evening's moment of maximum expectation and concentration, 2,700 people collectively holding their breaths. The moment breaks into applause as the conductor, Jaap van Zweden, strides out, sweeps his arm over the audience and orchestra, then takes his perch. Another moment of expectant silence and the baton falls. A shiver and crescendo from the percussion swells into brass and strings, and Steven Stucky's "Elegy" commences.

From the moment the oboe sounds, forests and wetlands come alive on stage. In this place of high human culture, we are lifted into joy and beauty partly by the sounds of other beings, our senses immersed in the physicality of plants and animals.

The oboe's sound is rooted in plants from the coastal wetlands of Spain and France. The reeds that impart vibrations to the musicians' breath are parings of a giant cane indigenous to the brackish, sandy shores of the western Mediterranean. Growing more than six meters tall, the hollow stems of this grass grow only two to three centimeters wide. This seemingly preposterous architecture—plants taller than houses, on stems narrower than my thumb—endows the reeds with their sonic properties. Tough fibers made from interconnected plant cell walls run lengthwise through the canes. This dense, uniform array of microscopic filaments stiffens the canes, allowing just a little flex in strong winds. It takes tools as sharp as surgeons' knives to excise thin slivers to make reeds for wind

instruments. Only after blades have shaved the reed to translucent thin-
ness can human hands or lips feel any springiness. In the sound of the
woodwinds—oboes, clarinets, bassoons, saxophones, and others—we
therefore hear one of the more extreme plant architectures, a skinny giant
that yields material uncommonly lightweight yet very hard and stiff. Reed
instruments in India, Southeast Asia, and China use plants with similar
qualities, either giant canes, palm fronds, or bamboo. Reeds made from
more diminutive grasses or from shaved tree wood produce soft or coarse
sounds with inconsistent tone. The northern European whithorn and
bramevac, for example, use willow bark reeds to evoke squeals from coni-
cal wooden horns, sounds that lack the fine control and predictability of
cane- and bamboo-reeded instruments. Oboists play with the finest reeds
of all. When I spoke with Sherry Sylar about her work, she told me that the
oboist's relationship with reeds is like woodworking, a precise craft of manipulating plant material. The oboist is both luthier of cane and musician.

The oboe's bore and its finger holes sculpt the pressure waves within the
instrument, pulsations that then push sound into the hall. It is the bore's
smoothness and taper, the bell's flare, and the dimensions and sharpness of
the finger holes' many openings and edges that combine with the resonant
properties of wood to give the instruments' bodies their acoustic signature. Any warps, pits, cracks, uneven surfaces, or irregularities in proportions degrade the sound. Oboes and other wind instruments, then, need to
hold their shape, surface gloss, edges, and proportions, even when bathed
in the warm moisture of human breath. This calls for dense, smooth-
grained wood. The predecessors of modern oboes and clarinets, shawms
and hautboys, were made of boxwood, fruitwoods such as apple and pear,
or tight-grained maple. These trees grow slowly, layering wood into
themselves in thin yearly accretions. Similarly dense and smooth apricot
wood is favored for the surnāy of western and central Asia and bamboo for
Japan's hichiriki.

Before the nineteenth century, the music of reed instruments flowed from the woods of their homelands. Now we often hear materials that have been transported from other continents. Most oboes and clarinets used by professional musicians, for example, are made from mpingo, also known as East African blackwood or grenadilla, or other tropical woods such as cocobolo or rosewood. These materials became available to European instrument makers after colonial occupations of Africa, South America, and Asia. The superior stability, density, and smoothness of these woods were ideal for instruments that are repeatedly bathed in human breath then dried, a process that cracks or warps other woods. Along with nineteenth-century innovations in metal sound-hole keys and levers, forest products shipped to Europe from tropical forests produced many of the instrument-making traditions that prevail today.

The Musical Instruments Collection at the Metropolitan Museum of Art, a short walk across Central Park from Lincoln Center, reveals the tangled relationships among local ecologies, colonial trade, and the craft of instrument making. At first, the galleries seem like mausoleums for sound. Silent instruments sit illuminated behind sheets of plateglass, reliquaries for the remains of music whose spirits have flown. The glass, polished wooden floors, and long, narrow dimensions of the galleries give the sound of footfalls and voices a lively, clattery feel, unlike the expansive warmth of concert halls, reinforcing the sense of isolation from musical sound. This initial impression evaporates, though, when I let go of the idea that this is a space for direct experience of sound. Instead, we can marvel here at stories of materiality, human ingenuity, and the relationships among cultures.

Like the Paleolithic mammoth-ivory flutes whose construction relied on the most sophisticated craft of their era, the instruments on display at the Met show how, across cultures and time, people have drawn on their highest forms of technology to create music. Trumpets and whistling jars from the precolonial South American Moche civilization reveal mastery of ceramics. Pipe organs were, for centuries, among the most complex machines in Western Europe. An Algerian rebab bowed lute and Ugandan ennanga harp show precise engineering of wood, skin, and string. The technologies of silk production, wood carving, lacquer, and ornamented inlay converge in a Chinese guqin, a long stringed instrument played on a tabletop or lap. In the twentieth century, industrial innovations appear, from electric guitars to plastic vuvuzela horns.

Precolonial instruments often used indigenous materials. Walking through the galleries is an education in the many ways that humans have sonified matter from their surroundings. Clay, shaped then fired, turns human breath and lip vibrations into amplified tones. Rocks turned to bells and strings reveal metallurgical connections to land. Plant matter is given voice in carved wood, stretched palm frond, and spun fiber. A bestiary of animals sings through taut skins and reshaped teeth and tusks. Each instrument is rooted in local ecological context. Condor feathers in South American pipes. Kapok wood, snake skins, antelope horn, and porcupine quills on African drums, harps, and lutes. Boxwood and brass in European oboes. Wood, silk, bronze, and stone in se, shiqing, and yunluo, Chinese percussive and stringed instruments. Music emerged from human relationship with the beyond-human world, its varied sounds around the world revealing not only the many forms of human culture but the diverse sonorous, reverberant properties of rock, soil, and living beings.

But for all its magnificent and often fine-grained ecological and cultural rootedness, human music is not narrowly provincial. Music's power to connect stretches far beyond its unifying effects on listeners in the present moment. Music making binds the ecological, creative, and technological histories of seemingly distant cultures. Ideas and materials have moved from one place to another since the dawn of instrumental music. The swans whose bones gave Paleolithic artisans material for flutes were not part of the fauna of the tundra around the caves. Transport or trade brought the swan's wing bones into the places where they became musical instruments. Human desires have driven trade for instrument making ever since. Listeners seek sound that pleases and moves them. Musicians demand stability and consistency from their instruments. Our eyes delight in the form, hue, and surface ornamentation of instruments, a visual complement to sonic beauty. All these qualities demand the best materials, stimuli for trade.

The extensive trade network that connected China, India, western Asia, North Africa, and Europe—the "silk road" of the first millennium CE, carried ivory east from Africa to Asia, silk strings west from China to Persia, and southern Asian tropical woods to temperate regions. Ideas about the forms of instruments moved alongside materials used in instrument making. Double-reed instruments and bowed stringed instruments came to Europe from Africa and western Asia. Lutes, drums, harps, and trumpets arrived in China from central and western Asia.

Eighteenth- and nineteenth-century colonial land seizures, forced labor, and rail and shipping networks brought new materials to European instrument makers. When a modern orchestra, folk group, or rock band takes the stage, the air comes alive with the sounds of vibrating plant and animal parts, the voices of forests and fields reanimated through human art. But we also hear the legacy of forced occupation and resource extraction, now turned to modern globalized trade. Melodies soar from hollowed mpingo wood in oboes and clarinets, a voice from East African savannahs. Electric guitarists press their hips into the mahogany bodies of their instruments and slide their fingers over Madagascan rosewood fingerboards, playing with slices of giant rain forest trees. String players bow with horsehair tensioned by South American Pernambuco wood. Many bows are tipped with ivory or tortoiseshell. All of these European instruments had long precolonial histories, grounded in local soils and materials, but were transformed into their modern forms, in part, by the export to Europe of materials from colonized lands. The changes wrought by colonialism create striking visual differences among the European instruments of different ages in the Met galleries. In the eighteenth and nineteenth centuries, dark tropical woods and abundant use of ivory replaced much of the lighter box-wood, maple, and brass of earlier European instruments.

Eighteenth- and nineteenth-century European colonizers picked out the material most pleasing to their ears and most useful to instrument-making workshops. A few European materials made the grade and were retained, even as "exotic" woods and animal parts became more readily available. Spruce and maple, especially, remained the favored wood for the bodies of stringed instruments and the soundboards of pianos. Calfskin topped tympani. These European materials were joined by ivory, favored for its workability and stability, and tropical woods whose density, smoothness, elasticity, and tones met musical needs: mpingo's tight, silky grain; Pernambuco's extraordinary strength, elasticity, and responsiveness; rosewood's warmth and stability; and padauk's resonance. These tropical woods all belong to the same taxonomic family, tree cousins to the beans, and have tight-grained, dense wood from slow-growing trees. Most take seventy or more years to reach harvestable age. On a concert stage, we hear the voices of tree elders.

The industrial economy continues the same path, plucking materials and energy from around the world. Long-buried algae drilled from oil wells are distilled and polymerized into plastic keyboards. Amplifiers are plugged into an electric grid powered by the incineration of mined coal, the flow of water through dammed rivers, or the decay of mined uranium rock.

The tropical woods and ivory most favored for instrument making are now mostly threatened or endangered. Nineteenth-century exploitation has turned to twenty-first-century ruination. Demand for materials for musical instruments, though, was not the primary cause of many of these losses. The volume of ivory used for violin bows and bassoon rings was dwarfed by exports for tableware handles, billiard balls, religious carvings, and ornaments, although piano keys consumed hundreds of thousands of pounds of tusks in the late nineteenth and early twentieth centuries. Pernambuco was extirpated from most of its range not by violin bow makers, but through overharvesting for dye made from its crimson heartwood. The country Brazil gets its name from *brasa*, "ember" in Portuguese, for the glowing-coal color of the wood whose trade was so important in the founding of the country.

Mpingo woodlands are in decline, driven by export for instruments and flooring, and by local uses for carving. Compounding the problem of overharvesting is the twisting, gnarled form of mpingo trunks. Carving straight billets for oboes and clarinets from such wood is challenging, and often less than 10 percent of the cut log is usable. Rosewoods, often used for guitar fingerboards, are mostly exported for furniture, with more wood in one bed frame or cabinet than in any guitar shop. Although trade in many rosewood species is restricted by international law, the wood is now so valuable that financial speculators and luxury goods manufacturers drive an illegal market worth billions of dollars yearly.

The sound of contemporary music is therefore a product of past colonialism and present-day trade, but, with very few exceptions, it is not a driver of species endangerment. Indeed, the relationships between musicians and their instruments—often built over decades of daily bodily connection—serve as an inspiring example of how we might live in better relationship to forests. An oboe or violin contains less wood than a chair or stack of magazines, yet this single instrument yields beauty and utility for decades, sometimes centuries. Contrast this with the culture of overexploitation and disposability that pervades so much of our relationship to material objects and their sources. For example, we threw out more than twelve million tons of furniture in the United States in 2018, 80 percent of it buried in landfills, most of the rest burned, and only one-third of 1 percent recycled. Much of this furniture was sourced from tropical forests, often supplied to the United States through manufacturing hubs in Asia. Such trade is increasing and the World Wildlife Fund states that the "world's natural forests cannot sustainably meet the soaring global demand for timber products." If the rest of our economy took as much care of wood products as musicians do of their instruments, the deforestation crisis would be greatly eased.

Driven to action by a desire to honor the materials with which they work, some musicians and luthiers are now at the forefront of seeking alternatives to the exploitative use of wood, ivory, and other materials from threatened species. This is especially important work because musical instruments are now far more numerous than in past centuries. More than ten million guitars and hundreds of thousands of violins are made annually. Such volume of trade cannot be built on rare woods. It is therefore now possible, with some searching, to find instruments made from wood certified to come from sustainable logging operations. The Forest Stewardship Council, for example, puts its stamp of approval on several new lines of instruments. The Mpingo Conservation & Development Initiative in southeastern Tanzania promotes community-based forest management where local residents own, manage, and benefit from mpingo and other woodland species, managing forests sustainably to help the local economy. Instrument makers are also introducing new materials, relieving pressure on endangered woods. Until the late twentieth century, only twenty tree species provided most of the wood for guitars, violins, violas, cellos, mandolins, and other Western stringed instruments. Today the variety of wood sources for instrument making has increased to more than one hundred species. Alongside this diversification of natural products, manufactured materials like carbon fiber and wood laminate are substituting for solid wood.

In the decades that come, unless our path changes, it will not be the overharvesting of particularly valuable species that challenges our sources of wood and animal parts for instruments. Instead, the loss of entire forest ecosystems will remake the relationship between human music and the land. The forests from which we now draw our most precious musical raw materials are in decline. In the first dozen years of this century, forest loss exceeded gain by nearly three times, a global net reduction of more than 1.5 million square kilometers. Tropical forests fared worst, followed by the spruce and other boreal forests of the north. Increasing fire, forest clearing for commodity crops, and changing climate will likely accelerate these changes in coming decades. Music will, in future, still give voice to the Earth, just as it always has. It will tell of the ancient bond between ecosystems and human artistry but also of extinction, technological change, and the subjugation of forests by human appetites.

A few old instruments—carefully tended by musicians—now evoke the memory of the departed or degraded forests. On the stage at Lincoln Center, we hear woods from past decades and centuries. Sherry Sylar plays on oboes whose woods were harvested decades ago in the early twentieth century. Each one has a "passport" documenting the wood's provenance, showing that it was not obtained through recent cutting of now-endangered trees. When we talked, she described how some colleagues scour the country for sales of older oboes, hoping to find instruments with good wood from ages past. The music of Sylar's violinist colleague, Sheryl Staples, comes from a Guarneri violin. Its woods are at least three hundred years old, harvested from spruce and maple forests that grew on a preindustrial Earth. Although wood for instruments still comes from the Fiemme Valley forests in northern Italy that supplied Guarneri and Stradivarius, springtime there now comes earlier, summer is hotter, and winter snowpack is diminished compared with that of previous centuries. This yields wood with a looser, less sonorous grain than the tight woods of past centuries. In another hundred years, it is likely that heat, droughts, and changed rainfall will push alpine forests off these mountain slopes. Music often now speaks of the Earth as it was, not as it is, a memory carried in wood grain.

Sitting in my seat at the Lincoln Center, I arrive in intimate contact with

the world's forests—their past and future—and the history of human
trade. The sounds of the orchestra are worldly, immersing me in the beauty
and brokenness of both biodiversity and human history. Music is not tran-
scendent or abstracted, it is immanent and embodied. In a time where forests are in crisis and mass extinction is underway within life's community,
it is perhaps time to unshroud and honor these relationships from which
music blooms.

I first held a violin in my late forties. Placing it under my chin, I let go an
impious expletive, astonished by the instrument's connection to mammalian evolution. In my ignorance, I had not realized that violinists not
only tuck instruments against their necks, but they also gently press them
against their lower jawbones. Twenty-five years of teaching biology
primed me, or perhaps produced a strange bias in me, to experience holding the instrument as a zoological wonder. Under the jaw, only skin covers
the bone. The fleshiness of our cheeks and the chewing muscle of the jaw
start higher, leaving the bottom edge open. Sound flows through air, of
course, but waves also stream from the violin's body, through the chin
rest, directly to the jawbone and thence into our skull and inner ears.

Music from an instrument pressed into our jaw: these sounds take us
directly back to the dawn of mammalian hearing and beyond. Violinists
and violists transport their bodies—and listeners along with them—into
the deep past of our identity as mammals, an atavistic recapitulation of
evolution.

The first vertebrate animals to crawl onto land were relatives of the modern lungfish. Over 30 million years, starting 375 million years ago, these
animals turned fleshy fins into limbs with digits and air-sucking bladders
into lungs. In water, the inner ear and the lateral line system on fish's skin
detected pressure waves and the motion of water molecules. But on land the

lateral line system was useless. Sound waves in air bounced off the solid bodies of animals, instead of flowing into them as they did underwater. In water, these animals were immersed in sound. On land, they were mostly deaf.

Mostly deaf, but not totally. The first land vertebrates inherited from their fishy forebears inner ears, fluid-filled sacs or tubes filled with sensitive hair cells for balance and hearing. Unlike the elongate, coiled tubes in our inner ears, these early versions were stubby and populated only with cells sensitive to low-frequency sounds. Loud sounds in air—the growl of thunder or crash of a falling tree—would have been powerful enough to penetrate the skull and stimulate the inner ear. Quieter sounds—footfalls, wind-stirred tree movements, the motions of companions—arrived not in air, but up from the ground, through bone. The jaws and finlike legs of these first terrestrial vertebrates served as bony pathways from the outside world to the inner ear.

One bone became particularly useful as a hearing device, the hyomandibular bone, a strut that, in fish, controls the gills and gill flaps. In the first land vertebrates, the bone jutted downward, toward the ground, and ran upward deep into the head, connecting to the bony capsule around the ear. Over time, freed from its role as a regulator of gills, the hyomandibula took on a new role as a conduit for sound, evolving into the stapes, the middle ear bone now found in all land vertebrates (save for a few frogs that secondarily lost the stapes). At first, the stapes was a stout shaft, both conveying groundborne vibrations to the ear and strengthening the skull. Later, it connected to the newly evolved eardrum and became a slender rod. We now hear, in part, with the help of a repurposed fish gill bone.

After the evolution of the stapes, innovations in hearing unfolded independently in multiple vertebrate groups, each taking its own path, but all using some form of eardrum and middle ear bones to transmit sounds in air to the fluid-filled inner ear. The amphibians, turtles, lizards, and birds each came up with their own arrangements, all using the stapes as a single middle ear bone. Mammals took a more elaborate route. Two bones from the lower jaw migrated to the middle ear and joined the stapes, forming a chain of three bones. This triplet of middle ear bones gives mammals sensitive hearing compared with many other land vertebrates, especially in the high frequencies. For early mammals, palm-sized creatures living 200 million to 100 million years ago, a sensitivity to high-pitched sounds would have revealed the presence of singing crickets and the rustles of other small prey, giving them an advantage in the search for food. But before this, in the 150 million years between their emergence onto land and their evolution of the mammalian middle ear, our ancestors remained deaf to the sounds of insects and other high frequencies, just as we, today, cannot hear the calls and songs of "ultrasonic" bats, mice, and singing insects.

The evolutionary transformation of parts of the lower jaw of premammalian reptiles into the modern mammal middle ear is recorded in a sequence of fossilized bones, stony memories from hundreds of millions of years ago. As embryos, we each also relive the journey. During our development, our lower jaw first appears as a string of interconnected small bones. But these bones do not fuse into a single lower jaw as they do in living or ancient reptiles. Instead, the connections among them dissolve. One bone becomes the malleus of the middle ear. Another becomes the incus bone that connects the malleus to the stapes. A third curls into the ring that holds our eardrum. And one elongates into our single lower jawbone.

When I lifted the violin to my neck and felt its touch on my jawbone, my mind filled with imaginings of ancient vertebrates. These ancestors heard through their lower jaws as vibrations flowed from the ground, to jaw and gill bones, to the inner ear. The violin drew me into a reenactment of this pivotal moment in the evolution of hearing, without the indignity of prostrating myself. High art meets deep time? Not in my incapable hands, but certainly in the artistry of accomplished musicians.

Bone conduction of sound gives violinists a different experience of sound than their listeners. Most of the sound flows through air, joining player and audience. But sound waves also flow up through the jaw, turning the bones of the head into resonators that fatten the experience, especially for low notes. These vibrations also run down through the shoulder, into the chest. Playing the violin without such bodily contact—resting it on a spongy cloth against the shoulder and forgoing jaw contact—yields an insipid experience. The instrument feels distant, even though it sounds loudly in our ears.

The violin's form gives it a special connection to the far recesses of our evolution, but this is just one of the many ways the human body is intimately connected to the materiality of instruments.

From our seats in the hushed auditorium, we listen and watch: Fingertips brush, press, and slide along strings. Cellos stir the skin and muscles of inner thighs. Reeds tremble between wet lips. Breath flows across the open mouths of flutes. Hands, arms, and shoulders pound tympani and send shudders through maracas. Lungs cry out through trembling lips, their agitations shaped and amplified by brass coils soaked through on the inside with the moisture of human breath.

Through the orchestra, we experience a direct connection not only to the distant stories of ear bone evolution but also to the living presence of animal sensuality. The groin-thrusting and guitar-neck stroking of rock musicians is the most unsubtle example, but these antics pale in comparison to the diverse bodily intimacies on display at an orchestral concert. The composition of music often tells of desire, passion, or heartbreak, stories or emotions all the more powerful for being evoked not as abstractions, but as products of moving lips, flowing blood, activated nerves, and animated breath, the bodily homes of love and erotic desire.

But music's relation to the human body is far more than this. A catalog of the many ways that musicians' bodies connect to their instruments sounds racy partly because we live in a culture where sensuality is equated with sexuality. Music, though, gives voice to the diverse ways that the body can give us sensual experience. Sexual, sometimes yes. But the body also grieves, exults, bonds, explores, strives, hungers, builds, and rests. An accomplished musician—through their intimate relationship with their instrument or voice, built through years of muscular, sensory, intellectual, and aesthetic training—invites us into these experiences. Every note is an extension of bodily movement, a sonic pathway from the interior of one person to another. Nerve to nerve we connect, sound wiring us into "the other." Even the tempi of music are manifestations of our body, beats that often reflect the one-two rhythm of bipedal walking, ticking within a range that exactly spans the pumping rate of the human heart.

If you play an instrument, you understand. My own amateur relationships with the violin and guitar connect me back into my body. The guitar's sound waves leap into my chest, up into the throat, a centering flow. Singing with the guitar is a matter of unifying vocal folds with the vibratory tones of wood. The song is breath, flesh, and forest. The violin takes me deeper into chimeric union. Every knot or strap of muscular tension reveals itself through the bow and its rosined passage over strings. A hairbreadth's difference in the position and angle of fingers on fingerboard levers tone up, down, or into blurry hesitation. I ease my neck and shoulders, and the sound clarifies, like a gleam from sunlight on clear water. But my experiences are shallow compared with those immersed in the discipline and artistry of instrumental music. Sherry Sylar told me, "Playing the oboe is an addiction for me, I feel grounded when I play it, the sound resonates throughout my body. It is an organic experience that nothing else quite replicates." A live concert invites listeners into the simultaneous and unified experience of dozens or hundreds of such bodily exultations.

The experience of music, then, embeds us not only in the ecology and history of the world, but in the particular qualities of the human body. One of these qualities is our special human ability to wield tools and craft ivory, wood, metal, and other earthly materials into instruments. Another is the musicians' ability to animate these mergers within listeners' bodies, through sound. Music incarnates us, literally "making us flesh."

Might the internal, subjective experience of human music also ground us in the earth and unite us with the experiences of other species? Our culture mostly says, no, music is uniquely human. Philosopher of music Andrew Kania tells us, for example, that the vocalizations of "non-human animals" are "examples of organized sound that are not music." Further, because singing creatures like birds and whales "do not have the capacity to improvise or invent new melodies or rhythms," they "should no more count as music than the yowling of cats." Musicologist Irwin Godt concurs, writing that "the birds and bees may make pretty sounds . . . but despite the effusions of the poets, such sounds are not music by definition. . . . It makes no sense to muddy the waters with non-human sounds. This is a fundamental axiom."

When I step outside the walls of the performance hall or seminar room, spaces whose "fundamental axiom" is the sensory exclusion of the beyond-human world, these ideas seem to me hard to defend.

If music is sensitivity and responsiveness to the vibratory energies of the world, then it dates back nearly four billion years to the first cells. When sound moves us, we are also united to bacteria and protists. Indeed, the cellular basis of hearing in humans is rooted in the same structures, cilia, possessed by many single-celled creatures, a fundamental property of much cellular life.

If music is sonic communication from one being to another, using elements that are ordered and repetitive, then music started with the insects, three hundred million years ago, then flourished and diversified in other animal groups, especially other arthropods and the vertebrates. From the katydids animating the night air in a city park, to the songbirds that greet the dawn, to the thumping fish and caroling whales of the oceans, to the musical works of humans, animal sound combines themes and variations, reiteration and hierarchical structure. To argue that music is sound organized only by "persons" and not "unthinking Nature," as philosopher Jerrold Levinson has done, is akin to claiming that tools are material objects modified for particular use only by humans, thereby excluding the artisanal achievements of nonhumans like chimpanzees and crows. If personhood and the ability to think are the criteria by which to judge whether a sound is music, then music is a multiplicity encompassing the many forms of personhood and cognition in the living world. Erecting a human barrier around music in this way is artificial, not a reflection of the diversity of sound making and animal intelligences in the world.

If music is organized sound whose intent is wholly or partly to evoke aesthetic or emotional responses in listeners, as Godt and others claim, then the sounds of nonhuman animals must surely be included. This criterion aims, in part, to separate music from speech or emotional cries, a challenging line to draw even in humans where lyrical prose and poetry erode the division from one side and highly intellectualized forms of music chip away at the other. All animals live within their own subjective experiences of the world. Nervous systems are diverse, and so the aesthetics and emotions that are part of these experiences no doubt take on multifarious textures across the animal kingdom. To deny that other animals have such subjective experiences is to ignore both our intuitions from lived experience (we understand that our pet dog is not a Cartesian machine) and the last fifty years of research into neurobiology, which now can map within the brains of nonhuman animals the sites from which emerge intention, motivation, thought, emotion, and even sensory consciousness. Laboratory and field studies show that nonhuman animals, from insects to birds, integrate sensory information with memory, hormonal states, inherited predispositions, and, in some, cultural preferences, producing changes in their physiology and behavior. We experience this rich confluence as aesthetics, emotion, and thought. All the biological evidence to date suggests that nonhuman animals do the same, each in their own way. For the cat, then, "yowling" is music if it stimulates aesthetic reactions in feline listeners. The subjective responses of other cats are the relevant criteria by which to judge the sound's musicality. That we presently find it hard to access the sentient experience of cats demonstrates human technological and imaginative limitations, not the absence of music in their caterwauling. Further, the current models of the evolution of animal communication strongly suggest that the coevolution of aesthetics and sonic display explains much of the diversity of sound that we hear in other species. Sonic evolution without aesthetic experience has little diversifying power. Aesthetic definitions of music, then, are biologically pluralistic, unless we make the unsupported and improbable assumption that experiences of beauty are uniquely human.

If music is sound whose meaning and aesthetic value emerge from culture, and whose form changes through time by innovations that arise from creativity, then we share music with other vocal learners, especially whales and birds. In these species, as in humans, the reaction of individuals to sounds is largely mediated by social learning and culture. When a sparrow hears a mate or rival sing, the bird's response depends on what it has learned of local sonic customs that have been passed down culturally. When a whale calls, it reveals to others its individual identity, clan affiliation, and, in some species, whether it is up-to-date on the latest song variants. These responses are aesthetic: subjective evaluation of sensory experience in the context of culture. Often this results in richly textured patterns of sonic variations across the species' range. Cultural evolution in these species also changes sound through time, at a pace that is swift in some and leisurely in others, depending on their social dynamics. New sonic variations arise through diverse means: selecting sounds best suited to changing social and physical context, mimicking and modifying sounds from other individuals and species, and the invention of entirely new twists on old patterns. These diverse forms of animal music combine tradition and innovation, just as human music does.

If music is sound produced through modification of materials to make instruments and performance spaces in which to listen, then humans are nearly unique. Other animals use materials external to their bodies such as nibbled leaves or shaped burrows to make or amplify sounds but none make specially modified sound-producing tools, even the skilled toolmaking primates and birds. Music, then, separates us from other beings in the sophistication of our tools and architecture, but not in other regards. We are, as other musical animals are, sensing, feeling, thinking, and innovating beings, but we make our music with tools in a built environment of unique complexity and specialization.

As human musical sounds flow into us and move us, we are embedded in nested forms of music: the experience of themes and variations within the piece; the tension between novelty and tradition within the musical genre we are hearing; the cultural particularity and interconnectivity of the style of music we're hearing; and the special form of music in the human species, an art form emerging from and living in relationship with the diversity of music in other species.

Walking into the august spaces of Lincoln Center, I felt that the dominant narrative of our age was being forced onto me, an alienating falsehood: that we live apart from and above all other earthly beings. But, as the orchestra filled the hall with sound, I was plunged back into reality, a joyful return.

Animality. Connection. Belonging. No wonder we feel music so deeply. We have come home. Home to the nature of our bodies, both in the sensory present and through evolutionary history. Home to the ecological connections that give us life. Home to the beauty and fissures in our relationships with other cultures, lands, and species.

On the program that night were three compositions that told stories of belonging, connection, and fracture: Steven Stucky's "Elegy," from his longer work *August 4, 1964*; Aaron Copland's *Clarinet Concerto*; and Julia Wolfe's *Fire in My Mouth*. Copland's piece draws North American jazz and South American popular music into twentieth-century North American orchestral music. Instead of looking back, resurrecting the sounds of eighteenth- and nineteenth-century European concert halls, the work seeks to interweave American musical ideas with European orchestral traditions. Stucky and Wolfe explore pivotal moments in US histories of war, and of civil and workers' rights. Wolfe also draws our imaginations into the materiality of instruments and everyday objects. In her evocations of the sounds of the Triangle Shirtwaist Factory and the terrible fire that killed many of its workers, she calls on violin bows whishing through air, fingernails on the varnish of wooden instruments, books thrown to the ground, and the coordinated snap of hundreds of scissors. This music—beautiful, troubling, opening—deepens our capacity to feel the injustices of past and present, and to understand how protest and societal change rise out of grief, offering an invitation to connect with the wounds of the past and the questions of the present. Art here is not an anesthetizing ornament but part of the human quest for meaning. I walked out of the soundproofed hall and onto the plaza moved and inspired.

Music wakens or deepens within us the capacity to experience beauty through connection to others. This has been sound's role in the animal kingdom for hundreds of millions of years, now expressed in our species as one of the most powerful experiences we can have of our own bodies, emotions, and thoughts, and those of others. This is why we make music at moments of importance in our lives and at times of significant transition: in civic and religious gatherings, and in the lives of communities joining couples and burying the dead.

Now our power, greed, ignorance, and insouciance have ignited global crises of mass extinction, climate, and injustice. We need more than ever to listen to others with our bodies, emotions, and minds. Can we expand the circle of who and what is included in this "other" that we come to know through music? Because music is both fully human and entirely of this earth, music embodies interconnection and belonging. This remains true even when we wrap ourselves in architectures and cultural practices that evince separation and superiority. The belief in a *maestro* species, "he who is greater," can be dissolved by music's unifying powers. Experiences of musical beauty can knit us back into life's community. But we must first choose to listen.

PART V

Diminishment, Crisis, and Injustice

Forests

The spicy aroma of bruised sassafras leaves envelops me as I stride under a canopy of oak. Thorny greenbriars snatch my legs. I dodge the nastiest tangles in the understory, but mostly I try to walk a straight line. A pedometer on my hip counts paces: 260, equal to 200 meters from the last survey point. I swing my backpack to the ground and retrieve a clipboard. A tick clambers over the tape that I've used to seal socks to trouser legs, a defense against the dozens, sometimes hundreds, of blood seekers I encounter each day. Pluck, pinch, flick. Gone.

I jab the stopwatch and pour my attention into my ears, keeping eyes on the forest canopy.

Husky voice, phrases of four up and down notes. Scarlet tanager, twenty meters away.

Chippy-chup, a flutter of high sound. Two American goldfinches, twenty-five meters away.

Slurred, bright phrases, alternating inflections up and down, a question and an answer. He sings, *Where are you? There you are.* Red-eyed vireo, close, only five meters away, above me on a maple branch.

Two crows fly over, *caw caw-CAW*.

In the distance, fifty meters away, a rapidly modulated whistle, building to an emphatic end, *we-a-we-a-WEE-TEE-EE*. Hooded warbler.

Click. Five minutes are up. Scrawl on the datasheet: "Transect V, point 2. Time: 0610. Wind: Beaufort 2. Temperature: 25°C. Vegetation: white oak and red maple canopy; sourwood, blueberry, and sassafras understory." I pull out a range finder and turn its dial as I gaze through the two eyepieces, checking my distance estimates. Stow the gear. Sip of water. Two hundred and sixty paces to the next five-minute count. Repeat five hundred times.

From mid-May to mid-June, over two years, I threaded survey lines across the forests, tree plantations, and rural settlements of the southern Cumberland Plateau in Tennessee, on land that forced removals took from its Cherokee citizens in the 1830s. A satellite photograph of the region now shows a swath of green tree canopy running from Kentucky to Alabama through a landscape otherwise dominated by agriculture and urban areas. The region is one of the largest blocks of forest in the eastern United States. Unlike the National Forest and National Park lands to the east, the forests here are mostly privately owned. As the largest temperate forested plateau in the world, the region is a biodiversity hotspot, especially for salamanders, migrant birds, land snails, and flowering plants. The Natural Resources Defense Council calls the region a threatened biogem. The Open Space Institute has three funds dedicated to land conservation in the region.

At the time of my surveys, in 2000 and 2001, the diverse oak and hickory forests of the region were being leveled and turned to monoculture plantations of loblolly pine trees, a species native farther south and much favored by the pulp industry for its rapid growth. At the time, timber corporations and state agencies either denied that conversion of forest to plantation was underway or claimed that the change was of little conse- quence to biodiversity, pointing at housing development as the main threat to the region's forests. Aerial photographs refuted the denial, showing an accelerating rate of forest clearing and plantation establishment. The ef- fects of forest loss on biodiversity were harder to pin down. These changes cannot be seen from aerial photographs. But they can be heard, and so I set out into the woods with a clipboard to listen.

A complete inventory of all the species in any landscape is impossible. We don't know the identity of most microbes and many small inverte- brates. Among known species, enumerating each one could occupy doz- ens of scientists for years. Conservationists therefore focus their efforts, hoping that samples of a few species will reveal patterns relevant to all. In forests, surveys of birds are the most commonly used technique to rapidly assess biological diversity. Birds are sensitive to changes in vegetation, in- sect abundance, and the physical structure of habitats. Their populations are like probes into the hidden properties of habitats. Many species could serve this role, but birds have a special advantage. They sing. A few min- utes of listening can reveal the outlines of a bird community. Sampling other species requires hours of sifting through soil, setting out traps, ex- amining specimens in the hand or under a microscope, or sequencing DNA. Birdsong also entrances human senses, and many naturalists have spent years learning and appreciating their sounds. Finding skilled birders is easier than finding qualified nematode, fungus, plant, or insect taxono- mists. Birds also stimulate human concern more than many other animals. Compared with studies of less charismatic creatures, studies of birds yield information more immediately appealing to human aesthetics and ethics. Song, evolved to mediate social interactions within species, is now a con- duit for humans to listen across species boundaries.

Clearing the land for a pine plantation is a brutal assault. First, every tree is cut. Sometimes the trees—oaks, hickories, maples, and a dozen more species—are taken to mills to be pulverized into cardboard or, for the larger logs, sawn into lumber. Much of the forest is stacked in piles the size of churches and burned. Any remaining saplings and understory are then bulldozed. Trucks or helicopters finish the job of "suppression" with herbicide. Without poison, many of the forest plants would resprout. Millennia of fires and windstorms have taught the vegetation to rebound. But the plantation demands not resilience from the former forest, but near annihilation. Rivulets and forest wetlands were often bulldozed along with the forest. Downstream, what were clear mountain streams ran like chocolate milk, so opaque that I could not see my skin through the water in my cupped hands.

Clearing complete, immigrant laborers, mostly teenagers and young men, plant rows of pine saplings from nurseries. The pay, according to a 2003 study in Alabama, is between $0.015 and $0.06 per tree. A fast planter can make $80 a day, ten times the rate of pay for agricultural work in Mexico. The work is hard and the pace unrelenting. In the words of one planting contractor from Alabama, "We have offered up to $9/hour without a single American worker lasting more than 3 days. . . . It's not a good job. Without the migrant workers, agriculture and forestry would die in this country." The newsprint and toilet papers that come from these plantations exact a heavy toll on both the land and human bodies. They also contribute little to local economies. Local government officials complained that logging trucks do not even buy their fuel in the counties where the plantations grow.

Short of a layer of asphalt, it is hard to imagine a more thorough transformation of the forest. The change is evident to any resident or visitor. But human testimony from these lands is rare. The timber company owns tens of thousands of acres. There are no settlements on the land, few public roads into the heart of these operations, and the surrounding rural counties are sparsely populated. Stories of the forest seldom leave these places. Scientific measurement can be a missive from an otherwise unheard landscape. Science is not only a process of study and discovery, it is also a way to bear witness, albeit via human ears listening to a tiny portion of the forest community's many inhabitants.

In the indigenous oak forests of the region, I heard, on average, six bird species at each survey point. As I moved from point to point, the species changed, revealing variations in habitat. In all, I encountered forty-three species in these forests. Some were very common. I heard the singsong warbling of the red-eyed vireo at nearly every point. Others, the whiny scolding of the blue-gray gnatcatcher, for example, I found only occasionally. But overall, the bird community had an even mix of species, a community with many voices, not dominated by a small number of species. In older pine plantations, this diverse weave of sound was thinned to frayed muslin. Each survey point averaged four species. Across all survey points, I found twenty species. The birds tended to be the same from one place to another, dominated by red-eyed vireos and pine warblers. Younger plantations, those whose trees were just a few years old and grew ankle to shoulder high, were similarly simplified, but inhabited by birds that prefer thickets and forest edges, like indigo buntings and field sparrows.

My surveys showed not only that plantations were depauperate places for bird diversity, but also that the rest of the rural landscape was, contrary to the claims of the plantation apologists, home to rich communities of birds. Both rural residential areas and forests that had been logged but then left to regenerate without herbicides or bulldozing had bird diversity as high as or higher than that of mature oak forests. These lands retain large patches of forest, and thus many bird species, but also include brushy areas and fields that attract sparrows, buntings, wrens, and others. From the front porch of houses in these wooded areas, you can hear ten or more species singing at any one time. In all, rural settlements were home to more than sixty species in my surveys.

My surveys were possible only because of birdsong. At least 90 percent of the birds I detected I heard but did not see. Of course, such a survey misses all the silent birds—those sitting on nests, occupied with feeding during my visit, or whose singing peaked earlier in the spring—but nonetheless, aural surveys give an index by which to compare habitats. In all, I noted 4,700 individual birds across the five hundred survey points. Fed into graphs and statistical analyses, my experience of the presence of these animals was given legitimacy by the language of science and thus communicative power within human institutions. In the end, my surveys, and the extensive work of habitat mapping and analysis by a dozen colleagues, persuaded a national conservation group to successfully pressure timber corporations to stop converting native forests to plantations and to work with the state to set up conservation lands. A victory, of sorts, although by then most of the corporate-owned land had already been converted and would soon be spun off to private investment firms as part of a continent-wide divestment of land. To this day, local economies receive little economic benefit from these forests and plantations.

Maps demonstrated the extent of forest changes—from 1981 to 2000, 14 percent of the oak forest was converted, mostly to pine plantations—and analyses of bird surveys showed how these changes affected wildlife. Such graphs and statistics help us to understand and communicate. But they also serve as a substitute for lived experience by decision makers. In the Manhattan lawyers' offices where the fate of the forests was decided in a meeting of besuited corporate CEOs, forest managers, scientists, and conservation advocates, few people had spent more than a few hours on the land they controlled. There were no representatives of local communities present. In the absence of the scent of trees, the varied songs of birds, the sight of running water, and the feel of soil and tree roots in fingers, a handful of graphs had to suffice.

The sustained direct sensory experience that is the root of human aesthetics, understanding, and ethics has almost no place in our corporate structures. For large businesses and nonprofits, and for many parts of government, listening is present only in highly mediated forms.

The pine plantations I surveyed were not silent, but their soundscapes are impoverished compared with the forests they replaced. This method of growing and harvesting wood pulp directly suppresses sonic diversity. And so it is across much of Earth. Worldwide, human needs and desires are curtailing and extinguishing the voices of other species. We live in a time of rapid diminishment of sonic diversity, both in the direct extinction of other species and through the shrinkage of habitat.

Humans, especially those of us in industrialized societies, now use 25 percent of all the energy captured and made available by plants across the world, a percentage that doubled during the twentieth century and is still increasing. One species among millions takes one-quarter of the available energy and matter at the base of the food chain. In regions where agriculture dominates, our take is much higher.

Areas free from the yoke of human management are shrinking. Earth lost nearly twelve million hectares of tree cover in 2019, nearly four million of which was primary forest in the tropics, a continuation of a decades-long pattern. The loss is not evenly spread, however, with forest losses concentrated in the tropics and gains in many temperate regions such as abandoned agricultural land in Eastern Europe. Yet even in places such as North America and Europe, where tree cover is in some regions expanding, old-growth forests are still being cut, as in the Pacific Northwest and Poland's Białowieża Forest. Other terrestrial habitats are also in decline worldwide. The area of cultivated pasture has increased, but natural grasslands have declined by up to 80 percent. The area of coastal and inland natural wetlands has halved globally. We are narrowing the foundation of the rest of the biosphere. No wonder biological diversity in all its forms—genes, species, sounds, cultures, communities—is in retreat.

Sonic decline is a symptom of the loss of biological diversity. But sonic diminishment is not only an indicator of loss. Sound connects animals in the present moment, sustaining their vitality by uniting them into fruitful communicative webs. The silencing of ecosystems isolates individuals, fragments communities, and weakens the ecological resilience and evolutionary creativity of life.

Sound might also guide us to be better members of life's community. Listening connects us directly to Earth's living communities, grounding ethics and action. Lately our ears have received technological help from computerized recording devices. Unlike my bird surveys in Tennessee, these electronic ears hear the entire soundscape and can discern patterns across vast troves of sonic data. This promises deeper awareness of the voices of thousands of animal species, perhaps guiding more effective conservation action.

A diesel truck idles in the street outside, a thin plume of dark smoke wafting over the curb and across small suburban lawns. Its rumble penetrates the house and settles in my chest. The air is dry, prickly with smoke from Rocky Mountain wildfires and ozone from traffic and oil drilling. Underfoot, tufts of plastic fiber sprout wall to wall, years of wear evident in their uneven thatch. More than three months into the COVID-19 lockdown, the honey locust tree poking between concrete driveway and lawn has been my spring and summer forest. The tree is a transplant from forests to the east, planted among Austrian pines, Japanese maples, and native cottonwoods in what was shortgrass prairie, now part of the vast spill of suburbia across the Colorado Front Range. Often there are no bird or insect sounds here, or their voices are few: house finches nesting in gutters and field crickets chirping from the grass around irrigation nozzles. Instead, the soundscape largely comprises a seethe of traffic, droning heating and air-conditioning systems, hiss and spatter from lawn sprinkler nozzles, mowers and leaf blowers, and a smeared canopy of airplane noise from flights headed from Denver to the West Coast. On the edges of town, in the protected areas set aside by town planners, the traffic sounds blend with animal voices indigenous to the region: the whistled songs of meadowlarks, yips of prairie dogs, and the gruff cries of patrolling ravens.

Headphones on. Borneo: a forest in East Kalimantan Province, Indonesia, just two hundred kilometers north of the equator. I pull up a two-day continuous recording from a site in a lowland rain forest that, as far as anyone knows, has never been logged. The microphone sat in a weatherproof box hung from a tree. Researchers set up and retrieved the device but otherwise left it unattended. The eavesdropper turned the moment-by-moment life of the forest into accretions of data in a memory chip. Later this sediment of zeros and ones was copied to a laptop computer in the field, then to a server in a lab in Queensland. I press play, and the sounds of the tropical forest reawaken in the miniature magnetic coils and paper cones in my headphones in Colorado. The sound is an obedient phantom, a presence removed by human technologies from its living sylvan bodies, resurrecting on our command.

The sound is disembodied but still powerful. I cue the digital sound file to midnight in the forest and drop into shimmering insect sound. At least fifteen species are singing, and their voices cover almost all the audible range, except for the very lowest frequencies. The singers differ in the texture of their sounds, some silky, others raspy or bristly, but they are so tightly packed that I feel as if I'm suspended in a dense, lustrous cloud. The snap of falling water on waxy leaves adds an irregular beat. This is not rain, but the fat drops that fall from the tree canopy in a downpour's aftermath. A distant croak pops into the lower registers, perhaps a tree frog in the canopy. I drift in the sound, letting the insects carry me through the

Bornean night. A few voices hold steady, a bright drone. Some pulse second by second or rasp in short bursts. Others swell then recede like ocean swell, cresting every fifteen seconds, then easing back.

I wake at 1:30 a.m., Borneo time, ninety minutes after starting the playback. The forest's sounds lulled me to sleep. My ears, perhaps starved by suburbia for the diverse voices of forest life, reached into me and dialed back my consciousness. My sleep had a familiar texture, not groggy or fogged, but clear, like immersion in the refractions of water. The only other time I sleep this way is under trees when I'm taking a break from hiking or when I'm in a tent in the forest. For fourteen million years our great ape ancestors slept in tree nests. This dip into sylvan sleep might be a hazy remembrance, wakened by my ears, of a long ancestry.

Refreshed, I return to the soundscape of the Bornean forest. As the night progresses, insects continue to dominate, peppered with some thumps and twangs that I take to be frogs. Birds and primates are silent. At 3:00 a.m., the fat, even weave of trills and burrs has spun into two thick cords of trilling. Many of midnight's insects have dropped out and now half a dozen species dominate the air. By 4:45 a.m., new insect sounds, zips and chirps, take over from the steady trillers. One katydid's rasps are so soft and low that they are almost like bleats. Then, six minutes later, the first sound from a bird, a rapidly repeated *tut*, like water dripping fast from a faucet, the predawn call of the Bornean barbet, a jay-sized bird that lives in the forest canopy, its green plumage blending with the foliage as it hunts small animals and gobbles fruit. Many trees in these forests rely on the barbet and its kin to disperse their seeds. Distant whistled bird cries follow a minute later. Then, close to the microphone, rough, vigorous croaks, coming first alone, then in twos and threes, *crac crac-CRAca cra-CRA*. Rhinoceros hornbills—giant fruit-eating birds of the primary forest—are waking and sharing their morning greetings. Bird whistles and fluty notes from half a dozen species build over the next ten minutes. As the sun rises and the day unfolds, cicadas emerge, buzzing like those I am familiar with in temperate forests. A few screech like the whine of a drill or scrape like a knife on a sharpening stone. At dusk, dawn's crescendo of bird sound returns, then gives way to crickets and katydids.

I delight in these sounds, imagining the rich forest around me. But I also feel an uneasy sense of dislocation, especially if I listen for more than a few minutes at a time. My ears are fully immersed in one of the most diverse places known on the planet, but the rest of my body, including all my other senses, is in a rental house in North American suburbia. The rain forest is spiced with thousands of leafy, fungal, and microbial smells. Every tree has its bouquet, and the soil rewards nasal explorations with striking aromatic variegations. I breathe only truck fumes and the exhalations of a house interior, backed by haze from tens of thousands of fracking wells east and north of town, and a dense network of busy roads. Ants, beetles, and leeches swarm the forest floor, necessitating regular plucking from human ankles and legs. My feet now feel only the scratch of carpet fibers on bare soles. The humidity and warmth of the rain forest air blur the boundary between forest and human. There, human sweat and the dripping moisture of leaves merge, as if tree sap and human blood were one. Suburban heat, though, rises lifelessly from asphalt and is walled out of house interiors. My eyes see three plant species from my desk and, if I'm lucky, a couple of birds, not the hundreds of the rain forest. Even my gut is in a different sensory world than the sounds in my ears, well fed with nutritious food, but dissociated from the flavors and textures of food traditions around and in the forest.

Is this what human musicians felt when wax cylinders first played their music back to them? The music is there, faithfully recorded, but is removed from the contexts of place, sensory presence, and living connection. Is this what the first readers of the written word felt when language that formerly lived only in the breath was encoded on a page? I have spent my life immersed in recorded music and the written word. Through the motion sickness that I feel in this extended listening to the rain forest, a queasiness I have never felt in a living forest, am I tasting what we lost when we forsook aural culture for written words and recorded sound? For our ancestors, listening and speaking were entirely embedded within all the senses and in a singular place and time. Now music and words arrive through ears or eyes only—ears in headphones, eyes on books—and are deracinated from their place of origin. I love my records and, yes, books, but wonder how their abstractions (from the Latin *abstrahere*, "to drag away or divert") have shaped me.

I dive back in. Despite the undertow of unease, I revel in these marvelous records of one of Earth's most diverse and striking soundscapes. I click and listen again to the hornbills' waking and the cicadas' saws. Then I upload other sites from the same forests, some unlogged and others growing back after a selective commercial cut of their trees. These recordings were made as part of a research study led by Zuzana Burivalova, a professor at the University of Wisconsin, with colleagues from conservation groups and universities in Indonesia and Australia. Through multiple sound recordings across seventy-five different sites, they hoped to assess how the animal diversity of the forest was faring and make recommendations for future conservation in the region.

These recordings are staggeringly varied. At each site, hundreds of voices come and go over twenty-four hours. As I skip around in the digital sound archive, I land each time in what, to my ears, is a different sonic world. These sonic patterns are nothing like those of the cities and forests of the temperate world. Midnight in New York City is a little noisier than two in the morning, but the types of sounds are the same: sirens, airplanes, cars, and chatter on the street. Dawn in an old-growth forest in Tennessee has many more voices than noon but is largely composed of the same singers. The timbres and rhythms of sounds cycle over days and nights in these places, but not with the same granularity as in the Bornean forest. Time is denser and more finely textured in tropical forests than elsewhere. The same is true of space. As I click from one site to another, I hear contrasts matched only by the most extreme differences in the temperate world, as if I were walking from a deeply shaded forest into a swamp or open meadow, or from a busy street into an urban park. Every site in these recordings has a vigorous character of its own, defined by many layers of insect sound and hundreds of different bird, frog, and mammal calls.

As I think about the researchers, I feel a pulse of anxiety at the thought of trying to quantify the differences among these sites. These recordings comprise more than three thousand hours of digital sound files. It would take more than a year of full-time work just to listen to every recording.

Enter big data for sound. Thanks to software developed by a team at the Queensland University of Technology and Burivalova's coding and statistical analyses, we can listen for acoustic patterns in lengthy sound recordings. The software slices every recording into one-minute segments, then dices each one-minute sliver into more than two hundred frequency segments. In this way, the continuous stream of sound is cut into countable pieces. The software then looks for patterns across the whole soundscape. How, for example, do the loudness and the frequency of sound differ among sites? Is the sound saturated at some sites, with every frequency and minute filled, but more threadbare in others? How do these patterns change over the day and night?

As we'd expect from experience in the forest, the computer found a peak in the saturation of the soundscape at dawn and dusk. These are the clamorous choruses of birds, frogs, primates, and insects that mark the rising and the setting of the sun in tropical forests worldwide. Both logged and unlogged forests showed these peaks. Night was less saturated with sound in the unlogged than the logged areas, likely because some night-singing animals like katydids and some frogs are especially abundant in the open areas left by selective tree felling. In the day, unlogged forest was more saturated, a reflection of the more diverse animal communities in these forests. These are the kinds of patterns that human observers readily notice and have, over many decades, quantified with clipboard-in-hand field surveys. Selectively logged forests are home to many species, but their living communities are usually less diverse than in unlogged forests.

The analysis also found patterns that time-limited traditional surveys would likely miss. In particular, the logged forest was more acoustically homogenous than unlogged areas. As my naive ears listen to a fraction of the sound recordings, all the sites sound fabulously different. But the software was able to hear beyond such human limitations and precisely measure how similar the sites are to one another.

This work is on the leading edge of a revolution in how scientists listen to the world. In 2000 and 2001, I tromped through the forests, marking off each bird as I heard it, much as thousands of other field biologists have done all over the world in our attempts to measure, understand, and ameliorate the many effects of humans on other species. But these surveys are time-consuming and sample a minuscule portion of the soundscape.

Extended sound recordings, analyzed by computer, offer a complement to more traditional forms of field study. In addition to larger time samples and greater statistical power, the recordings solve problems inherent in surveys that rely on field observers. We all differ in our hearing abilities and identification skills, adding variability to the quality of observations. Naturalists and scientists also have taxonomic biases. It is not hard to find people who can name every bird sound in their region. But few people can identify by ear all the insects, especially in tropical regions. In addition, not all tropical species sing at once in a narrow breeding season as they do in temperate regions, necessitating surveys over many months. Scientific study quickly hits limits of human capacity and knowledge.

By processing vast troves of digital sound data, algorithms extract patterns and trends unknowable by older scientific methods. In the last ten years, the price of recording devices with capacious memories has dropped. AudioMoth, for example, is smaller than a pack of cards and can record continuously for several days or, if programmed to record only a few hours each day, more than a month. The device and its supporting software are open-source, its blueprints and code freely available to all, and, for those who prefer not to solder their own gear, costs only seventy dollars premade.

These technological advances have led to thousands of research projects that typically fall into one of two categories, reflecting two different types of software analysis. Some software is programmed to sift through the recordings, plucking out specific sounds. Managers in Cameroon's Korup National Park used a grid of recording devices to measure gunshots and the effectiveness of antipoaching patrols. Hydrophones in Massachusetts Bay have tracked cod-spawning aggregations using recordings of the fish's mating chorus, pinpointing the most productive sites and uncovering declines. Rare and threatened species such as elephants in dense African rain forests, fish in tropical wetlands, and birds in Puerto Rican forests have all been studied with the help of electronic ears deployed throughout their habitats. Species like bats and insects whose sounds are too high for human ears are also readily tracked with electronic sound recorders. Once detected and classified by the software's algorithms, sounds from multiple recorders can estimate changes in behavior and population size, or be compared with the data on other recorders to estimate the location of the animal.

The other approach is the one taken by Burivalova and her colleagues. Instead of picking out the identity of individual species, the software scans and analyzes the entire canvas of sound, measuring saturation, loudness, and frequency to find patterns across space and time.

No software can yet identify all the singing species in one place and thus dissect all the component parts of a soundscape, although some can simultaneously pick out a couple dozen voices. When I stand in a Tennessee forest and name all the birds, frogs, squirrels, and insects singing around me, and recognize the meaning and emotion in a human companion's voice, I'm outperforming the most powerful "artificial intelligence." Perhaps future technologies will surpass us, but for now, humans can still beat a computer in a contest over sonic pattern recognition. This is a reminder of the potential cost of listening through computers. As is true in so many parts of our lives, our time and attention are drawn by these new technologies inward into the human world of electronica, rather than outward into direct sensory experience of the living Earth. Even the name of the new technique, passive acoustic monitoring, suggests a withdrawal of active human senses.

In addition to their potential utility for researchers and managers today, soundscape recordings also create an archive for the future, digital memories of how Earth sounds today. Generations to come will listen with questions we cannot imagine. Every stored recording is a gift to tomorrow.

The soundscapes of years to come will be missing some of Earth's voices. Part of what we record is thus a preamble to extinction. Digital sound files will help us to grieve. They will also partly inoculate against the problem of the "shifting baseline," the diminishing expectations of each generation as it gets used to a world less filled with song. My grandfather told me how he missed the bird- and insect-rich sounds of the fields and towns of his youth in northern England. Without his story, I'd encounter modern soundscapes as "normal." Each sound recording is an anchor against this tide of forgetting.

Most automated soundscape recordings to date have been short-term and focused on particular questions and regions. But larger-scale archival work is also commencing. The Australian Acoustic Observatory, for example, is installing recorders at one hundred sites across the continent, with the aim of recording continuously, initially for five years, and making the stored sound freely available. These electronic remembrances are a technological complement to the stories we must tell one another. Data needs accompanying narrative. Hopefully, if we act now, our legacies will convey not only loss but evidence of renewed flourishing in years to come.

Despite their utility as time capsules for the future, I was skeptical about whether these technologies could help to conserve forests. Another gadget, I thought, great for naturalists and academics in pursuit of new projects, but of little relevance to slowing the ravages of deforestation. After all, we know the nature of the problem: millions of hectares of tropical forests are lost annually to fire, saws, and bulldozers. A bleeding, fading patient needs immediate practical help, not an ever more precise and technologically sophisticated diagnosis.

Conversations with Burivalova, the project's leader, and one of her coauthors, Eddie Game, lead scientist for The Nature Conservancy's Asia Pacific region, showed me otherwise. They explained how extended field recordings and computer analysis of large acoustic datasets could both guide on-the-ground conservation and attract more funding for this work. With other researchers, Burivalova and Game have also deployed recording devices to help people in Papua New Guinea to track biodiversity in forests and agricultural areas, information that then informs local decision-making about future uses of the land.

"It worked out better than I thought it would," Eddie told me. "In Borneo, the recordings are more sensitive to differences in forests than I had expected. . . . We know from our previous work and research by others that well-managed logged forests can have about the same gross biodiversity as protected forests. But this masked local differences and uniqueness in protected forests. Previous research mostly used field surveys of birds and mammals and missed these fine-grained differences. In Papua New Guinea, the sound recordings give local people a powerful and relatively cheap way of monitoring their forests.

"We're an organization that prides itself on having evidence that what we're doing is effective. When we speak to academics, they see this work as really boring science, but for us it is very meaningful to know that what we consider to be better land management practices result in richer soundscapes." He explained that the variable sonic texture and local differentiation of unlogged forests suggest that dividing logging into several small areas rather than one large one could have a lower impact, allowing local differences to persist.

"How can we help the logging industry be more friendly to biodiversity?" Burivalova asked. "Even companies that are interested in being more environmentally friendly cannot do much in terms of biodiversity monitoring. It's too expensive and difficult. Acoustic recordings could give them an easier way to gauge how they are doing."

To those steeped in the antilogging rhetoric from parts of the environmental movement, talk of conservationists working with timber companies in Bornean rain forests might seem wrongheaded. In the United States, the ravenous excesses of the timber industry have provoked a strong counterreaction. The Sierra Club, for example, opposes commercial logging on federally owned lands, even those expressly created with the aim of supporting public oversight of forestry. Loggers reliably appear as villains in North American forest-infused fiction and nonfiction alike.

But the chainsaw can paradoxically be the salvation of the forest. In Borneo, selective logging removes large, commercially valuable trees. The rest are left in place because they are either too small, or not valuable, or legally protected. These "secondary" forests—those that have been logged, often two or three times—harbor many of the same species as the primary forest. Such logging certainly has ecological costs. Some species are lost, especially those like woodpeckers and fruit-eating birds that specialize on the largest trees. Logging roads can increase erosion and serve as conduits for the arrival of people seeking land to clear for smallholdings. But if done right, logged forests regrow. Four hundred million years of evolution has taught forests resilience. Given a chance, biodiversity surges back. In Tennessee, selectively logged forests have high bird diversity, but monoculture plantations do not. In Borneo, secondary forests are havens for indigenous species when compared with industrial-scale oil palm and pulpwood plantations. Replicated bird surveys in Malaysian Borneo, for example, found that the numbers of threatened bird species were two hundred times lower in oil palm plantations than in selectively logged forests. Even in the "wildlife friendly" plantations that included fragments of remnant forest within the plantation, the abundance of these birds was sixty times lower. Plantations are also poor habitat for frogs and insects.

Both Burivalova and Game also emphasized in our conversations that the wider context of surrounding land is very important. A secondary forest ringed with plantations is biologically impoverished compared with one embedded in a forested landscape. A primary forest in a sea of secondary forests likely has a more thriving ecological community than one hemmed in by plantations.

Logging provides livelihood for local communities, work and income rooted in the regenerative power of soil and trees. Oil palm plantations and mines also provide income, but they do so at a greater cost to the fertility and diversity of the land.

We are not creatures disembodied from needs for food, energy, and shelter. Wood can be renewable. Fossil fuels, steel, plastic, and concrete generally are not. To lock up many forests in "protected areas" free of human use, then, is to exile ourselves from the community of life, forcing us deeper into unsustainable relationship with synthetic materials or forest products shipped in from elsewhere, imposing the costs of our consumption on people and forests out of range of our senses. The question should not be whether we cut trees, but where and how we should do so. We certainly need extensive areas set aside, free from the saw. We also need policies and on-the-ground enforcement to bar rapacious cutting that degrades the land. But a thriving future also requires that we participate in the forest community as all other animal species do, as consumers. This is a matter of ecological and economic realism. Our lifeblood is drawn from the Earth. People need work. The oft-cited alternative to extractive use of forest products—ecotourism by wealthy foreigners flying in from overseas—helps in some areas but spurs an increase in deforestation rates in others, is not a viable source of income for local people in most parts of the tropics, and assumes that ever-increasing international travel by the wealthy is sustainable.

In the future, sound recordings might also serve to strengthen monitoring by governments, local communities, companies, and organizations that try to monitor and "certify" the ecological soundness of wood and other products.

At present, forest certification schemes use crude measures of "sustainability" and "responsibility." Inspectors spend limited time on the ground and check off relatively easily observed indicators: are roads built to minimize erosion, are workers wearing safety gear, does the map pinned to the supervisor's office wall accord with the management plan, is land tenure clear, are known special areas like streams and wetlands protected, and does the written plan seek long-term viability of the forest? These are important questions, but they do not assess the presence of most forest species, let alone their well-being or changing fortunes. Soundscape recordings could, through the intermediaries of technology and statistics, elevate the voice of the living Earth community. The thunderous diversity of rain forest sound would then meet silent piles of human paperwork. Out of this incongruous union, a more vibrant future for all might grow.

Beyond their practical importance in land management, recordings can also spirit the forests' voices up over the Bornean forest canopy—south across the Java Sea, north across the South China Sea, east across the Pacific Ocean—into the ears of those of us who need to hear. Donors, policy makers, and grant makers hear the unearthed sound and are moved to act. The rest of us, those without the leverage of unimaginable wealth or political power, also understand through these sounds: we are connected. One-third of the plant photosynthesis that supports life on land happens in tropical forests. The wood in our houses, paper, and furniture are often rooted in Southeast Asia. The palm oil in cosmetics, processed food, biodiesel, and farm animal feed is grown on former rain forest land. But we have broken all direct sensory connection to these forests that sustain us. Sound can bring us partway back to embodied sensory understanding. We might then make wiser choices about how and whether to use the products of forests far over the horizon rather than the materials and energies close at hand.

Eddie leans forward. "People really get that sound is linked to biodiversity. I've had more substantive conversations about forest monitoring with this sound data than with anything else. They experience the forest. The thing that blows their minds is how noisy it is, constantly."

He pauses, eyes flicking upward as he searches for words.

"Through the sound, they get *close* to this almost undefinable property, 'biodiversity,' closer than any metric, or graph, or photo."

There is another "algorithm" that can "process" thousands of hours of "data" about the changing forest: lived human experience. Almost all tropical regions are home to people whose ancestors have lived within the forest for centuries or millennia. Many of these cultures are now besieged. Forest conservation is therefore a matter of human rights.

In the Western tradition, forests are often seen as places of darkness, home to brigands and exiles. Wolves of all kinds. The edges of civilization. The forest is umbral, full of confusion. Dante lost the right road in a dark and savage forest. Children become disoriented in the forests of the Brothers Grimm. Ever since the Neolithic agricultural revolution, we have cleared trees to make way for pasture, crops, and towns. Even when Western cultures desire to manage land for wood or forest conservation, they usually do so as enterprises that exclude people from the land. In the United States, for example, national forests and national parks were established by expelling from within their boundaries every single human inhabitant, save for those who retained private "inholdings" or employees in park compound housing. Contemporary US state tax incentives for keeping land in "forest" often disappear if people live within the forest. In the official statistics of both the US government and the Food and Agriculture Organization of the United Nations, forests that have houses built among them or in which people grow food count as "lost" forest, but tree plantations and barren ground left after clear-cutting count as "forest."

When this Western mindset meets tropical forest, human calamity often ensues. Governments declare forests *terra nullius*, empty land, opening a "frontier" to colonization of land that is home to people whose cultures have lived there for centuries or millennia. Corporations—both for-profit extractive industries and nonprofit conservation organizations—take title to land and drive out its human inhabitants. These are not only injustices of yesterday, enacted in the age of wooden ships, muskets, and disease-poisoned blankets. Indigenous cultures today are under sustained attack, their lands and lives taken by force and murder, and by the violence of the laws of nation-states and the global economy.

In Kalimantan, the Indonesian portion of Borneo, an alliance of fifteen organizations representing indigenous communities submitted an urgent appeal to the United Nations Committee on the Elimination of Racial Discrimination in 2020, stating that "vast encroachment on and takings of indigenous lands for road-building, plantations and mining" were underway, "all of which threatens to cause imminent, gross and irreparable harm to the Dayak and other indigenous peoples." In Brazil, also in 2020, representatives of dozens of indigenous groups vigorously opposed new laws that would further "open indigenous lands up to exploitation." Deforestation rates are now rapidly increasing after years of decline in Brazil, reaching their highest levels in a decade in 2020, more than eleven thousand square kilometers lost. Indigenous leader Célia Xakriabá of Brazil says, "I can hear the song of the birds now, but it's also a song of misery, of sadness, because most of them, they are alone. They have lost their partners. . . . And we, the indigenous are becoming more alone, because they [miners, loggers, ranchers] are taking people from us." In the Democratic Republic of the Congo, Rainforest Foundation UK found in 2019 that people "living around Central Africa's largest national park have been subjected to murder, gang-rape and torture at the hands of park rangers." The "widespread physical and sexual abuse being inflicted by 'ecoguards'" occurred in conservation parks that were first established by driving indigenous peoples from the forests.

The nonprofit organization Global Witness recorded 212 murdered defenders of the land in 2019, violence disproportionately directed at indigenous peoples, an underestimate because many deaths happen out of the media's gaze. Conflicts over tropical forest lands in Colombia, the Philippines, and Brazil topped the list. Amazon Watch reported an "unprecedented wave of violence and intimidation" in 2019: more than two dozen murders, seven indigenous leaders assassinated, and multiple instances of violence directed at the person and property of people defending forest land from mining, logging, and clearing for agriculture. "If we don't stand before the world and say, 'This is happening,' we will be exterminated," said Ermes Pete, an indigenous leader from Colombia, during a protest against rising violence and the murders in 2020 of more than 200 civic leaders.

Not only are the voices of indigenous peoples in tropical forests often not being heard, they are, in many places, being actively suppressed. Deafness to these people and the knowledge that they have of the forest is not merely a by-product of expanding industrial activity and land colonization: silencing is strategy. To listen would acknowledge the presence and the rights of indigenous people, and open the door to ways of being that are a threat to short-term extractive economies, the theft of land, and the transfer of control to outsiders.

To speak and to listen, then, are acts of resistance that can inform action. Listening can restore life-giving flows of knowledge among peoples and between people and the community of life. But not all forms of listening are equally open to the voices of the oppressed. Our modes of listening must remedy injustices and not reinforce them.

As science expands its ability to remove the ears of local people from assessment of forests—first in the tradition of foreign field naturalists flying in to "sample" biodiversity and now through electronic ears hooked to "artificial intelligence"—we often bypass the senses and intelligence of people who not only have been listening to and understanding the forests' many rhythms and cadences for centuries but whose cultures were born and now belong within the ecology of the forest. The soils and biodiversity of these forests are, in part, a product of thousands of years of care by indigenous peoples. Because many listening technologies now circumvent the need for the human senses, they carry with them the danger that such lived human experience in the forest will become irrelevant within the processes of science and policy making.

Technologies and the methods of science do not necessarily lead to injustice, but they distance us from subjective, embodied knowledge, sliding without friction into the oppressors' dehumanizing tool kits. It does not have to be so. The indigenous communities in Kalimantan appealing for help from the United Nations decried the recent removal of "environmental and social impact assessments as prerequisites for business permits." These changes to the law will allow timber and oil palm corporations to further displace indigenous communities from their lands and despoil the forest. "Environmental and social impact assessments" need, in many cases, the methods and insights of science. Eddie Game's plan to get sound recorders to local communities in Papua New Guinea, for example, now funded through a partnership with the United States Agency for International Development, aims not to usurp control but to give local people access to information they can use as they see fit to manage their land.

Listening technologies are most likely to yield positive results when they restore imbalances of power. At present, control in forests is mostly in the hands of resource-extraction corporations, governments, and, in some places, large aid agencies and conservation groups. If the many voices of the forest—human and beyond human—could penetrate these organizations, all might benefit, especially if listening is not just relegated to superficial consultation with local communities as plans from elsewhere are implemented. But a surer path to right the relationship between people and the forest is to change the power dynamic at its root by restoring indigenous peoples' control over their lands and futures.

We are a long way from such justice. A 2015 study by the Rights and Resources Initiative found that in half of the sixty-four countries they studied, indigenous communities had no legal path to obtain title to their lands. In Indonesia, less than one-quarter of 1 percent of the land is community owned or controlled, although the Indonesian Constitutional Court ruled in favor of communities' customary forest tenure rights, so there is some hope of an increase. In the United States, indigenous communities own or control 2 percent of the land area. In Australia, 20 percent. In Colombia, Peru, and Bolivia, about one-third. In Papua New Guinea, 97 percent. These figures illustrate wide differences among countries, but they also gloss over many nuances and imperfections in indigenous communities' tenure of land, including violations by governments and corporations seeking minerals and timber. In general, though, these percentages are increasing as dozens of countries decentralize control of forests. Activism by local communities, pressure from foreign donors and agencies, and limited administrative capacity of central governments have driven these changes.

Where land title and control have been returned to indigenous communities, rates of deforestation often decline. In the Peruvian Amazon, for example, eleven million hectares of land have been titled to more than one thousand indigenous communities since the 1970s. Rates of forest clearing on these lands, as assessed from satellites in the 2000s, dropped by three-quarters. During the boom of forest clearing in the 1990s in the northern Ecuadorian Amazon, deforestation rates in indigenous territories that overlapped protected areas were low. But indigenous territories that lacked these formal protections had much higher rates of forest loss, partly because local communities could not prevent incursion by mining and logging and partly because some communities chose to clear land for agriculture. A 2021 United Nations report found that Latin American forests controlled by indigenous communities were better protected than others, but that there was a pressing need to compensate these communities for the benefits such as carbon storage and biodiversity that their forests provide. In Nepal, when local communities control forest management, both poverty and deforestation decrease, especially in larger forests that have been under community control for some time. Honoring the needs and rights of local communities is an end in its own right and a necessary precondition to the work of habitat protection and restoration.

The "unlogged" areas in Burivalova and colleagues' acoustic monitoring study were located in a thirty-eight-thousand-hectare forest managed by the Wehea, a Dayak culture indigenous to the area. Ledjie Taq, chief of the Wehea, recounted in a 2017 interview with journalist Yovanda how, in the 1970s and 1980s, illegal logging, then palm oil plantations impoverished much of the forest and drove people from their land, giving them no choice but to become laborers for industry. But, he said, "The Dayak people cannot be far from the forest. The forest is a storehouse of life. . . . We gathered strength and put up a statue of our ancestors. We announced that Wehea is a customary forest [forest belonging to indigenous peoples]. We made rules for everyone, especially the local people." These rules govern hunting, tree cutting, clearing of land for agriculture, and access by outsiders.

In 2004, with the help of researchers at Mulawarman University, The Nature Conservancy, and the regional government, the forest became the largest, and one of the few, indigenous-community-controlled forests in Indonesia. In their publications, Burivalova and her colleagues called the Wehea forest "unlogged" and the areas where commercial timber harvest had occurred "selectively logged." Another categorization might name the sites "land controlled by indigenous communities" and "land controlled by central government and corporations" (the Indonesian government grants logging concessions).

Around the forest protected by the Wehea, palm oil farms, timber plantations, and mines continue their expansion at the expense of forests, feeding the global economy. Fire also takes its toll, driven by climate change and the more than four and a half thousand kilometers of drainage canals dug into the wet soil of Borneo's peat forests. In 2015, one of the worst years, twenty-two thousand square kilometers of forest burned in Kalimantan. Forty million people in Southeast Asia swam for weeks in a pall of smoke as thick as murky water. In cities hundreds of kilometers away, every breath brought into the body the vaporous, toxic ghosts of the burned forest and all its inhabitants. Chemical analyses of carbon in the smoke showed that the burned forest peat had lain buried in the soil for a thousand years or more. Urbanization is about to be added to the dire effects of land clearing for commodities and fire. In the next decade, more than a million people will leave sinking Jakarta and move to Indonesia's soon-to-be-built new capital city in East Kalimantan, about two hundred kilometers from the Wehea forest.

The magnificent diversity of a rain forest's sounds is not only the product of millions of years of past biological evolution. It is a sonic manifestation of the work of traditional custodians of the land, people whose own often endangered languages are part of this diversity of sound. Where these people's human rights are honored, life and sound often flourish. The future vitality of these richest soundscapes on the planet depends, in large part, on whether we restore the rights and agency of forest peoples. This is not a reincarnation of the "noble savage" idea from the Romantic movement in Western Europe, where indigenous people and cultures were presumed to be primitively uncorrupted by the hand of civilization, childlike in their harmony with "nature." Rather, might those of us in colonial cultures recognize that many forms of civilization have developed across the world, all of them entitled to freedom from murder, land theft, and disenfranchisement?

In a world where colonial and industrial cultures are manifestly failing to protect forests, oceans, and air—the foundation of life on Earth—it seems especially provident to allow cultures with better track records to, at the very least, have control over the lands they and their ancestors have lived on for centuries. These are not "pristine" lands. No human culture lives without effect on other species. As humans spread around the world, our arrival coincides with the decline or extinction of the tastiest and easiest to hunt animals. But some cultures have found more effective and fruitful ways to guide and contain human appetites and thus be responsible members of life's community. In an era of ecological collapse, these are the voices that should lead and advise us. Instead, many are crying out for their lives as colonialism and resource extraction continue to plunder, kill, and displace. In 2019, nearly four million hectares of primary tropical forest were lost from our planet. We've lost about the same amount every year for the last two decades. These forests are home to hundreds of indigenous cultures. Tropical forests also house most of the world's terrestrial species and huge stores of carbon. The loss of these forests is accelerating the climate crisis. The present system of governance and trade is failing in its most basic tasks: to protect the rights and homes of people, and to ensure that we bequeath undiminished the diverse marvels and life-giving properties of the living Earth.

"Culture and nature are the main wealth owned by the Wehea Dayak people," says Ledjie Taq. "If we don't look after them and pass them on to our children and grandchildren from an early age, then we won't be able to pass on anything."

The dignity and value of human cultures. The riches of nature. Look after them and pass them on. To do so, we need to listen to our animal cousins through bird surveys and recordings of the combined voices of the forest animals. But alongside these studies rooted in Western science, we also need to hear our human sisters and brothers. They have news for us about their forest homes. To listen is to honor those who are speaking. We cannot do so while also denying them agency and removing their source of life, the forest. To listen in tropical forests is to hear the need for justice.

A great silencing is underway in tropical forests. As they are lost or degraded, so too are the diverse voices within them, both human and nonhuman. In these imperiled forests, it is not only the sounds of insects, birds, amphibians, and nonhuman mammals that are in sharp decline but also the acoustic richness of our own species. Because linguistic diversity is especially high in tropical forests, deforestation is a leading cause of endangerment of human languages. The fate of sound in tropical forests, then, reveals the impoverishment and homogenization of human and nonhuman life.

Headphones off. Outside my window, a European starling lets tumble a stream of whistles and clicks, mixed with *ki-ki-ki*, an imitation of the kestrel that patrols these suburban streets. One of the five lawn service companies that tend to the turf of neighboring houses is leaf blowing grass clippings from a concrete sidewalk. A garbage truck with a bin-grabbing pincer like a stag beetle's mandibles wheezes and clatters on its rounds. Mostly, though, the house interior is quiet, an unchanging soundscape of fridge compressors and laptop fans.

These are sounds that unify the suburbs. In a world in tumult, our senses here are soothed by familiarity and predictability. It is a universal human desire to make homes that buffer us from the sensory extremes and vagaries of the outside. From Paleolithic caves to modern apartment buildings, human dwellings cocoon us and keep us safe from the threats and discomforts of cold, wind, noise, or attack from others. Industrial power has now made this buffering so complete that it imposes a disconnection that undermines the powerful relationship between sensory experience and human ethics.

Many of us now live in almost complete sensory isolation from the people, other species, and land that sustain us. Buildings cut us off with their walls, but more severe are the fractures imposed by supply chains for material goods, pipelines and wires for energy, and land-use plans that exclude native habitat from much of the suburbs and cities. Click-and-deliver internet shopping now removes us from contact even with traders and shopkeepers. The cardboard box delivered to my door is the apotheosis of colonial trade: goods shorn of any trace of living relationship to people or land.

Users, like me, of paper pulp from pine plantations or timber from Bornean forests almost never know where our goods came from. I look around at the objects in my house. With the exception of some garden vegetables, the provenance of everything I own bears no relation to my body and senses. This ignorance and isolation not only are the products of globalized trade, but are the source of the sensory alienation needed to sustain a destructive economy. With our senses cut off from the information and relationships that root and orient ethics, we are adrift. Ecological despoliation and human injustice can thus continue unconstrained by lived relationship. It was these sensory connections that mediated human environmental ethics until the colonial and industrial era.

When I first listened to Borneo's forests in the suburban room, I felt I was wrenched from one world to another. But these are the same world, deeply linked. The unruffled peace of the suburb is the corollary of the storms underway in forests and other habitats. From ruined ecologies and human societies, we extract the resources to build and sustain the calm. Manufactured quiet and predictability provide the conditions needed for continued despoliation, over the horizon, beyond the senses.

Oceans

I drop the needle onto a vinyl album. Industrial diamond meets sound ensnared in polyvinyl chloride. The claw of the record player's stylus follows the spiral furrow. The jewel follows the wavy plastic groove, every microscopic side-to-side motion conveyed to magnets and wire coils in the stylus's head. Burned coal and methane, arriving on wires strung across the sky, electrify my amplifier.

The powers of factories, oil wells, and mines converge. A humpback whale's song awakens, leaping out of the sea into air, breaching out of the 1950s into an experience of the moment.

Two long introductory cries, a pause, then a string of rumbles and throbbing pulses. The first cry is more than three seconds long, an interlacing of dozens of frequencies, each one swelling and receding at a different pace. The higher registers sweep down, a moan. The lower tones hold steady, droning, then twirl up, accenting the end. Echoes from undersea canyon walls or from the sea's surface add reverberance. The second cry is a little shorter, simpler. Its stack of frequencies flow in concert, a downward inflection that leads into a steady wail, then an up-down bump, *weeEEow*, before fading into echoes. A growl undergirds these sounds, builds in vigor, then resolves into a string of percussive jabs, a trill made of low, fleshy twangs that snake through variations of pitch and tempo.

The Cold War captured this whale's song. The work of zoologists and musicians then propelled it into public imagination, awakening human ethical concern for our sea cousins. Later, the song returned to the oceans in the form of whaling bans. The album is a triumph of interspecies listening.

But the vinyl disk spinning on my turntable is also a record of how far the soundscapes of the oceans have been degraded in our lifetimes. The oceans of the 1950s were orders of magnitude quieter than they are now. If there is an acoustic hell, it is in today's oceans. We have turned the homes of the most acoustically sophisticated and sensitive animals into a bedlam, an inescapable tumult of human sound.

The humpback that kicks off the first track of the album was recorded by Francis Watlington, a descendant of whalers who had emigrated from the United Kingdom to Bermuda in the 1600s. Watlington worked in Bermuda for the United States Navy in the 1950s and 1960s, inventing, installing, and monitoring hydrophones that eavesdropped on the Atlantic Ocean. Several patents for underwater listening devices bear his name. Archival photographs show him in cramped rooms surrounded by wires and monitors, at home in the habitat of an inventive electronic engineer.

Watlington and his colleagues ran a cable from their onshore lab to a hydrophone three kilometers offshore and down seven hundred meters to the seafloor. At this depth, they hit the "deep sound channel," the lens formed by pressure and temperature gradients that transmits sounds thousands of kilometers through the ocean. The electronic ear sought the thrum of engines and the squeal of sonar signals from enemy ships or submarines. Alongside this military intelligence, the hydrophone caught the sounds of humpback whales as they moved in springtime from the Caribbean to northern feeding grounds. From shore, Watlington could see the whales blowing and breaching over his hydrophone. The signals arriving in his laboratory revealed their sounds. Few human ears had listened at such depths before, let alone recorded the sound. Intrigued by what he heard, Watlington kept the magnetic tape in whose tiny flecks of iron oxide were contained the mark of the whales' songs, a collection that spanned the years 1953 to 1964. In 1968, he shared the by-then declassified tapes with Katharine and Roger Payne, zoologists who were visiting Bermuda to make their own recordings of humpback whales.

The Paynes, working with mathematician and scientist Hella and Scott McVay, fed the magnetic tapes into a sonograph printer, a Second World War technology that turns sound recordings into inked glyphs on long scrolls of paper. On each scroll, time passes lengthwise and the sound's frequencies are represented by up-and-down lines and smudges across the short dimension of the paper. The whale's cries look like the scratch marks of a clawed paw, parallel striations showing the many layers of harmonics. Where the cries resolve into drones or whistles, only a single line is visible, one frequency. Thumps are bold vertical stripes of charcoal. Clicks are light touches of pen. Like a musical score, the scrolls reveal to the eye both the form of each sound and the relationships among the sequences of cries, whistles, bangs, and rattles.

On scraps of paper, the internal structure of the whales' sounds became apparent. Long sequences of sound are repeated every few minutes. The Paynes and McVays discerned five different levels at which sounds were grouped and repeated: single pulses or tones, more complex cries or whistles, phrase-like clusters of these shorter elements, strings of phrases, and finally long unbroken sessions. The shortest elements lasted a few seconds. Some sessions lasted many hours. Because the sounds contained repeated structure, like human and bird sounds, they called the sounds song.

Roger Payne gathered some of the best recordings and, in 1970, put out the *Songs of the Humpback Whale* album that now spins on my turntable. These whale sounds are likely the most widely heard sounds of any individual nonhuman animal. The album sold more than a million copies. An excerpt, a flexible plastic disk included in a 1979 *National Geographic* magazine, reached ten million more, the largest single pressing in the history of the recording industry. Today digital downloads, CDs, and pirated copies continue to carry these whales' songs to millions of human ears.

In the 1970s, the recordings made the pages of the journal *Science* and were mixed into Judy Collins's song "Farewell to Tarwathie," inspired music by composer Alan Hovhaness performed by the New York Philharmonic, and were etched onto NASA's gold-plated copper album of Earth sounds on the Voyager satellites. The last was packaged with a cartridge and a needle, in case the turntable and vinyl revival have yet to reach beyond our solar system. They were also played from Greenpeace's boats as they harassed whaling vessels, and in the US Congress as testimony in debates about whale conservation. The songs of whales became both a rallying cry for the growing environmental movement and a bridge for human imagination into both the mysteries of the sea and the personhood of whales.

Watlington's ancestors hunted whales, then sent the animals' abundant oil to the cities of Europe and North America. There, whale meat and oil fed, lit, and lubricated the bodies and industrial apparatus of growing human populations. We often think of whaling through the lens of Herman Melville's accounts of sailing ships and hand-powered chases. But the kill of nearly three hundred thousand sperm whales from 1900 to 1960 equaled the entire haul of the previous two centuries. In the 1960s, another three hundred thousand sperm whales were killed. Twentieth-century industrialization—fast ships, exploding harpoon guns, and floating and onshore factories—turned whaling into an activity more like war than fishing. In the first decade of the twentieth century, whalers killed fifty-two thousand animals. In the 1960s, the decadal kill had increased to more than seven hundred thousand. In all, whalers killed about three million animals in the twentieth century. Some whale populations, such as the Antarctic blue whales, were reduced to one-thousandth of their former abundance (now edged up to about one hundredth). Most others by 90 percent or more. The voices of hundreds of thousands of singing beings were erased from the oceans.

By the 1970s, crashing whale populations and the rise of plastics, industrial animal farming, and synthetic lubricants made whale bone, meat, and oil mostly obsolete. Our physical hungers sated from other sources, we no longer needed the material substance of whales. Watlington became a different kind of whaler, capturing and storing not whale bodies, but sounds, and his haul arrived in the same markets that his forebears supplied. Watlington's and the Payne's recordings fed, lit, and lubricated pathos, curiosity, and a slow transformation of morality. After providing bodily sustenance for many human generations, in the 1970s whales turned, especially in industrialized English-speaking cultures, into ethical goad, muse, and metaphor.

The humpback's songs fell on ears ready for an emotionally charged expression of both despair at destruction and hope for the future. In the United States, the Environmental Protection Agency and Earth Day were both founded in the same year as the album's release, the result of years of activism. At the same time, the United Nations was planning its first environmental conference. It helped that humpback whales sound mournful to human ears. Moan, wail, cry. A lamentation and dirge from below the waves. As Pete Seeger sang, "the passionate wail / That came from the heart / Of the world's last whale." Had Payne put out an album of other whale sounds, the project might have flopped, disks remaining unsold in a warehouse. Sperm whales use streams and clusters of clicks both to communicate with one another and to explore their world through echolocation. They creak like old door hinges, clack like metronomes, and, when gathered in groups, hammer and peck like dozens of frenzied woodpeckers. Played at the right volume, they could also blow out your ears, the loudest animal sound known. Minke whales boing, pulse, thump, and twang, their calls rubbery and percussive. Fin whales *oop* and grunt mostly too low for human ears to hear. The North Atlantic right whales' groans sound as though they've arrived through a long resonant drainpipe. They also "gunshot," like a large-caliber rifle. The gray whales' wavering complaints are croaks and bellows like those of distressed bulls or fierce growling cats. Most of these sounds miss the sweet spot for human sonic perception and emotional reaction. Their complexities come in forms alien to our ears and neural processing. The sperm whales' clicks, for example, are richly endowed with meaning, conveying individuality, clan, and familial identity, and, it seems, continually changing social and behavioral intentions. But we humans hear them as mechanical clacking. The tempo, frequencies, cadences, and timbres of the humpback whales' sounds overlap enough with those of human speech and music that their sounds evoke empathy.

Our senses bias us toward feelings of kinship with species whose communicative sounds most closely resemble ours. Because concern follows closely on the heels of empathic connection, our senses shape our ethics. Without sensory connection, we fail to enter into the embodied relationships that are the foundation of ethical deliberation and right action. But these senses can also bias and prune our regard for others, elevating some species and obscuring the rest.

Now that human action is the dominant force shaping the future of the planet, our sensory biases and bodily hungers remake the form of the world, preserving the parts that grab our senses and, often, discarding or abusing the rest.

Our senses, and thus our ethics, now face two challenges in relation to the seas. The first is that the ocean lives almost entirely outside of the reach of our senses. A visit to the ocean shore reveals little of what lives below. This is the barrier that early recordings of whale sound partly breached. The second challenge is that the few sensory connections we have to the undersea world do not faithfully represent the present state of the ocean.

The recordings of whales from the 1950s and 1960s come to us from another world, a time when suboceanic noise had only just begun. Contemporary "whale sound" albums and nature documentaries deliver soundtracks carefully recorded and edited to avoid and remove the clamor. A search for "whale sounds" in online music stores yields hundreds of albums that promise relaxation, sleep, meditative calm, and help with tinnitus, stress, and "holistic" healing. Unsurprisingly, humpback whales are the stars—few people de-stress by having their bodies zapped and muscles paralyzed by blasts of sperm whale echolocation pulses. The "authentic nature sounds" delivered by these albums omit the blare and cacophony of the lived experience of real whales. When the 9/11 attacks reduced large-vessel traffic in the Bay of Fundy, the levels of stress hormones in North Atlantic right whales dropped. These hormone samples were extracted from whale poop found by trained sniffer dogs leaning over the prow of small boats, their noses guiding human scientists to the floating records of whale stress. To be *authentic*, a whale soundtrack ought to suffuse our blood with alarm chemicals and steep our mind in anxiety and dread, distress rooted in the infernal noise that we pump into the whales' world. Instead, we feed ourselves the aural equivalent of synthetic tranquilizers, manufactured anodynes for the senses and soporifics for ethical discernment and action.

In the 1970s and 1980s, activists succeeded in preventing the total extermination of the whales. Numbers of some species have grown back. A few populations, gray and humpback whales in the North Pacific, for example, may have now recovered to prehunting levels or even higher. But most whale populations are still far lower than prehunting numbers. These are metrics of whole populations. For some of these populations, the prognosis for survival has bettered. For others, the apocalypse is still imminent. For individual whales, though, the present world is severely degraded. Many are trapped or wounded by plastic, including entanglement in discarded rope. They can no longer sleep or cruise the surface without injury because ship strikes are a major source of mortality. And even at the height of the whale hunt, the sounds of the ocean were largely as the whales' ancestors had experienced for millions of years. That world is now gone.

Ah, the aromas of the sea. Sulfurous seaweeds. Ammoniac stench of gull roost. Lung-constricting acidity of diesel fumes. Bilge water's oily sheen, high in the nasal passages. A fresh nip of forest scent blows in from the Douglas-fir trees that throng low, rocky hills behind the marina, a dark breath of moss and wet ferns.

All aboard! We clatter along the metal gangway, banging backpacks, coolers, and cameras against its railings. Our tourist cruise is scheduled for six hours, but we've brought enough freight for days. No problem with ballast. I wedge onto a plastic bench, facing the port railing. The two dozen other passengers array themselves along the benches or prop against the small wheelhouse. As we cast off, bags of potato chips crackle and a vinegary aroma mixes with the engine exhaust.

Vibrations from the boat engine thrum in our chests, a sound so deep that most of it escapes our ears, perceived instead through nerves in muscles and organs. The drone is calming at first, perhaps a bodily memory of the hum of blood and heartbeat in the womb. As the day passes, calm will turn to exhaustion from relentless inner shaking.

As we get underway, I feel a surge of delight to be on the water, away from conference rooms and computers. The low hummocks of the San Juan Islands slide past as we thread the waterways. The bow cleaves a gray-blue sea, stirring rafts of common murres and pigeon guillemots into skittering flight. Knots of floating giant kelp and eelgrass drift past, some with crabs atop the dislodged tangles. Waifs of sea fog linger in the island coves. The boat's speed delivers a rush of tangy sea aroma: algal iodine and briny mud.

We're whaling with cameras, joining a flotilla of a dozen other boats from harbors all around the Salish Sea. The fuzz and beep of ship radios stitch a net over the water, a blurry facsimile of the wide-ranging sonic connections of the whales themselves. Every skipper hears the voices of the others, relayed by electromagnetic waves. The quarry cannot escape. WHALES GUARANTEED shout the billboards on shore.

We motor on, cutting a sinuous path as we weave around island headlands. A sighting . . . close . . . off the southwest shore of San Juan Island. Through binoculars: a dorsal fin scythes the water, then dips. Another, with a punch of mist as the animal exhales. Then, no sign. But the whales' location is easy to spot. A dozen boats cluster, most slowly motoring west, away from the shore. We power closer, slowing the engine until we travel without raising a wake and take our place on the outer edge of the gaggle of yachts and cruisers.

A sheet of marble skates just under the water surface. Oily smooth. A spill of black ink sheeting under the hazed bottle glass of the water's surface. Only when the notch between the animal's flukes guns past and disappears does my conscious mind catch on. The whale's approach was a gesture made entirely of muscle. Power like a liquefied kick from a draft horse. Frictionless motion, a river-smoothed stone flung across ice. *Praaf*! Surfacing fifteen meters ahead of the boat, the exhalation is plosive and rough.

The pod of about ten animals comes to the surface. Part of the L pod of orcas, our captain says, one of three pods that form the "southern residents" in the waters of the Salish Sea between Seattle and Vancouver, often seen hunting salmon around the San Juan Islands. Others—"transients" that ply coastal waters and "offshores" that feed mostly in the Pacific—also visit regularly. The L pod continues west, heading toward the Haro Strait. They move like waves: head up, blow a puff of breath, back and dorsal fin arch up, head plunges down, tail rises then slaps at the water. The undulation seems leisurely, easy, but the whales' purchase on the water is evident in their speed. No human kayaker could keep up this pace. Our engines purr as the U-shaped arc of boats tracks the pod, leaving open water ahead of the whales.

What to call them? Killer whales? But every animal kills to live, save for the few—corals and the North American spotted salamander—that have invited photosynthetic algae under their skins. A humpback slaughters more animals in a single swallow of plankton than these whales manage in months of hunting fish or seals. Orca? From the Roman god Orcus, overlord of the underworld and of broken oaths. A name that carries within it a memory of a severed bond. Or *qwe'lhol'mechen*, the Lummi Nation name, "our relatives under the waves." Each moniker holds a mirror, perhaps, to the culture that speaks it: killer, promise breaker, or cousin.

We drop a hydrophone over the boat's gunwale, its cord feeding a small speaker in a plastic casing. Whale sounds! And engine noise, lots of engine noise.

Clicks, like taps on a metal can, come in squalls. These sounds are the whales' echolocating search beams. Air from storage sacs under the blowhole punches across "phonic lips" that squeeze together and vibrate. The sound shoots forward through the head, where it passes through a fatty lens whose layers of different viscosities focus the sound waves before they stream out of the forehead. When these sonic bullets hit solid objects, they ricochet back at the whale. Fatty tissues and elongate bones in the lower jaw receive the sound and shunt it to the middle ear, acting like sponges and reflectors for sound waves. Every object reflects sound in a different way, and the whales use the echoes not only to see through the murky water but to understand how soft, taut, fast, or tremulous matter is around them, using sound as we use a sense of touch. Because sound waves in water pass readily into flesh, this tactile sense also penetrates other animals. X-ray touch, delivered by sound. This ability is shared by all seventy-two species of toothed whales—dolphins, porpoises, narwals, sperm whales, and beaked whales—but is absent in the fifteen species of baleen whales, such as humpbacks, blues, rights, and minkes, although these animals are also highly sensitive to sound, orienting in the tenebrous depths by listening to the three-dimensional structure of sound around them. Vocalization and hearing for whales are as if the human senses of touch, kinesthesia, sight, and hearing were united, drawing into our bodies the motions of trees around us, the inner forms of animal companions, and the textures of distant rocks and buildings.

Mixed with the staccato of the whales' clicks are whistles and high squeaks, sounds that undulate, dart, inflect up, and spiral down. These whistles are the sounds of whale conviviality, given most often when the animals are socializing at close range. When the pod is more widely spaced during searches for food, the whales whistle less and communicate with bursts of shorter sound pulses. These sonic bonds not only connect the members of each pod, but distinguish the pod from others. Pods are matrilineal. Shared lingo—the distinctive tonal quality and patterns of whistles and pulses—marks affiliation with a group of mothers and grandmothers. All seventy individuals in "southern resident" pods share call types, including rich warbles and harsh honks, whereas pods of "northern residents" among the islands and inlets north of Vancouver Island are more screechy. The "transient" and "offshore" groups that also swim these waters have their own sonic cultures and mix only with whales of their own kind. These differences are conservative, lasting decades and perhaps longer, and mark firm boundaries among groups. "Our relatives under the waves" live in societies whose hierarchical structures are both mediated and preserved through sound.

Every group has a particular hunting behavior. "Southern residents" feed primarily on Chinook salmon, along with some other fish and squid. "Northern residents" are also fish specialists. "Transients" feed on marine mammals, with a particular fondness for seals and porpoises, and will also chomp at seabirds. Compared with "southern residents," these mammal hunters are very quiet, especially as they stalk, listening without echolocating or chattering, although they erupt in sound after a kill. The "offshore" community hunts a wide range of fish, as well as blue and Pacific sleeper sharks. Our names for these cultures are misleading: "residents" also make long offshore journeys, to California for the southern group and Alaska for the northern whales. "Transients" are no more nomadic than the others. These animals all belong to the same species, but they live in communities mostly walled off from one another through cultures of sound and hunting styles. The same is true in other parts of the species' nearly worldwide range. In Antarctica, five different communities live together but seldom mix, variously specializing in hunting different species of whales, seals, sea lions, penguins, or fish. These communities have diverged genetically from one another, especially in the far northern and southern edges of the species range.

Here off the coast of San Juan Island, the whales' voices are like fine silk stitched into a thick denim of propeller and motor sound, clicks and whistles sometimes audible but often disappearing into the tight weave of engines. Through the hydrophone, our boat sounds like an unbalanced fan, a wobbly churn. Pistons merge into a low grind. The dozen other boats, all creeping along under engine power as they track the whales, interweave their throbs, whirs, and shudders. Combustion engines swaddle the whales in an inescapable, constricting wrap.

As the U-flotilla follows the whales, a rigid inflatable boat with SOUNDWATCH (for Soundwatch Boater Education Program) blazoned on its side weaves among the other boats. The three people on board wave to the gaggles of tourists gathered at the boats' railings. Then a cruiser cuts in front of the path of the whales. The inflatable guns its outboard and arcs so that its path meets the miscreant. Some friendly hand gestures. A long pole delivers a leaflet. Boater education achieved. The inflatable returns to the cluster and bounces among the private motor boats, delivering more leaflets.

Since the early 1990s, Soundwatch has deployed small vessels in the areas most favored by the whales and their boating watchers, averaging more than four hundred hours of patrols per year. Over that time, the number of private and commercial boats seeking whales has increased, although the number coming close to the pods has decreased, perhaps as a result of regulations and volunteer guidelines that now reduce boat speed and closeness of approach. Unlike the in-your-face tactics of Greenpeace activists zipping in inflatables around whaling ships in the 1980s, Soundwatch aims to "politely initiate communication" and inform boaters about how to minimize disturbance of whales. They also collect data on boater behavior. Over the years, the most common violators of the "no-go" zones and boat speed limits are the skippers of private vessels, often those passing by on their way to go fishing or cruise the islands.

Feeling the engine throb through the soles of my feet on the boat's deck, I sense that the chorus of chugging engines that partly encircle these whales, even if it meets "guidelines," is hardly a neighborly welcome. Every turn of the propeller blade, no matter that we're going slow and avoiding close approach, is a tap delivered to the vibration-sensitive fat-filled lower jaw of the whales. I "politely initiate communication" and query our genial captain about sound and the whales. "Naw, we're not bothering them. If we keep our distance and go slow, no problem. Look at them, they're playing now."

In the distance, I see two huge ships, a container ship and an oil tanker headed north through the Haro Strait, likely bound for Vancouver, the largest port in the region. Our hydrophone's portable speaker is too small to relay most of these ships' low noise, but with heaphones on, I hear a continuous background rumble. These are two of the more than seven thousand large vessels that combined make more than twelve thousand transits through the strait every year. These range from bulk carriers, to container ships, to tankers, many of which are two to three hundred meters long. Large vessels also ply the waters west of the Haro Strait, headed to ports and refineries in and around Seattle and Tacoma. Each one of these vessels makes sound audible underwater from tens, sometimes hundreds, of kilometers. Unlike small pleasure boats that are usually moored at sundown, these large vessels make noise all night and day and are often most active and loudest at night. The largest container ships blast at around 190 underwater decibels or more, the equivalent on land of a thunderclap or the takeoff of a jet. In contrast, pleasure boats and passenger ships sound at about 160 and 170 underwater decibels, respectively. The decibel scale is logarithmic, and so the largest ships release thousands of times more sonic energy than small boats. The racket comes from many parts of the ships. Hulls stir the water into a low roar as they slice through. Fuel explodes in pistons and animates the metallic thrashing of engines as large as office buildings. Propellers spin so fast that water cavitates at the blades' tips, creating air bubbles that blast as they then implode, smearing into a rumble and hiss. These sounds block both the echolocation and communication of the whales.

The "southern resident" whale community whose life centers on these waters cannot bear the noise, especially in the long run. Their population is in decline, likely headed to extinction unless the world gets more hospitable. In the 1990s, the community numbered in the nineties. Now they've dropped to the low seventies, losing one or two more animals every year without raising new calves. In 2005, they were listed under the Endangered Species Act. No single factor is responsible, but the interaction of shipping sounds, dwindling food supply, and chemical pollution is, for now, closing the door on their future.

These whales are the falcons of the ocean, rocketing down one hundred meters or more in pursuit of their nimble and speedy prey, the Chinook salmon. In the gloamy, silty depths, visibility is poor, but the fish's swim bladders are bright in the echolocating beam, bubbles of sound-reflective air. Sound frequencies of boat noise overlap with the clicks that the animals use to echolocate and find their prey. Noise raises a fog, blinding the hunters. If a whale is within two hundred meters of a container ship or one hundred meters of a smaller boat with an outboard engine, its echolocation range is reduced by 95 percent. This is true worldwide but is an especially acute problem in and around the Haro Strait. Models of shipping traffic suggest that in this region, large ships account for two-thirds of the noise that mars the whales' hunting. The remainder of the noise comes from smaller vessels, including the whale watchers that swarm the animals. Worldwide, small-boat traffic is only a sonic problem for whales close to shore and near busy ports. Over most of the oceans, it is the noise from large ships that fogs their hearing.

In air, we hear only a low groan from passing vessels. The sound is mostly transmitted down, below the waves, and the aerial portion is quickly dissipated. Under the surface, the sonic violence of powered boats travels fast and far through the pulse and heave of water molecules. These movements flow directly into aquatic living beings. Sound in air mostly bounces off terrestrial animals, reflected back by the uncooperative border of air to skin. Our middle ear bones and eardrum are specifically designed to overcome this barrier, gathering aerial sound and delivering it to the aquatic medium of the inner ear. Sound, for us, is focused mostly on a few organs in our heads. But aquatic animals are immersed in sound. Sound flows almost unimpeded from watery surrounds to watery innards. "Hearing" is a full body experience. For toothed whales, the embrace of sound is deeper. Ship noise envelops their sense of echolocating "sight" or "touch," as if the noisy trucks rumbling past my window were pressing their tarry sounds into my eyes and skin. For most whales, and for many fish and invertebrate animals, eyes are only occasionally useful. In the abyssal depths, the animals swim in ink. Along coasts, the water is so turbid that animals see, at most, a body length ahead. Sound reveals the shapes, energies, boundaries, and other inhabitants of the sea. Sound is also a communicative bond. In the ocean, as is true in the rain forest where dense foliage occludes vision, sound connects you to unseen mates, kin, and rivals, and it alerts you to nearby prey and predators. In much of the ocean today, though, it is as if every rain forest tree had a ship's engine blaring from its trunk.

If salmon were abundant, all this noise might not be a problem. Even a blinded falcon can snatch quarry from a teeming flock. But the Chinook salmon that compose most of the whales' diet here are in crisis. Dams, urbanization, agriculture, and logging have cut off or degraded most of the freshwater rivers and streams in which the fish spawn and live out their first months. Pollution, fishing, and a warming ocean kill the smolts and adults as they complete their journey from fresh water, to estuary, to ocean, and back, a loop that takes three or more years. Chinook salmon numbers in this region have declined 60 percent since the 1980s, likely more than 90 percent since the early twentieth century. Pollutants add to the burden. The bodies of the whales in this region are among the most toxified of any animal. PCBs are a legacy of industry. DDT lingers from past agriculture. Flame retardants from people's homes vaporize, glom on to dust, and are washed downstream. Partly because of this toxic burden, few calves are born to these pods and neonates usually die shortly after birth.

The combination of noise, declining prey, and contaminants is deadly. Under current conditions, models forecast, at best a fragile "southern resident" population. Any additional stress will send them to extinction. To increase the whales to their former abundance, Chinook salmon would have to be sustained at or above the highest levels known since the 1970s. Instead, salmon are dwindling. Robust mitigation of noise and contaminants could nudge the population up, but only if shipping were greatly slowed and a century of pollution were to be reversed. Hope lies in a confluence of actions. Models suggest that if we could reduce acoustic disturbance by half and increase Chinook salmon populations by one-sixth, the whale population could once again be viable. The "northern resident" population lives, for now, in quieter, less polluted waters, preying on more abundant fish, and are faring much better.

From 2017 to 2020, the Port of Vancouver enacted a voluntary slow-down of shipping traffic headed through the Haro Strait. For thirty nautical miles, large vessels slowed, adding about twenty minutes to the ships' voyages. Ship noise increases with speed, and so dialing back the throttle lessens the cacophony in a place where the southern resident whales often feed. More than 80 percent of vessels complied with the project and hydrophones deployed around the strait found a reduction in noise levels.

Yet traffic increases yearly in the region, more than eliminating the quiet gained by shaving some noise from each passing ship. In 2018, crude oil exports from Vancouver jumped by two-thirds, mostly headed to China and South Korea. In 2019, the Canadian government approved an expansion that would nearly triple the capacity of the pipeline that supplies much of the oil from the tar sands region of Alberta. Vancouver's port is expanding and, in 2020, was waiting for approval and funding for a 50 percent increase in size. In 2019, the nonprofit Friends of the San Juan cataloged more than twenty other proposals to build new or expanded shipping terminals for containers, oil, liquefied gas, grain, potash, cruise ships, coal, and car carriers in the region. If approved, these would increase traffic by 35 percent, not counting increased tugboat, barge, or ferry traffic. If increased shipping is blocked in Vancouver and if demand for shipped goods does not decline, traffic will be displaced to other ports, some in regions so far protected from heavy industry. For example, although proposals for new liquefied natural gas shipping terminals in and around Vancouver have been withdrawn or blocked, new gas pipelines and shipping routes are being developed in places with less opposition. Seven hundred kilometers north of Vancouver, the fjords that lead to the port of Kitimat are home to several species of whales living in relatively unpolluted and quiet waters. A liquefied natural gas terminal is under construction there that is slated to add seven hundred new large-vessel transits, a more than thirteenfold increase, not counting the powerful tugs that would accompany the tankers as they navigate rocky fjords.

The United States Navy also plans expanded exercises in the region, including the use of explosives and loud sonar. By its own estimates, across the Pacific Northwest coast, navy "acoustic and explosive" exercises, including those in the favored waters of the "southern residents," will kill or injure nearly three thousand marine mammals and disrupt the feeding, breeding, movements, and nursing of 1.75 million more. The ocean falcons face both thickening fog and a navy that proposes to permanently cloud their eyes.

The whales in and around the San Juan Islands and the Haro Strait live in a constriction point for much of the trade that passes between Asia and North America, supplemented with some shipping from the Middle East and Europe. The vast majority of the consumer goods and bulk commodities that move between the continents do so on ships. I look around at my material possessions. Whales, either in the Haro Strait or perhaps off the coast of Los Angeles, heard the arrival of every item made in a country on the Pacific Rim: laptop, silverware, watering can, furniture, and car. Whales living along the Atlantic coast were immersed in the sounds of deliveries from Europe and North Africa: office chair, books, wine, and olive oil. Having lived most of my life inland, many hours' drive from the sea, I have seldom seen or heard whales. But the whales hear me. They are immersed in the sounds of my purchases from over the horizon every day of their lives.

The converging shipping lanes around major seaports are focal points for a noise problem that extends across the oceans. In the 1950s, when Watlington recorded the humpback whales off Bermuda, about thirty thousand merchant vessels plied the world's oceans. Now about one hundred thousand do, many of them with much larger engines. Tonnage of cargo has increased by ten times.

Ambient noise captured by hydrophones on the Pacific coast of North America has increased by about ten or more decibels since the 1960s, when the measurements started. By some estimates, the energies of noise pollution in the world's oceans have doubled every decade since the mid-twentieth century. The noise is worse around the major shipping lanes that connect major ports across the northern Pacific and Atlantic, for example, but because sound propagates readily in water, the rumble reaches for hundreds of kilometers. When a large ocean-bound ship crosses the continental shelf, its sound shoots to the deep ocean floor, several kilometers down, then bounces up off the sediment and into the deep sound channel. This channel carries the noise thousands of kilometers. Like smoke in a room, the haze is worse close to the smokers, but spreads out from its sources to fill the entire room. Across much of the world, it is now impossible to measure the wind-stirred "background" levels of ocean sound. In a few places, shipping noise is less pronounced, especially in the southern oceans around Antarctica and in places where islands and seamounts act as sound shields.

Near to shore, small-boat traffic adds another, higher-pitched, layer of sound, as I discovered on the deck of the whale-watching boat. The number of recreational boats in the United States increased by 1 percent per year for the last three decades. In coastal Australia, the annual rate of increase in the number of small boats has recently reached up to 3 percent. The sound from these smaller vessels does not travel as far, but is the dominant sound source for many animals living in coastal waters. At close range, sonar—sounds emitted from shipboard devices to detect the seafloor, schools of fish, and enemy submarines—can add to these higher-pitched noises. Some naval sonar is loud enough to permanently damage the hearing of marine animals at close range.

Into this global mire of noise comes the loudest human noise of all, the percussive beat of our industrialized search for buried sunlight.

Like whales seeking their prey with echolocating clicks, human prospectors blast sound into the ocean, seeking oil and gas buried under ocean sediments. Ships drag arrays of air guns that shoot bubbles of pressurized air into the water, a replacement for the dynamite that was formerly tossed overboard for the same purpose. As the bubbles expand and collapse, they punch sound waves into the water, an industrial version of the fizz of snapping shrimp claws I heard on Saint Catherines Island. These waves spread in all directions underwater; those that go down penetrate the seafloor, then bounce back when they hit reflective surfaces. By measuring these reflections from the ship, geologists can not only see through the water column but also build a three-dimensional image of the varied layers of mud, sand, rock, and oil tens or even hundreds of kilometers under the seabed. Like a whale guided by the reflective ping of a Chinook salmon, oil and gas companies use sound to find their quarry. But unlike the click of a whale, these seismic surveys can be heard up to four thousand kilometers away.

The blast of an air gun emerges from a meter-long missile-shaped canister towed behind the survey ship. The sound can be as loud as 260 underwater decibels, six to seven orders of magnitude more intense than the loudest ship. The guns are typically deployed in arrays of up to four dozen. These batteries go off about once every ten to twenty seconds. The ship tracks methodically back and forth through the ocean, like a lawn mower, in surveys that can run continuously for months, covering tens of thousands of square kilometers. When the surveys encompass the open ocean, beyond the edge of the continental shelf, as they frequently do in this era of expanding numbers of deepwater oil rigs, the sound flows into the deepwater channel and, like shipping noise, spreads across ocean basins. In some years in the North Atlantic, dozens of surveys run at once and a single hydrophone can pick up the relentless sound of seismic surveys off the coasts of Brazil, the United States, Canada, parts of northern Europe, and the west coast of Africa. Seismic surveys are widely used wherever unctuous treasure might be buried under the sea, including Australia, the North Sea, Southeast Asia, the Middle East, and South Africa.

Underwater seismic pounding feeds every one of us who use oil and gas. Yet we have no shred of sensory experience of the consequences of our hunger for these fossils. Stand on an ocean shore, and you will not hear the sound of seismic surveys. Take a ship into deep water and, even there, water's reflective boundary and our air-adapted ears shield us. Analogy fails too. A pile driver in your house, running without stop for months? That gives an approximation of the loudness and relentlessness, but we can walk away from the house, and even when we stand next to the machine, the assault mostly affects only our ears. For aquatic creatures, sound is sight, touch, proprioception, and hearing. They cannot leave the water. Few can swim the hundreds of kilometers necessary to escape. The pile driver is coupled, minute by minute, to every nerve ending and cell, suffusing them for months on end with the violence of explosions.

Ocean life, especially near to shore or along busy trade routes, now lives in a din previously unknown except near underwater volcanoes or during an earthquake. Wind-stirred waves, breaking ice, earthquakes, the motion of bubbles in water columns, and the sounds of whales and snapping shrimp are the sounds to which marine life is adapted. But the blast of air guns, the needling and stab of sonar, and the throb of engines are new and, in most places, far louder than just a few decades ago.

The worst places in the ocean are now intolerable for much ocean life. Whales flee areas in which seismic testing is underway. A study off the southwest coast of Ireland found nearly a 90 percent decrease in sightings of baleen whales and a halving of sightings of toothed whales during active seismic surveys compared with "control" surveys with no blasting. Air guns also decimate the base of the ocean food web, the plankton and larvae of marine invertebrate animals. In an experiment off the coast of Tasmania, a single air gun killed every krill larva—a key prey animal in the food web of southern oceans—within more than a kilometer and wiped out most other plankton. The sound waves from the blast may have shaken the animals to death and, for the survivors of the initial shock, so ripped up the sensory hairs that cover the animals' bodies that the plankton soon died, stripped of any ability to hear or feel their world. The sensory systems of larger invertebrate animals like lobsters can also be permanently destroyed by exposure to seismic surveys. Yet trade groups for the oil exploration industry continue to lobby for relaxed regulation of seismic testing, claiming that there is "no known detrimental impact to marine life" of large-scale surveys. They also claim that because the blasts go off every ten seconds and each impulse lasts one-tenth of a second, "sound is only produced for one percent of the entire survey period." By this logic, a boxing match is not a violent affair and a beeping smoke alarm is mostly silent.

Naval sonar—the high-amplitude blasts of sound used to "see" below water through reflected echoes—can cause whales to dive and surface so fast that their veins bloat with nitrogen bubbles, connective tissues disintegrate, and organs hemorrhage. Sound bleeds them to death from within. Under assault from sonar, some whales come into the surf, try to hide behind rocks, or beach themselves in a bid to escape the torturous whine. These strandings and frenzied bids to escape the water bring the whales into the human visual realm, a rare sign accessible to the human senses of the crisis below the waves.

Even when sound is not immediately lethal, it exacts a toll. A recent review of more than 150 scientific studies of whales, dolphins, seals, and other marine mammals found that noise reduces feeding, cuts off echolocation, increases time spent traveling, decreases rest, changes the rhythms of diving, and drains energy reserves. Some species respond to ship noise by increasing the loudness and rate of their calls, others go silent.

Whales are social animals, living in continuous acoustic contact with their families and cultural groups. Whaling greatly reduced the complexity and abundance of these societies. Noise further degrades and severs social bonds. In highly social terrestrial animals, we know that reducing or eliminating connection to others injures, and in extreme cases kills, individuals. Less is known about the physiology and psychology of whales than about land-dwelling animals, but it is likely that noise increases distress and, in the long term, narrows the sonic pathways through which whale cultures thrive and evolve.

Noise also changes both the behavior and physiology of fish. In a noisy environment, they often become agitated, darting about as if a predator were close. But when a real predator shows up, they seem unable to defend themselves, not startling and speeding away as they should. For fish that use sounds during their breeding displays, noise has variable effects. Some species ramp up their calling, perhaps to shout over the background, but others go quiet. For many, noise either blocks or greatly reduces the range over which they can be heard. Some fish obsessively clean their nests and tend to their fry when noise rises, an increase in effort that, like their increased swimming, costs them both energy and time. When feeding, fish exposed to noise catch fewer prey, are less efficient, and find it harder to discriminate between good and bad food. Fish in noisy places have higher levels of stress hormones, and the development of their hearing suffers. In some species, mortality rates double from the combined effects of these changes.

The negative effects of noise even penetrate the ocean sediments. A study of burrowing clams, shrimp, and brittle stars showed that they change their behaviors in noisy conditions, reducing their movement and feeding. These changes to seemingly obscure creatures in ocean mud have consequences that ramify throughout the ecosystem. The burrowing and mud-filtering activities of these animals partly control the movement of nutrients in the ecosystem, including how fast these chemicals are recycled into the web of life or buried in deep layers. If this study represents a general finding, the noisiness of our oceans may leave its impression even in the stone that is left behind from our era, discernible to future geologists as a changed chemical signature in mud and rock, alongside the plastics, pollutants, and acidity that we have cast into the waves.

Off the west coast of San Juan Island, our whale-watching boat leaves the flotilla, the allotted time for our jaunt completed. The whales had swum north and circled back toward the island, their retinue following at a distance. We saw no more close approaches, but we gazed on their harlequin backs and flukes as the whales dallied at the surface.

Returned to shore, I feel wobbly on the unmoving asphalt. In a few hours, my muscles and inner ears came to know and expect the motions of water. When I feel steady enough, I get in the car and turn on the ignition. Gasoline squirts into pistons. It was likely barged here through the Puget Sound. Tree latex and fossil oil in my tires whirl over the road, flaking rubbery dust onto the impervious surface, a silt destined to wash to the sea. Back at a hotel, I plug my laptop, imported from across the Pacific via ship, into the wall socket. The screen's glow and microchips' warmth are fed by turbines in dams across formerly salmon-filled rivers, supplemented by the splitting of uranium atoms and the combustion of coal and gas. I lie down on a mattress permeated with flame retardant.

Headphones on. Click: Orcasound.net. Click: listen live. As the sky fades from dusky gray to security-light pearl, I drift with the sounds of water ticking and sloshing against a hydrophone thirty meters off the west coast of San Juan Island. Gentle knocking sounds. A crab moving around the kelp? A high whine, like an electric motor, runs for two minutes, then cuts off. A few outboard engines pass, atonal whirs. Through the night, the sound threads in and out of my sleep, waking me into confusion before dawn with the burr and slash of propellers gunning a boat through the water.

The noise in the ocean today is infernal, but not hopeless. The acoustic devilry that we daily stream into the subsurface world can be stopped. Unlike chemical pollution that lingers sometimes for centuries, or plastics that will persist for millennia, or the death of the coral reefs that will not be reversed for millions of years, sound pollution can be shut off in an instant.

Silence from humans is unlikely, though. Whether or not we are aware of our dependence on the sea, we are a maritime species. The energy and materials that supply our bodies and economies move largely by ship. Most of our oil, gas, and food travels among continents by sea. There is little chance, therefore, that the noise will cease entirely. But quieter oceans are within reach.

It is possible to build almost silent ships. Navies have been doing so for decades. Some submarines are so stealthy that their presence can only be revealed by pumping into the water sonar loud enough to deafen any passing dolphin. Fisheries researchers seeking to measure fish abundance and behaviors do so from vessels with engines, gears, and propellers engineered to reduce noise and thus not alarm fish. The hush from these ships comes at the cost of efficiency and speed. Yet even for large commercial vessels, noise can be greatly reduced through careful design. Regular propeller repair and polishing reduce the formation of cavitation bubbles that are the main source of noise. Further reductions come from changing how engines are mounted, adjusting the shape of propeller blades, modifying propeller caps, sculpting the flow of wakes, adjusting how propellers interact with rudders, and operating propellers so that they spin at rates that reduce cavitation. Slowing the vessel, even by 10 or 20 percent, also cuts noise, sometimes by up to half. Many of these changes save fuel, giving a direct benefit to the ship operators, although not always enough to offset the costs of expensive reengineering. More than half of the noise in the oceans comes from a minority—between one-tenth and one-sixth—of the vessels, often older and less efficient craft. Quieting this clamorous minority could significantly reduce noise.

But without a reduction in the volume of traffic, quieter ships might lead to more ship strikes if whales cannot hear approaching danger. For millions of years whales have safely traveled and rested at the water surface. Now blows from hulls and slashes from propellers are significant risks for whales in ocean shipping lanes and around busy ports. Technological tweaks have unintended consequences, especially if the movement of goods around the world continues to grow.

The most harmful effects of sonar can also be reduced, at least for large marine mammals, by locating navy exercises away from known feeding and calving grounds, tracking whales and shutting down war games when they are close, gradually ramping up sound levels so that animals have time to escape, and reducing long-term exposure by not repeatedly subjecting the same animals to high-amplitude sonar. As with shipping noise, reducing the overall number of ships conducting exercises would have the most significant effect.

Even seismic surveys can be hushed. Were we to wean ourselves from the black milk of Earth, we would have little need to rake the oceans with sonic death rays. Failing that, other methods now exist to map the subsurface. Machines that send low-frequency vibrations down into the water column yield excellent maps of buried geology while making less noise than air guns. This "vibroseis" technology is regularly used on land but has yet to be widely adopted in the ocean. Marine vibroseis produces sounds that overlap with animal senses and communicative signals but does so over smaller areas and in a narrower frequency range.

These changes are mostly now only experimental, hypothetical, or enacted in small corners of the oceans. Regulation of marine noise happens piecemeal by country, with no binding international standards or goals. The noise in the oceans continues to worsen. The United States Navy's 2020 plans for sonar in the waters around Washington State were so aggressive that the governor and five leaders of state agencies wrote to the National Marine Fisheries Service demanding changes, including the use of preexisting real-time whale alerts systems and expanded buffers around high-energy sonar buoys. A 2016 estimate of global shipping noise projected a near doubling by 2030. A review in 2013 found that expenditures on seismic surveys were increasing at nearly 20 percent per year, more than ten billion dollars annually, capping two decades of rapid growth. Decreasing oil prices and the COVID-19 pandemic have now slowed this rise, but demand for more surveys will likely surge if prices rise. The US military plans to soon start broadcasting continuous noise into all ocean basins to guide underwater vehicles.

The ever-increasing noise in the ocean is directly tied to the extinction and diminishment of life's diversity elsewhere, especially in tropical forests. In Borneo, forest-based local communities are being extinguished in favor of logging, mining, and tree plantations. These commodities all serve the global economy and are transported by ship. The worldwide decline of local economies, caused by ever-growing volumes of international trade, results in deforestation, loss of land rights to local communities, and ocean pollution of all kinds, including noise. The impoverishment of sonic diversity on land and water, then, is part of the same crisis. Were we to re-create vibrant local economies, we'd have less need to transport materials and energy across oceans. We would also directly sense the human and ecological costs of our actions, a stronger foundation for wise ethical discernment. Such a reformation of the economy would not resolve the many problems we create, but it would better position us to find solutions and answers.

We possess the technology and economic mechanisms needed to reduce our noise. But we lack sensory and imaginative connection to the problem and thus the will to act in solidarity with "our relatives under the waves."

My turntable spins. The humpback whales' songs are alive again in my headphones. I try to imagine where these animals are now. Watlington and the Paynes recorded them in the 1950s and 1960s, and so the whales were likely born sometime in the first decades of the twentieth century. These animals lived through and within the peak of our slaughter of their species. More than two hundred thousand humpbacks were killed between 1900 and 1959. Nearly forty thousand were killed in the 1960s. The singers I hear from the album on my turntable may, if they were the unlucky ones in the 1960s, have been killed and turned into soap, transmission fluid, textile mill lubricant, rust-resistant paint, and, after their oils were hydrogenated, margarine. Certainly, many of their kin met these fates.

If they survived, the singers on the album may still be with us. These animals would recall the sonic magnificence of the oceans before the mid-twentieth century. For bowhead whales that can live for centuries, the sonic revolution in their world is more drastic. In their younger years, some of these animals knew the oceans before engines, air guns, or sonar penetrated the waters. In those days, and for millions of years before, whales filled the oceans with sounds. Whales were up to one hundred times more abundant then than now, a total population of millions of individuals. Today a single whale can sometimes be heard from across an entire ocean basin. Imagine millions of these animals giving voice. Every water molecule in the oceans continually thrummed with the sound of whales. Vociferous fish, now decimated, formerly sang by the billions on their breeding grounds and added their sounds to the whales' calls. The ocean world pulsed, shimmered, and seethed with song. Unlike air guns, sonar, and shipping noise, these sounds did not kill, deafen, and fragment the community of life. Instead, as sound does in all living communities, they connected animals into fruitful and creative networks. Given a chance, this could return.

The work of Roger Payne and other apostles of whale song in the mid-twentieth century drew our imaginations into the oceans. What we heard compelled us to act. Now the seas are riven with new crises, yet our cultural imagination is mostly disengaged from the sonic tumult we create. Networks of hydrophones along coasts, feeding sound into homes, classrooms, and museums, are now healing this disconnection. Journalists, such as Lynda Mapes and her colleagues at *The Seattle Times*, are creating stunning multimedia evocations of coastal whales and their environments. These are inspiring catalysts. But most who profit from the sonic destruction of the oceans—almost all of us in industrial societies, from consumers and shareholders, to regulators and corporate heads—seldom feel the appalling nature of the world we create. Even ocean activists mostly rely on visual tools in their campaigns, hanging banners and writing screeds, rather than bringing seismic pounding, the shriek of sonar, or the roar of propeller cavitation home to their sources.

I watch the needle track its furrow through the plastic disk. Sound—coming to me through the ocean water of my inner ears—joins me to whale bodies, nerve to nerve, cousin to cousin. We loved you enough to rocket your voices into space. We curtailed our ravening appetites just in time to save the last of your kind. Can we now listen and act, saving you from sonic nightmare?

Cities

From the apartment's open window come two seconds of whistled melody, then a quiet chitter like an afterthought. A pause for another couple of seconds, then the song repeats, a new arrangement of fluted warbles, crowned with a soft squeak. The song continues for ten minutes, each phrase a variation of whistles and short trills.

A Eurasian blackbird perches on the apartment building's gutter and casts his song into the courtyard. The paved space is enclosed on all four sides by high walls, and so the sound is trapped, reverberating and bouncing back to me in lush, vigorous tones as I listen from my window on the fifth floor. As he sings, the bare walls are gilded and the cool, dewy air of this May morning glows. Usually, this center courtyard in a Parisian apartment block is an acoustic annoyance, catching and transmitting to every window the clatter of rubbish bins on concrete pavers and the chatter of passing residents. But the blackbird uses this space to his advantage, posting himself at its rim and pouring in his song. This modern open-topped cavern gives a richer, longer reverberation than the one I heard at Geißenklösterle cave, listening to the blackcap. I'm astonished to hear such sonic beauty from a bird in an unexpected place. There are no trees in the courtyard, yet the song blooms here as if in a wooded valley. The French name for the bird, *merle*, captures some of the spirit of the sound, rolling on the tongue like his introductory whistles. The English name is accurate enough for the charcoal feathers of the male, although he also sports a golden, sometimes amber, beak and a yolk-yellow eye-ring, and the female is dusky brown.

I rented this small apartment in Paris for a few days, expecting nothing more than a convenient place to stay while I visited family. But the blackbird's song woke one of my earliest memories. The whistled melody and the rich tone imparted by the courtyard unearthed a long-buried sensory remembrance, a fragment of childhood experience. Without my understanding why, the sound felt deeply familiar, in the same way that the aromas of foods from early youth can evoke memories of belonging. As a child, I lived in a similar apartment in Paris, but until this moment I had no conscious memory of any bird there. Later, my mother confirmed that, yes, a blackbird sang every spring from the courtyard and small roof garden behind our apartment on Rue Tiphaine. She said that for her, the blackbird's song was a reminder of the richness of the dawn chorus of birds in the English countryside of her youth. The song was a welcome sign of spring but also melancholy in its aloneness, missing the dozens of other species that sing alongside the blackbird outside the city.

Nearly half a century had passed since I last heard a blackbird singing in a courtyard, but the melody and timbre of the sound somehow traveled with me all those years, held in the sparkle of electrical charges on the fatty membranes of my nerve cells. When the sound came to me again, years later, these energies woke and pushed feelings of delight and warmth into consciousness. Thank you, memory. Impressive.

Our human experience of long-term auditory memory sets us apart from our close relatives, the other primates, but likely not from other vocal learners like birds and whales. Nonhuman apes and monkeys have excellent memory for visual and tactile experience, but these powers seem not to extend to sound, especially in the long term. Humans, though, readily recall the nuances of sounds. Most of these memories are short-lived, but some can last a lifetime: The sound of a loved one's voice. A melody from childhood or adolescence. The pronunciation and meanings of words, even those unused and unheard for decades. The soundscapes of city streets and backyards. The inflections and textures of the voices of other species. These dwell inside us, acting not as static archives but as living guides to the meanings of sensory experience, activating in an instant.

Our sonic memory differs from that of other primates because evolution has reshaped our brains to make us better participants in aural culture. Like many singing birds, human culture is transmitted by sound, as well as by sight and touch. But the cultures of monkeys and nonhuman apes are almost entirely visual and tactile. As a result, humans and birds have well-developed connections between the areas of the brain involved in the perception and comprehension of sound, links that are much weaker in other primates. Brain scans show that these neural pathways are required for long-term auditory memory. My decades-long memory of the blackbird was indirectly made possible by human language.

Aural memory, then, allows us to understand and navigate both the human and beyond-human worlds. The human talent for long-term memory of sound may have helped us as we explored new regions. Recall of both individual sounds and of the feel of soundscapes gave our ancestors points of reference through which to assess and understand new environments. In some cultures—notably among some aboriginal nations in Australia— human song becomes part of this geography of sound. Song lines meld human and beyond-human sound and stories into memories that travel through time across many generations. The computers that scientists use to analyze thousands of hours of digital sound in Borneo and elsewhere are an extension of an ancient human capacity for place reading through sound.

Listening to the blackbird's voice bloom, I have a strong feeling that he is using the space to his advantage, like a human singer finding a favorable performance spot. Acquaintances tell me of blackbirds holding forth at the rim of courtyards in Berlin and London, creating stunning aural displays. Intention is hard to prove, though. Perhaps the birds perch at random throughout their territories, sometimes happening upon reverberant spaces. But such insensitivity seems unlikely for a bird whose energies are largely devoted to song for much of the year, starting in earnest in January, peaking in April and May, then declining through the summer and autumn. Surely, he is a connoisseur of the qualities of his voice in the world, listening, remembering, and adjusting just as he did as a youngster when he learned his song through attentive listening and studious refinement through practice?

Such improvisational and flexible use of the city accords with the rest of the bird's biology. There are no records of free-living blackbirds in Paris before the 1850s, although some were kept as caged singing ornaments. The birds' captors used whistles and hand-cranked miniature organs— devices called *merlines* for blackbirds and *serinettes* for finches such as canaries—to tutor the birds' songs. Now blackbirds are common wherever trees are scattered among buildings or in the city's many parks, large and small. The same is true across much of Western Europe. Before the nineteenth century, blackbirds were forest specialists, living only in wooded countryside. As they colonized cities, their voices, behavior, and physiology changed. The song that I remember from my childhood has within it the imprint of the city.

Urban colonization started in winter. A few adventurous nineteenth-century blackbirds lingered in the city instead of decamping to southern Europe and North Africa like most of their kind. These birds were likely drawn to both heat and food. The city is usually several degrees warmer than the countryside. Seeds and fruits in gardens and parks, along with spilled and discarded food from domestic animals and humans, increased the allure. Blackbirds were joined in this winter move to the city by other birds like greenfinches, blue tits, and mallards. These innovators thrived and soon started breeding within the city, abandoning their ancestral woodlands and marshes and becoming urban creatures. Birds on other continents have made similar adaptations to urban life, often breeding at higher densities in the city than in rural areas. House sparrows, European starlings, and rock pigeons are among the most widely distributed animals on the planet. They are joined by a wide taxonomic variety of other species, including lorikeets and ibises in Australia; night-herons and monk parakeets in North America; bulbuls and mynas in Asia; mousebirds, kites, and martins in Africa; and various crows and magpies worldwide.

In Paris, the blackbirds' colonization of the city was helped by the construction of parks and wide, tree-lined avenues in the middle of the nineteenth century. Georges-Eugène Haussmann, at the direction of Napoleon III, razed much of Paris, transforming a tangle of narrow streets into an ordered web of grand boulevards, linked to parks and public squares. To accommodate the hundreds of thousands of displaced people and to provide a larger canvas for his work, Napoleon annexed surrounding towns in 1859 and 1860, expanding Paris to its present borders. The street where I heard the blackbird as a child, now in the fifteenth arrondissement, was a small independent town in the 1850s, between the marshes along the Seine and the wall and tollgate that defined the southern edge of the city. Likely no blackbirds sang in narrow streets that ran, without sidewalks or trees, between closely packed building facades. After Haussmann was finished with it, a tree-lined thoroughfare ran across its north edge, connecting small parks and apartment buildings, some with vegetated courtyards. The bird that I heard in the 1970s was likely a descendant, a century later, of avian colonists to this new Paris. Haussmann's project, which turned the city center and surrounding towns into a modern urban space, paradoxically coincided with and facilitated the arrival of a bird that formerly sang only in the forest.

In cities, blackbirds sing higher, louder, and at a faster pace than in the countryside. This amped up vibe has many causes, each an adaptation to the new urban habitat.

Traffic noise is the most obvious acoustic difference between the city and its surrounds. Engines, tires on asphalt, and the throb of road construction erect a wall of mostly low-frequency noise. When I'm in the city, I usually don't notice this background growl. My attention is drawn instead to the intermittent punches of sirens, horns, and shouts. But a microphone fed into a computer reveals what our minds usually filter out: in the city we swim, always, within a sea of low-pitched noise.

The rumble of cities is so pervasive that it penetrates a kilometer or more into the earth. When COVID-19 lockdowns slowed human movement and industry, geologists recorded a global stillness previously unknown to their seismic instruments. No doubt animals like elephants and whales that sense low-frequency ground- and waterborne sound waves also noticed the difference, although how this affected their behavior is so far unknown. Stillness also fell on the world of aerial sound, but unlike our close attention to potentially catastrophic tremors in rock, we lack a standardized international network of sound monitors in air. Across the world, people suddenly became far more aware of the voices of the beyond-human world. These species were always there, but their sounds were masked by noise and our inattention.

Deep sounds have long wavelengths and are thus able to flow around obstacles. The low-pitched throb of the city carries far. Even in streets away from busy roads, rail lines, or construction, low-frequency noise pervades the air. In forests or prairies, away from the city, the overall sound level is softer and is often dominated by a bump in the midfrequencies, the sounds of wind in trees or grasses.

The higher-pitched sounds of urban birds punch and leap over the wall of low noise. Loudness lets them push through the clamor, like a human shouting over the noise of an engine. By singing higher—usually an increase equivalent to one or two human musical notes—the birds use frequencies less masked by the din of traffic. These adaptations to the city are not just transpositions of the song to a higher register; the birds change the composition of their songs too, using more high-pitched elements. Blackbirds also shrink the lower, introductory part of their song relative to the later, higher trills. The city has left its mark within the vigor, frequency, and form of the birds' songs.

For the white-crowned sparrows that I listened to in San Francisco, the background growl has increased over the last fifty years. This change has nudged their cultural evolution of song in new directions. Sparrows in noisy environments—whether near ocean surf or the roar of traffic—abandon the lower-pitched elements of their songs, either by dropping these syllables or by singing them higher. The ocean has always been present, but traffic noise has increased across the city, subjecting the sparrows in formerly quiet locales to higher levels of noise.

In the noisier parts of the Bay Area, sparrows now are shriller than those from the 1960s and 1970s. This change adapts them to the new soundscape, but the song is now, from a sparrow's perspective, less impressive. By cutting off what was the low end of their songs, the birds have lost one of the ways they demonstrated their vigor, by producing songs that zipped rapidly from low to high and back again. To compensate, urban white-crowned sparrows have found other ways to advertise their performance capabilities, increasing the complexity of individual song elements, adding ornaments and accents.

When the COVID-19 pandemic shut down much of San Francisco's traffic in the spring of 2020, background noise levels reverted to those of the 1950s. The sparrows responded by reverting to quieter and lower-pitched songs of a kind that had not been heard there in decades. We do not know whether these changes happened through the flexibility of individual birds or by cultural evolution as juveniles preferentially copied songs that worked well in the nearly car-free soundscape.

Experimental studies with randomized playbacks have confirmed that these responses to noise are not mere correlations. By blasting sounds of traffic or industry into some animal territories and not others, scientists have shown that birds assaulted with noise sing higher and louder. The effects start early. Even among nestlings, stress hormones are higher in noisy places. These youngsters raised in cacophony also have shorter telomeres, genetic markers of aging on chromosomes. Other species also feel the effects of noise. Reviews in 2016 and 2019 of more than two hundred scientific studies found that amphibians, reptiles, fish, mammals, arthropods, and molluscs were all affected. Noise variously affects feeding, movement, vocalization, and thus the fecundity and viability of animal populations. Excessive urban sounds can even interfere with other senses. Great tits find it harder to see camouflaged prey amid a din.

We intuitively understand these stories of noise. Our bodies have experienced the same. When a friend's voice is swamped by the wave of engine sound from a passing bus or the sonic assault of a busy restaurant, we feel the masking power of unwanted sound. We respond either with silence, waiting out the surge, or by cranking up our voices. When we try to talk over the noise, we instinctively do so by speaking both louder and at a higher pitch. We, too, get loud and shrill in the city. We also elongate our vowels, shoving them through the obstruction, and change the timbre of our voices, favoring higher harmonics. All this happens without conscious awareness, a process guided by brain stems that listen to the surroundings and adjust our voices. The increase in vocal loudness in noise was first described by French otolaryngologist Étienne Lombard as he researched hearing loss. The Lombard effect, because it is unconscious, cannot be faked. Patients who, for legal reasons, were pretending to be deaf had their deceptions unmasked. When Lombard played loud sounds into their ears, these bilkers spoke more vigorously, their attempts to defraud employers and the government betrayed by their brain stems. It is not only our voices that change in noise. We also add more spice and salt to food in loud environments, perhaps in a bid to push other important senses through domineering sound.

The Lombard effect is present in vertebrate animals from fish to birds and mammals, although in some species it seems to have been lost. The effect allows short-term compensation and accommodation of noise, a complement to longer-term genetic, cultural, or physiological adaptations. Because the effect changes so many aspects of sound—pitch, amplitude, timbre, emphasis on different syllables—disentangling which of these actually benefits wild animals is challenging. The energetics and anatomy of sound production underlie many of these tangles. For example, as every human toddler knows, high-pitched sounds require less effort to bawl than lower tones. To hammer your parents' ears: scream and squeal, don't roar or rumble. Although these shrill cries will not travel as far as low sounds, they give impressive volume for minimal effort. The same is true for non-human animals in noise. Because loud low sounds take more energy to make than high calls, it is most efficient to yell at high frequencies. The higher tones of animal sound in noise may be a secondary consequence of the animals punching more energy into each utterance.

A study of Eurasian blackbirds in and around Vienna found that in the forest, songs carried for 150 meters or more. In the loudest parts of the city, the songs carried only 60 meters. A higher-pitched song could partly vault over the noise and boost this reach in the city to 66 meters. But an extra five decibels yielded more benefit, pushing the reach of the song to 90 meters. Five decibels is about the extra amplitude generated by song-birds in city noise. The primary adaptation of blackbirds to urban sound-scapes, then, seems to be louder songs. The increased frequency emerges as a side effect of loudness, and as a bonus, gives an advantage by over-coming masking. The same is true for the changes in composition in the songs. City birds preferentially use high-amplitude elements of their songs, which also tend to be high-pitched.

It is not only noise that differs between city and the countryside. Urban blackbirds often live in denser populations, increasing the number of daily interactions with near neighbors. Their songs are partly a result of this changed social context. Even in the countryside, blackbirds with many close neighbors sing higher and faster. The city also seeps into the hormones of blackbirds. For unknown reasons, female blackbirds in cities lay eggs with fewer androgens such as testosterone than their forest-dwelling cousins. Adult male blackbirds have lower testosterone than birds in the countryside. City blackbirds also have higher levels of stress hormones, partly due to the burden of contamination with lead and cadmium in polluted cities, but their blood has a higher capacity to absorb and buffer chemical stress. Hormones are physiological stimuli for singing and social interaction, but exactly how they shape song and behavior in urban blackbirds is so far unknown.

Cities are like newly emerged volcanic islands in an ocean, akin to the earliest years of the Hawaiian or Galápagos Islands. Only a few species colonize these novel outposts. The islands are incubators of biological innovation: the new arrivals swiftly adapt, shaping their behaviors and bodies to the new world they have discovered. Eurasian blackbirds in Western European cities not only have songs different from their sylvan forebears, these urban birds also sing and forage at night under streetlights; breed three or more weeks earlier; tend not to migrate; have rounder wings well suited for short-distance flight and not migration; and have more cautious, neophobic personalities, yet they eat novel foods, feasting on seed at feeders, spilled human grain and rubbish, as well as enjoying exotic fruit on ornamental plants.

Although populations of urban blackbirds thrive, producing more than enough offspring in most years to sustain and even swell their numbers, individual birds bear a cost. Blackbirds age faster in the city than in rural forests, a decline revealed by their chromosomes. The ends of these chromosomes—the telomeres, markers of aging in animals from humans to birds—shorten rapidly in the city, perhaps because of the physiological stress of living under continual sensory and chemical bombardment. But predators and ticks are fewer in the city, as is the incidence of avian malaria, and so although urban blackbirds have ragged chromosomes, they often live longer than those in rural areas. They are perhaps like aged rock stars, their bodies ravaged by a loud, fast, and chemically sodden youth but persisting into a secure dotage.

So far, these differences have led to little genetic divergence between country and city blackbirds. The DNA of city birds tends to be less diverse than that of countryside birds, the mark of recent colonization by a few individuals, a similar signature to that in the genes of animals on oceanic islands. There is also some evidence that genes associated with risk-taking and anxiety are changed among urban blackbirds, although whether and how these subtle shifts in DNA change behaviors is unknown. The Eurasian blackbirds' transformation in the city seems driven not by genetic evolution, but by evolutionary changes that run parallel to genes. When mother birds provision their eggs with hormones, they shape the singing and behavior of their offspring. It is possible, then, that it is the physiology of egg laying that leads to different songs and other behaviors in the city. Cultural evolution may play a part, as it does in white-crowned sparrows, where the form of song adapts to place through listening, copying, and experimentation by young birds. Last, every individual bird molds its behavior to the moment, changing its song as the soundscape changes, preferentially singing when noise is low. Blackbirds using especially reverberant places to gild their songs are, perhaps, another example of this adaptation. The city provides not only acoustic difficulties but opportunities for sonic enhancement.

In just over one hundred years, some populations of the Eurasian blackbird have transformed themselves into city dwellers. In another century or two, genetic changes may catch up with and reinforce these differences. But just as Haussmann tore down and rebuilt Paris in the nineteenth century, it is likely that the next century will see changes just as radical, pushing the birds' behavior, physiology, and genetic evolution in new directions.

Paris and other cities will continue to heat up, driving out some species and inviting in new ones, including disease-carrying mosquitoes and ticks from the subtropics, some of which may prosper in the heat of the city, turning what is now a refuge from disease into an infectious trap. In the last twenty years, for example, Eurasian blackbirds in Germany have been cut back by 15 percent by the newly arrived Usutu virus from Africa, declines that are more severe in warmer years and locales. Human social demand for heat-mitigating street trees and parks will increase, a trend well underway in most major cities, expanding habitat for tree-loving urban animals. Human population density and resource use will change in unpredictable ways, just as it has for millennia. In the eighteenth century, no naturalist would have predicted that the Paris of the future would be an island of stone and concrete in a sea of leafy suburbs, filled with the song of a forest-specialist bird, a sound modified to fit the city. If blackbirds are present in another century or two, their songs will carry within them the now unknown nature of this future city.

By finding ways to thrive amid streets and parks, blackbirds and other urban colonizers have increased their breeding densities over time. Bird species that first colonized European cities in the 1800s now breed, on average, at densities 30 percent higher than those of their rural cousins, an impressive testament to life's adaptability. But most wild animals cannot live in the city. In the blackbirds' songs, I hear flexibility and resilience. My mother, on hearing the same song from a Paris apartment, also heard what was missing, the cadences of dozens of other bird species that she knew from the countryside. Yet the city also helps rural birds. By concentrating human activity, land use, and consumption, urban areas make possible the lives of nonhuman animals elsewhere. Were humanity to abandon its embrace of urban life and spread out more uniformly over the land, ecological calamity would unfold, a great silencing of the voices of other species. This is not a thought experiment. The suburbs have spread humanity's impact over the land, vastly increasing habitat destruction, energy use, and material needs compared with the "ecological footprint" of people living in cities. When I delight in the dawn chorus in rural woodland, I partly have the efficiency of the city to thank.

About 4 percent of the land surface of the world is urbanized, yet more than half of the human population lives in cities. The human density in the apartment building allows many hectares of forest or field to thrive unencumbered by suburban houses, roads, and lawns. City dwellers also use less fuel, metal, wood, and other material goods that must be mined or cut from the land.

In the blackbirds' song, I hear an animal finding its place within the city. In the silence around the song is implicit the possibility for ongoing life elsewhere. The city and the countryside live in reciprocity, not only in the human economy, but in the wider community of life.

Fifty years after my childhood in Paris, I listen from an apartment in New York City. Birds seldom sing on this street, although in the evenings I see night-herons flying over the crowd of apartment buildings, crossing Harlem as they leave their daytime roosts on the Hudson River to feed in the Bronx and East River to the east. The sounds of engines are ubiquitous and, in the hot summer, noise and fumes flow into the apartment through open windows.

As I child, my bedroom faced the street and I was captivated by the work of the *éboueurs*, bright-vested garbagemen, as they leaped on and off the footplates on the back of grumbling green trucks that seemed to me like hungry mammoths or dinosaurs, urban megafauna. Our apartment was one block back from a busy commercial street, and so noise and bright colors were spikes of stimulation on a street otherwise gentle in its bustle. Now, on a busier street in New York City, these enchantments have faded and life amid the sounds of the city's physiology—the feeding, blood flow, muscular contractions, and excretions of a huge metal and concrete organism—offer excitements often more trying than enthralling.

At two in the morning, a pickup truck parks under the fourth-floor window, open-doored, its radio cranked. The driver blasts the bus stop with pressurized water, a jet powered by a pump on the truck bed. The hose-down takes ten minutes, but the truck lingers with its speakers thumping for another fifteen. Buses pick up their pace before dawn, hissing brakes and bellowing engines as they stop then pull away onto a steep incline. Windowsills are black with the soot from their engines and those of the hundreds of delivery trucks that pass daily. Garbage trucks arrive just after sunrise to load the van-sized piles of trash from the curbside. Trash bags pound as they are slung into one another, workers shout, and hydraulics whir and gasp: a dawn chorus of plastics and discarded food on their way to landfills. All afternoon, a soft-serve ice-cream truck is parked across the street, its generator an atonal whine and chug, wheezing fumes from an exhaust pipe directed up at apartment windows. The six-lane Henry Hudson Parkway, a traffic artery only parklike in name, and Broadway, especially favored by late-night delivery trucks, undergird these sounds with a permanent drone from their paths just over one hundred meters away. Human voices mix into these sounds, but even the shouts are quiet compared with the engines. After dark, especially on the weekend, family groups pass with speakers on small trolleys, pumping out music on their way to and from this neighborhood's one small park.

These are mostly the sounds of good work and strong community: a city getting cleaned, public transit running, small businesses finding customers, food and other supplies arriving in the city, and people enjoying time together in public spaces. In aggregate, though, they produce a clamor vigorous and unpredictable enough to disrupt sleep and set nerves on edge. The anxiety twists higher when these wholesome sounds are spiced with occasional jolts of trouble: motorbikes modified to thunder so loud that their midnight passage sets off every car alarm on the block, an argument on the sidewalk that seems ready to tip into violence, or the un-settling crack and jangle of a window breaking.

Noise pollution is a grievance that dates to the first human cities. On clay tablets from Babylonia, in one of the earliest known written stories, we read of the gods' wrath at our din. Scholar Stephanie Dalley translates cuneiform from 1700 BCE: Ellil, the chief god, complained that the "noise of mankind has become too much. / I am losing sleep over their racket." To impose quiet on people "as noisy as a bellowing bull," the gods inflicted disease and famines. They also corrected their earlier omission and assigned a life span to humans, preventing endless population increase. Urban noise, by this account, brought us mortality and the yoke of disease. Perhaps urban scribes kept awake by the voices, music, and clatter of neighbors channeled their frustration into stories of revenge?

At the time these stories were impressed into clay, the global human population numbered fewer than 30 million, and Mesopotamian cities housed tens or hundreds of thousands of people. Now we number more than 7.5 billion, and our cities have populations in the tens of millions. Fifty-five percent of the human population now lives in cities. By 2050, the proportion is projected to be above two-thirds. The soundscape of the city is now the sonic context for most of humanity. Like blackbirds, we have adapted and thrived in this new sonic world but have also suffered.

On the A Line headed downtown from Harlem, four teenagers yell their conversation over the clatter and squeal of the decrepit subway car. One of them shushes the others, but they laugh in her face. "We're New Yorkers. Loud! That's what we do. We make noise." The machinery around them agrees. When I get out at Columbus Circle, I check the reading on a sound meter. Ninety-eight decibels when a through-train passes. These are sound pressures loud enough to damage inner ear hair cells. More than a few hours of exposure can permanently impair hearing. The teenagers' voices were loud but dwarfed by the power of wheels, brakes, and metal boxes jolting at speed along uneven tracks.

Cities are indeed noisy places, but it is not only loudness that distinguishes their soundscapes. The ambient sound level in many tropical and subtropical forests often approaches or exceeds seventy decibels. Some tropical cicadas are as loud as the subway, blasting at one hundred decibels. The late-summer nighttime chorus of katydids in Tennessee holds steady, for hours, at seventy-five decibels. When visitors from the city come to rural Tennessee in late summer, they complain that they cannot sleep for the insect racket, a reversal of the usual narrative about "noise" in cities and the countryside. A reasonably quiet apartment or office, even in a busy city, is more muted than this, usually between fifty-five and sixty-five decibels. The notion that "nature" is quiet is a product of expectations and experience in northern temperate regions. In Japan, Western Europe, or New England, the forest is indeed much quieter than the city, especially in the colder months of the year when insects, frogs, and birds are soft voiced or absent. The same is true in polar regions or the mountains where quiet reigns in the calm between windstorms. But it is often clamorous in places where plant life abounds and animal diversity is high.

City noise differs most markedly from other soundscapes in its tempi and unpredictable nature. I take a walk across Midtown Manhattan, sound pressure meter in hand. Just south of Columbus Circle, workers are cleaving the street's concrete. Like surgeons, they incise the skin to reach the arteries and nerves below. Their scalpel is a jackhammer, measuring ninety-four decibels from where I stand on the sidewalk, four meters away. Only two of the crew of five wear hearing protection. A young girl scrunches her face in pain and clamps her palms over her ears as she passes. Adults walk past, unflinching. A block north, a bus lets off its air brakes right as it draws level with me, startling a passing snowy canine puffball so that it jumps forward and yanks on the leash. Two blocks on, construction workers drop a metal pile of scaffolding tubes. The clatter breaks the stoicism of a couple of suited walkers, causing them to twitch, then dart their heads around. An ambulance punches its siren at a double-parked car. Someone shouts in my ear, trying to reach a friend on the other side of the traffic-filled avenue. Apart from the jackhammer that was plainly visible in a closed-off street lane, I could anticipate none of these sounds. Loudness is stressful and sometimes painful, but so, too, is immersion in a soundscape where explosions and poundings arrive seemingly at random. I feel as if I'm walking through a dark space where unseen hands sporadically reach out to slap and shake me.

In places where humans do not dominate, sudden loud sounds are rare and are usually cause for alarm. A tree falling. The sudden appearance of a stealthy predator. The yelp of pain from a bee-stung companion. Each sound stabs us with a surge of adrenaline. But most loud sounds in forests and other ecosystems arrive in more predictable ways and cause no distress. In the rain forest, the raucous cries of toucans and macaws flying in pairs over their vast domains taper in and out as the birds approach then depart. The chorus of cicadas and frogs also waxes and wanes in rhythms that, although sometimes overwhelming in their power, do not arrive as shocks to our ears. Vigorous ocean waves are soothing in their regularity. Even the bang and roar of thunder is usually predictable. We see, feel, or hear the storm approaching. It is the rare thunderclap that comes out of nowhere that is alarming. Now human nervous systems that evolved amid forest and savannah sounds find themselves unprepared for the city. In a day walking around Manhattan, I hear more unexpected bursts of loud sound than my ancestors likely experienced in a lifetime.

City noise—the unwanted, uncontrollable sounds of human activities— has well-known negative effects on our bodies and psyches. Loud sounds can lead to hearing loss, whether the immediate damage caused by jackhammers and other ear busters, or the slow erosion of inner ear hair cells brought on by years of exposure to subway stations, construction noise, or busy traffic. Hearing loss then leads to other problems, such as loss of social connections and an increased likelihood of accidents and falls. Noise not only assaults the hairs in our ears. When unwanted sound hits us,

whether from an airplane, passing trucks, or clatter in our homes, blood
pressure spikes, even when we are fast asleep. Noise also fragments sleep
and increases stress, anger, and exhaustion during waking hours. Our
hearts and blood vessels suffer. Heart disease and stroke increase with exposure to noise, likely because chronic exposure steeps us in stress hormones and high blood pressure. City noise can also disrupt levels of fats
and sugars in the blood. Children bear an especially high burden because
noise disrupts cognitive development. Exposure to chronic aircraft, traffic, or rail noise at schools leads to difficulties with focus, memory, reading, and test performance. Laboratory experiments on unfortunate rats
and mice confirm that noise both changes physiology and impairs brain
development. Sound's nature makes it an especially problematic source of
distress. Unwanted light is easy to block by closing our eyes or with a cur-
tain. Unwanted smell can usually be barred with a tight-fitting door.
Noise, though, moves through solid matter, finding ears that are always
open, always listening.

In Western Europe, where these effects have been well studied, the European Environment Agency estimates that noise is second only to fine
particulate matter pollution as an environmental cause of illness and premature death, annually causing twelve thousand premature deaths and
forty-eight thousand new cases of heart disease. An estimated 6.5 million
Western Europeans suffer from chronic sleep disruption and 22 million—
1 in 10 people—experience chronic high annoyance from noise. Few other
regions have measured these effects with as much precision, but the costs
of noise may be even more severe elsewhere than in Europe. Measurements of noise in African cities, for example, often exceed European urban
sound levels. Extrapolating from the European data—admittedly a coarse
approximation—suggests that, worldwide, noise in cities likely degrades
the health and quality of life of hundreds of millions of people, and annually kills hundreds of thousands. In general, these effects are worsening as
roads and skies get busier and industrial activities expand. For example,

between 1978 and 2008, air transport quadrupled, a trend that continued until the COVID-19 pandemic.

The burden of city noise is unequally shared. Sound pollution in cities is a form of injustice. Yet we are also a species that loves the soundscape of home. We not only adapt to and tolerate city noise, sometimes we bond to it as a signature of culture and place, the sonic vibe of our neighborhoods. City sounds, then, can paradoxically be both alienating and welcoming, sources of harm and of belonging.

After a summer staying in a friend's sublet in West Harlem, I move for a few weeks to another apartment, across the East River in Park Slope, Brooklyn. At this apartment, no expressway runs meters from the window. More than two hundred hectares of woodlands, lawns, and lakes in Prospect Park are minutes away on foot. Ice-cream vendors do not park all afternoon under our apartment windows. The buses in this new neighborhood run quiet and clean. I've ridden dozens of bus lines in New York City over the last twenty years, but until I arrived in Park Slope I had never ridden one that pulls away from the curb with a gentle sigh and unsmoky exhale, carrying its passengers on a WiFi-enabled glide. West Harlem is a mostly Latino and Black community; Park Slope is majority white, with double the median household income. More than 80 percent of homes in West Harlem are rentals, compared with just over 60 percent in Park Slope.

The unjust distribution of the harmful dimensions of city sound is the sensory manifestation of both the history of city planning and present-day policies. The expressways that run through many New York neighborhoods were routed deliberately to raze and fragment minority and low-income areas, displacing many people and, for residents who remained, increasing noise and air pollution. Robert Moses, the overlord of much of this work in New York, viewed such work as doubly beneficial, connecting mostly white suburbs to the city and destroying, in his words, "ghettos" and "slums." Moses's transformation of the city into a hub for private car traffic from outlying areas was repeated throughout the United States, underwritten by a 90 percent cost share from the federal government for urban freeway projects. By the late 1960s, so many minority neighborhoods had been ruined by the slash of freeways that activists pushed back. "No more white highways through black bedrooms" was one of their slogans.

Parks, on the other hand, were disproportionately built close to wealthy neighborhoods. At the time of Prospect Park's founding, in 1860, commissioners recommended seven park sites across Brooklyn. The city focused its attention on Prospect Park, despite the fact that it was remote from population centers at that time. Instead of providing the mass of people with easy access to green space, the park's planners chose a site next to the estate of Edwin Clark Litchfield, a railroad and real estate developer. At its founding, one of the explicit goals of Prospect Park was to attract more wealthy residents to the area and raise property values and tax revenue. West Harlem, on the other hand, has been repeatedly disenfranchised from park access. When Robert Moses rebuilt the West Side of Manhattan from 1937 to 1941, he added more than 130 acres of green, relatively quiet parkland to the riverfront, but this largess ended at the border of the Black neighborhoods in Harlem. Moses's projects were funded by every taxpayer in New York, but they benefited mostly whites, a form of robbery as well as exclusion. Later, in 1986, the city located the North River Sewage Treatment Plant on the riverfront in West Harlem, a billion-dollar project originally slated for construction farther south, closer to white neighborhoods. The plant exhales odorous and sometimes poisonous gases from sewage, along with fumes from the large engines that power the plant. In an attempt to offset some of these negative effects, a running track, pool, and other athletic facilities were built on the plant's roof alongside the smokestacks. The plant sits adjacent to the now-closed Marine Transfer Station, a twenty-four-hour conveyance point for garbage trucks unloading their trash onto boats. Instead of the wide terraced green spaces leading from city streets toward the Hudson River enjoyed by New Yorkers living a few dozen blocks to the south, West Harlem residents access the narrow strip of riverfront via either narrow stairwells from the roof of the wastewater plant or down 120 open steps leading to a dark tunnel. The elevator from the roof was, during my time in the neighborhood, out of service. A footbridge that provided easier access burned in the 1950s and was not replaced until 2016. Not only is parkland in short supply here, but accessing it requires significant effort.

Sound pollution intersects other forms of environmental injustice in the city. Old diesel buses cloud the air with both noise and particulate pollution. Seventy-five percent of New York City's bus depots are located in communities of color, neighborhoods that are also disproportionately affected by truck and car traffic, waste transfer facilities, and industrial sites. Latino and Black New Yorkers, on average, inhale nearly double the amount of vehicle particulate pollution than whites. In 2018, Eric Adams, Brooklyn borough president, joined by other elected officials, called the disproportionate use of old, polluting buses in low-income neighborhoods "unacceptable and intolerable." The Metropolitan Transportation Authority responded with a faster phaseout of some older buses and, by 2040, proposes to completely electrify the fleet. This would clear the air of bus noise and diesel exhaust but is contingent on funding. MTA's budget is mostly controlled not by the city but by New York State, which has, for decades, siphoned funds away from mass transit in the city, including using MTA funds for bailouts of struggling ski resorts. The growl and spew of buses in low-income areas of New York City has its origin partly in the snowy pleasures of a few mostly white upstate vacationers, a potent example of the twentieth-century American project of gutting the city in favor of the suburbs and exurbs. A comprehensive scientific review in 2020 of the ecology of cities worldwide found that patterns of pollution, treeless heat islands, access to healthy waterways, and other environmental dimensions of city life were "principally governed" by social inequities and structural racism and classism.

More traffic noise. Less parkland quiet. The contrast between the soundscape in West Harlem and Park Slope is a result of more than 150 years of unjust city planning.

In New York City, the sonic manifestations of power inequalities sometimes also extend to wealthier neighborhoods. The building demolition and construction industry can override all but the most powerful residents. In 2018, the city granted sixty-seven thousand exceptions to the rule that building construction should take place only between 7:00 a.m. and 6:00 p.m., more than double the number of permits in 2012. Each one of these exceptions took an already cacophonous process and expanded its disruption into the predawn and late-night hours. Fees from these permits added more than twenty million dollars to city coffers. Of the nearly three hundred million dollars spent on lobbying in New York State in 2019, real estate and construction account for the second-largest category, after lobbying for budget appropriations. The state comptroller's office reported in 2016 that noise complaints about construction in the city more than doubled from 2010 to 2015. Yet inspectors sent to building sites did not carry noise meters and almost never issued fines. The city departments charged with enforcing noise ordinances failed to use clusters of complaints to identify chronic problems. The more upscale parts of the city may be quieter than other neighborhoods, but they are not immune to the sonic assaults resulting from the unequal influence of well-connected developers. A city cannot function, of course, without building and renovation, but when jackhammers and trucks obliterate any hope of productive work or restful sleep, the city has failed in its basic task of providing livable habitat for humans.

Resistance comes from individuals, activist groups, and local elected officials. In West Harlem, a community-based nonprofit, WE ACT for Environmental Justice, has, for decades, fought for the rights and well-being of residents, winning settlements against the sewage plant, getting bus depots upgraded to be cleaner and quieter, fighting sources of asthma-inducing air pollution, and addressing the inequities of heat in urban areas. City Council members have lately pushed back at after-hours construction with bills that would, if enacted, more rigorously regulate noise. Individuals use small claims courts to enforce regulations that the city will not. These efforts build on a long history of attempts to reduce nuisance noises. In 1881, inventor Mary Walton, who lived near an infernally loud elevated rail line in Manhattan, patented noise-reducing supports for the rails, an innovation that was adopted in New York and other cities. In the first years of the twentieth century, physician and activist Julia Barnett Rice succeeded in limiting noise from boats and road traffic, especially around hospitals, and eventually won passage of federal noise control legislation. Horse-drawn milk delivery wagons in the early decades of the twentieth century were equipped with rubber wheels and the horses shod with rubber shoes to reduce clatter in the streets, an effort that seems charmingly quaint from the perspective of a city now ceilinged with helicopter and airplane noise and thrumming with construction. In 1935, Mayor Fiorello La Guardia declared October the month for "noiseless nights," calling on New Yorkers' "spirit of cooperation, courtesy and neighborliness" to reduce the clamor. Noise codes were enacted the next year. Eighty-five years later, the targets of those codes read like a description of a contemporary street: amplified music, engines, construction, unloading of trucks, nighttime revelry, vehicle-mounted loudspeakers, and "prolonged and unreasonable blowing of a [motor] horn."

Noise is one form of the lack of control over our sensory, social, and physical world. It often is the poor and the marginalized who experience the least control. Yet not all "noise" is bad and not all people experience the sounds of the city in the same way. In these differences are rooted bitter struggles over neighborhood identity and gentrification. When family and commercial life spills onto the street, as it does wherever homes are small and the summer is hot, the sound of voices, amplified music, and traffic becomes a defining feature of a sense of place, a signature of home.

But the sonic meaning of "home" is contested. When different expectations collide, conflict ensues. Sometimes these tensions are rooted in the inevitable frictions among neighbors living in close quarters. Because sound flows in wood, glass, and masonry, squeezes through cracked windows, and wraps its waves around corners and over rooftops, the voices and activities of our neighbors live inside us, in the motions of fluid in our inner ears. Such intimacy can disturb sleep and intrude or infuriate during the day. Sound entrains us in the lives of others and we must therefore surrender to them some control over our sensory experience. This is true everywhere, of course, in a forest or on the ocean shore, but there we find our inner agitation mellowed, perhaps because the sounds come in the foreign tongues of trees, insects, birds, and water on sand. If we heard in these sounds the droughted distress of the hissing pine needles, the lusty arrogance of the cicada, the clannish cursing of the crow, or the hurricane-born seethe of beach waves, might our minds add layers of judgment and analysis, complicating what was a soothing blanket? In a city, where we know the sources and meanings of sound all too well, neighbors can chafe or inflame our emotions, especially when we judge their noise to be a symptom of inconsideration. Bass-and-drum-heavy music, played late at night: put your hand on the wall and feel its throb. Predawn clatter of shoes on uncarpeted wood floors in the apartment above. Yet another shouted drama down the hallway. Kids gunning fireworks at midnight from the street corner, for the tenth night running. A small dog with Olympian stamina, flaying the neighborhood with its yapping for an entire afternoon.

In a neighborhood where bonds among neighbors are healthy, the flow of sound across the boundaries of one home to another is usually of little consequence. We tolerate and often enjoy the sounds of community. We resolve problems with a text message or neighborly talk the next day. But in neighborhoods riven by discord, sound can lead to further antagonism. One person's joyful expression of local culture is, for others, a noise nuisance. Where these fracture lines fall along lines of race, class, and wealth, different expectations of what a neighborhood should sound like become both symptoms and causes of gentrification.

The apartment where I stayed in West Harlem is in a neighborhood now mostly Latino. At night, especially at the weekends, life on the street is centered around music played from amplifiers on small handcarts or from tinny cell phone speakers. The ebb and flow of passing rhythms and melodies is the primary accompaniment to the traffic sounds of the city. Around the Fourth of July, fireworks set off nightly from the middle of the street added explosive ornamentation to the music. The detonations echo and reverberate in the canyons between tall buildings, adding lingering muscle to the display. As a white visitor to the neighborhood, I was a part of the process of gentrification, propping up housing prices and nudging retail toward whiteness. Had I dialed 311, the city government's clearinghouse number, and complained about "the noise," I would directly have called on the armed authority of the police to impose a culturally inappropriate preference on the local community. I enjoyed the music and felt no desire to call, but even if I had, as a guest and a cultural outsider, such an act would have been wrong.

Other white residents in the neighborhood do not feel the same way. As housing prices went up and whites moved in, noise complaints surged, especially after 2015. The decades-long practices of cranking up the radio while playing dominoes on folding tables on the sidewalk or kids setting off fireworks do not sit well with newly arrived white residents, many of whom are paying high rents in renovated or rebuilt apartment blocks.

The same dynamic plays out in other cities, reflecting class and racial tensions particular to each place. In New Orleans, white residents call the police to complain about Black second-line parades and street parties. New residential developments in Melbourne, Australia, elicit noise complaints from wealthier residents about long-standing live music venues, a fracture along lines of social class more than race. Near London's Chapel Market, newcomers to renovated apartments complain about the shouts— "Three apples for a pound!"—and the early-morning clatter of barrow wheels. At each place, it is not the sounds of the neighborhood that have changed but the desires and demands of listeners. Perceptions of "noise," weaponized through complaints to authority, serve to push out locals in favor of newcomers. In New York City, when a white hand dials 311 to complain about Black noise, the dialer does so with impunity (public records do not name the caller), and the subject of the complaint is exposed to an apparatus of law enforcement that is routinely violent and racist. Our judgments of what are appropriate and inappropriate levels of noise, and how we choose to act on these judgments, are therefore mediators in either tolerance or injustice. Housing prices drive gentrification but so, too, do cultural differences in sensory expression and expectation.

City life also teaches us that noise is gendered. The city plans that directed traffic and industrial noise into Black and other minority neighborhoods were penned by men's hands. The construction companies that push noise into the early morning and late night are run by men. The fireworks and car mufflers modified to sound like gunshots on the streets of New York are detonated mostly by young men. Men are the ones who sit with the car blaring its music to dozens of apartment windows or who strafe the narrow streets with motorbikes and cars retooled to maximize noise. City noise is often the sound of strident masculinity. Our culture encourages and tolerates men's violation of the sensory boundaries of others but actively silences the voices of women. In the roar of the city, then, we hear the same patriarchy that penned the biblical injunction "Let the woman learn in silence with all subjection," which caused Mary Ann Evans to publish under a man's name (George Eliot); empowers contemporary mansplainers; allows a misogynistic president to tell women journalists to "keep your voice down"; keeps women out of orchestras and off the conductor's podium; fills the Rock and Roll Hall of Fame with more than 90 percent male voices; and, to this day, shushes young women and admires garrulous boys. In every ecosystem, sound reveals fundamental energies and relationships. In the city, we hear human inequities of race, class, and gender.

Responses to noise, too, are gendered. Women have led the effort to reduce urban noise for centuries, especially in New York City. From Mary
Walton's nineteenth-century engineering; through Julia Barnett Rice's
work in the early twentieth century; to contemporary activism and policy
making by WE ACT for Environmental Justice, co-founded and led by
Peggy Shepard; and City Council legislation in New York City developed
by council members Helen Rosenthal and Carlina Rivera, women have
greatly improved the city's soundscape. This continues the much older
story of the role of female energies in shaping the sounds of the world.
Evolutionary elaboration and diversification of sounds, from crickets to
frogs to birds, have been driven, in many species, by the aesthetic choices
of females. It was mother's milk that gave mammals our muscular and
nimble throats, and thus allowed humans to speak and sing. The sounds of
our world are the product of all genders, but females have had a disproportionately large effect in producing much of what we admire and need in
soundscapes. The diversity of animal voices, the beauty of vocal expression, and the sonic livability of the city owe much of their existence to the
power of the feminine in biological evolution and human culture.

City noise also creates a hostile environment for those whose senses and
nervous systems differ from the norm. Many restaurants are now so loud
that anyone with even slight hearing loss is cut out of conversations, unable to discern patterns of speech amid the tumult. The noise in these
places is like a high step at the front door, impassable to wheelchairs, but
the barrier, in this case, is to those with ears that differ from the norm. Not
only are these restaurants excluding many people, but they are also subjecting workers to ear-damaging noise levels on a daily basis.

Neurotypicals and those who live unburdened by anxiety disorders often
thrive in the energies of noise. But clamor is often an unbearable assault to
those on the autism spectrum or for those for whom anxiety is a constant
companion. Noise can wall people out of participation in the life of the city,
a barrier no less real for being invisible to the eye. A few who cannot abide

the city's clamor have the privilege of being able to escape, but every child born into this soundscape and every adult whose job or family binds them to the city is locked into distress and, sometimes, terror. Noise is, in some parts of the city, oppression of the minority by the majority.

S tepping out of the subway station into Midtown Manhattan, I some- times feel buoyed by the vigor of the sounds around me, lifted by the sonic convergence of human work and society. But the same soundscape sometimes shoves me into the early stages of panic, a vise of sound that squeezes my heart and breath, and fills me with a frantic and despairing desire to escape. The city is a window into my autonomic nervous system, the unconscious tuning of my body and senses. Sound reveals not only the dynamics of our society but the texture of our psyches. My varying re- sponses to the city, then, are bodily symptoms of the city's sonic paradoxes.

The city draws me deeper into my humanity. My connections to others expand amid the city's confluence of cultures and its role as a hub for art and industry. I am fed by streets on which I hear dozens of languages, venues where both the leading edges and the canons of the world's music come alive, and theaters where the power of the living, spoken word is cel- ebrated. I am lifted by the sounds of urban birds manifesting life's adapt- ability and resilience: a kestrel peppering Broadway with its cries, ravens palavering from Brooklyn rooftops, and night-herons croaking as they wing over Harlem. We are a convivial species with curious, empathetic minds. The human qualities of imagination, creativity, and collaborative action flourish in cities' intensified social networks. I imagine that the in- habitants of the first cities in Mesopotamia felt the same surge of possibil- ity. In this new urban habitat, we can paradoxically become more fully ourselves, a homecoming for the human species.

Yet the city also ensnares us in the worst qualities of our species. Inside the trap, the city talks over us, constantly, with such vigor that the chemistry of our blood and the tone of our nerves revolt, sometimes to the point of sickness and death. No wonder we feel the need to be loud, to assert our presence and agency. But in doing so, we become part of the sonic distress for others. The assault is all the more powerful for its union of the senses. In the din and heave of sound, the bile of traffic fumes pervades our noses and mouths. We feel it in our lungs too, the tightness and empty clutching at air after a walk down a street clogged with honking SUVs, delivery trucks, and cars. Some drivers lean on the horn and will not let up. Others blast in triplets or in stuttering phrases, anger sonified. Then an ambulance tries to pass, its wail impotent in the logjam of metal. The cloud of exhaust hangs in the street canyon. At night, only one or two stars are visible, the rest veiled by the dome of light, the aura of particulate pollution reflecting the energies of billions of electric lights. Underfoot, the ground is unrelentingly hard. Footfall here is always martial, strident, and clipped, unlike the varied sounds of shoes and feet outside the city as they pass over leaf litter, rock, gravel, sand, and moss. The city grasps every sensory nerve ending and says: You cannot escape me.

In the sensory violations and dysphoria of the city, there is a door to empathic understanding of other species, "our relatives under the waves" and those terrestrial species who carry the sea only as memories in the fluids of their cells.

Submerged in sonic violence, I am whale, my entire body thrumming night and day with unwanted vibrations, energies alien to my flesh. My ancestors and their long experience of sound did not prepare me for this.

In a soundscape dominated by the noise of a single species, I am forest, stripped of the diversity of voices that took millions of years to evolve. I am now deep in the grief of extinction.

Reveling in the songs of the few remaining species, I am blackbird, a wild, broken singer. I feel myself propelled by life's joyful, improvisational imperative to find a voice in this strange new world.

The sounds of the city not only plunge us more deeply into our humanity. They are, if we attend to their effects, an immersion in bodily, sensory kinship with all speaking, listening beings. But unlike these other beings, we humans have a measure of control. We can choose a different sonic future. The whales, forests, and birds cannot.

{ PART VI }

Listening

In Community

A tone clear and warm as sunlight sounds from the giant bronze bell. The ring contains no hint of clang or jangle, just a single frequency, sweetened and fattened by overtones, pitched a few notes below middle C, exactly at the midpoint of the range of human speech. Although I stand two meters away from the bell, the sound seems to emerge from within me, a calming, centering glow that spreads from chest to extremities, then flows outward into my perception of the park in which I stand.

The barrel-shaped bell, a meter tall and more than half a meter wide at its mouth, is suspended from the domed roof of a pagoda. A horizontal wooden beam hangs from chains next to the bell. A child stands on tiptoe and reaches up to haul on a rope dangling from the beam. She pulls back, then releases, and the wooden striker swings onto the bell. The sound rings again. Pure and steady toned, with a slight pulsation, a swelling of amplitude that comes at a pace just slower than a calm heartbeat.

The sound is persimmon fruit in the mouth. The fading of red to orange in a sunset sky. The transience of all beings. So the Japanese literary tradition tells us, from the fourteenth-century epic *The Tale of the Heike*, to Masaoka Shiki's haiku, to the song lyrics of poet and teacher Ukō Nakamura. The sound of the temple bell, the bonshō, nourishes, lifts, and places us in right relationship.

This bell was made by the late Japanese Living National Treasure Masahiko Katori. Like other recipients of the honorific, Katori's artistry and craftsmanship are considered part of Japan's Important Intangible Cultural Properties, a government-sponsored system honoring practitioners of significant artisanal and artistic practices. Unlike other national schemes that identify and honor buildings, landscapes, or museum-worthy artifacts, these programs seek to elevate and protect not durable physical objects but the knowledge carried by people.

Like cultural knowledge, sound is unseen and ephemeral. When artisans die, the wisdom carried in their muscles and nerves goes with them. Likewise, a sound wave carries the meaning and memory imparted by its maker but soon disappears. If the artisan teaches others, knowledge passes on and is modified by students' interpretations and innovations. A sound wave, too, transmits its energies, sometimes only as the heat of friction as the wave dissipates, but sometimes when it is heard by living beings and changes them. The ring of the bell lives on in my memory, held in electrical gradients and a tracery of molecules, all sustained by the furnace of my metabolism. In writing these words, the bell's vibrations flow to the page and then into your mind and body. The sound of a single strike of wood on bronze lives on in human bodies, just as the cultural knowledge of Masahiko Katori is alive in the knowledge and work of contemporary Japanese artisans.

The sound of this particular bell—the Peace Bell in Hiroshima Peace Memorial Park—has, like the intangible cultural properties of Katori's work, received official government recognition. Along with other bells in the park, the ringing of this bell is Soundscape 76 of the 100 Soundscapes of Japan, a government program established to find and honor significant soundscapes, and to encourage deeper listening. This program, launched in 1996, is a rare example of government recognition of the value of soundscapes. The polity's typical relationship to ambient sound is through its attempts to regulate noise pollution, an important role but one focused on sound as a negative experience.

Globally, policies designed to preserve and honor valuable national or regional treasures are almost entirely focused on visible, tangible objects and physical spaces. From the point of view of preservation and curation, this focus is understandable. Objects can be sequestered into collections and viewed at will. The boundaries of parks and buildings can be marked and protected. But the wonders of human culture and the living world come to us through many senses. To honor only material objects and spaces is to exclude much of what gives life joy and meaning. Might we honor other manifestations of human culture and beyond-human life, as does the 100 Soundscapes of Japan project? The distinctive sounds of human neighborhoods and natural communities. The nuanced yearly cycles of aromas in forests and seashores. The taste of foods particular to a region. The feel on our skin of the wind blowing down a wintry street canyon or across a springtime park. The varied sensations of the ground under our feet. The shiver or glow of changing seasons. These, too, deserve attention, celebration, and, in some cases, preservation. Sounds can be recorded and archived, as can the chemical mixes of aromas, but these static records do not capture the living, changeable presence of the sensory environment.

The 100 Soundscapes of Japan were selected by a committee of the Ministry of the Environment from more than seven hundred nominations, some from local governments and businesses, some from individuals. The selection includes soundscapes from physical, biological, and cultural sources. This breadth is especially fitting because sound is always integrative, blurring boundaries as waves of energy meet, unify, and stimulate human perception. Some of the 100 Soundscapes are fleeting sounds, such as the sweet ringing of suzumushi crickets or the singing sands of Kotogahama Beach, and others are omnipresent, such as the rumble of waves on the shore of the Sea of Enshu. The collection attempts to capture some of the changing sonic qualities of human activity, including the anachronistic sound of steam engines alongside more contemporary sounds like whistles of ships and the ebullience of cultural festivals. The soundscapes are available to listeners regardless of wealth, class, or religion, although visiting all of them would require travel. Unlike some other forms of cultural and natural celebration, the sound of wind in the reed beds along the Kitakami River or the temple bells of Teramachi asks of listeners no admission ticket.

A survey in 2018 found that five of the original one hundred soundscapes were gone or inaccessible. Frogs had disappeared, trams no longer ran, or earthquake damage made access to sites impossible. The majority of those remaining had some form of local government or citizen group promotion or protection. The listing therefore provided a measure by which to monitor long-term change and has catalyzed local interest and awareness. Despite these successes, the 100 Soundscapes of Japan project has added no new sites since its founding. Yet the sounds of Japan changed significantly in the last quarter century. Bleeps, voices, and music from mobile phones are ubiquitous in cities; shipping traffic has risen in the oceans; private vehicle ownership increased then dropped; a pandemic temporarily silenced much industry; and the sounds of forests, wetlands, and shores shifted as species thrived or struggled. Regular additions to the national register of soundscapes would both record these changes for posterity and turn human ears back to the world, encouraging sonic curiosity.

Although the list is, for now, static, the project has stimulated new ways of relating to sound in Japan and overseas. Soundscape researcher Keiko Torigoe served on the selection committee and later visited some of the sites to understand how local communities responded to their designations as nationally significant soundscapes. In the Hamaoka sand dunes on the east coast of Japan's mainland, the local government commissioned and installed a statue of Namikozo, the "wave boy," an ocean spirit who announces the weather through the drumming of waves. Torigoe felt ambivalent about representing the intangible spirit of the waves in concrete form, although the sculpture does orient visitors to the soundscape and honors an important cultural story. River damming and tree plantations are threatening the shoreline here, and so the sound of water beating against sand is considered threatened by some residents. Farther south, in the subtropical forests of Iriomote Island, she found that tour boat operators had ceased using motorboats on a river whose bird and insect sounds are on the national soundscapes register. One of the goals of the 100 Soundscapes project was to draw attention to and protect vulnerable sonic communities. In this case, the river's soundscape directly benefited from a reduction in engine noise. In the far north, on Hokkaido Island, she found that the designation had provoked conversations about understandings of soundscapes. The listed soundscape here comprises the creaking, groaning, and hissing sounds of winter sea ice on the Sea of Okhotsk. But the most notable "sound" of ice for locals is the sudden silence that descends when the garrulous motions of the sea are quieted by a cap of weighty ice, a process that often happens over just a few hours. The cultural meaning of this silence has changed. Formerly, it was a sign of the arrival of the "white devil," an ice-imposed end to fishing that presaged months of hunger and poverty. But since the 1960s, scallop aquaculture has boomed, and ice sheets provide shelter for the bays in which the shellfish thrive. Now the ice's sounds and silence are marks of the productivity of the sea.

The 100 Soundscapes of Japan project has led to elevated sensory awareness in places outside the locales on the official list. The Soundscape Association of Japan, for example, now offers regular encouragements to deeper listening, both by sponsoring experiences such as walks where participants turn their attention to the soundscape and by hosting discussions about how best to appreciate, understand, and protect the sonic diversity of Japan.

In 2001, partly inspired by the success of the soundscapes list, the Ministry of the Environment expanded its work into the realm of aroma. Japan's 100 Sites of Good Fragrance lists those whose aromas have particular cultural or natural significance. These range from wisteria blossoms to grilled eel, sulfur springs to the scent of used books in Tokyo's Kanda district. As was true for the designation of soundscapes, the motivations for this project were both to honor the sensory richness of Japan and to underscore the need to control noise and odor pollution. Rather than focus government efforts exclusively on managing negative experiences, these projects remind us to seek out and embrace the positive too.

That Japan should be a global leader in the recognition and celebration of its sensory richness is not surprising. Japanese religious, literary, and aesthetic practices pay close attention to the nuances of sound, aroma, and light, and to the embeddedness of human culture among plants, other animals, water, and mountains. Matsuo Bashō's haikus, for example, are full of the sounds of frogs leaping into water, cuckoos singing, and cicadas trilling. Buddhist and Shinto temples draw our senses into the spiritual agency of trees, the life of water, and the insights offered by sand and stone. The "right to sunlight" is protected by law, forbidding building practices that cast too much shade on neighbors. These are cultural foundations of sensory attentiveness and respect.

The 100 Soundscapes of Japan project also drew inspiration from across the Pacific. In the 1970s, Canadian composers R. Murray Schafer and Barry Truax popularized the terms *soundscape* and *acoustic ecology*, and along with collaborating musicians and sound recordists, studied the varied textures of sound across Canadian and European landscapes. Schafer described this work as a "study of the total soundscape," whose aim was to encourage "aural culture" and reduce noise, asking of every community "which sounds do we want to preserve, encourage, multiply?" Keiko Torigoe and others integrated this Western approach into Japanese culture that was already, in her words, "open to the world of sound."

An official list of notable soundscapes draws private sensory experience into community. Just as we gather to eat, pray, sport, view visual art, and hear music, so, too, can we gather to listen to the sounds of Earth, the marvelously diverse interminglings of the voices of wind, water, and living beings, humans included. How else might we create a culture of listening?

We gather at a picnic shelter on the shore of Lake Cootharaba in Queensland, Australia. The Pacific Ocean rolls onto beaches only seven kilometers east, but here the water is calm, fed by the freshwater flows of the Noosa River. Underfoot, sand mixes with the shed leaves of eucalypt and *Casuarina* trees, a soft, aromatic duff. Under a high sheet of cloud, water and sky present a milky silver expanse, interrupted only by a narrow band of green from the trees on the opposite shore, more than four kilometers away.

But despite appearances, the waters are not uniform. The two dozen people assembled here have come to hear the multiplicity of the lake and its river, using our ears to connect with lives and stories either below the surface or in the water's relationships with people. Our guide, sound artist and researcher Leah Barclay, arrives with arms laden with wireless headphones. We each don a pair and flip a switch to set them to the right channel, tuned to a small transmitter in the bag of electronics that Barclay carries at her waist. This is the same setup used by DJs and dancers at "silent discos," but today this technology will evoke not human music but the many stories of the water.

We laugh awkwardly at the strangeness of breaking conversations as we encase our ears in headphones. Some ambient sounds flow through—human voices and wavelets on the sandy lake margin—but mostly we have entered an aural realm where we are all tethered to a single source, the soundtrack that Barclay creates and is beaming to our headsets. For the next ninety minutes, we take a slow walk along the shore. Our feet step on sand, boardwalk, and pavement, our eyes dwell among trees and people, but our ears plunge into layers of sound recordings and live hydrophone feeds from mostly below the water's surface.

At first, we are immersed in shimmering, squeaking, and popping. Barclay does not interpret but allows the sound to exist for what it is, an aural experience of the vitality of the river. From my own previous experiments with hydrophones, my imagination is drawn to the motions of gas bubbles rising through sediment and the clatter of swimming, crawling, and singing aquatic insects. As we move from the picnic area to a small beach and then through some woodland, other sounds emerge. The pulse of waves sucking on sand. The bass rumble of what might be thunder. Pops from snapping shrimp and clicks from dolphins, and the drumming and tapping of fish. Human voices dip into and out of these sounds, including the songs of Gubbi Gubbi people responding to the river, stories of bonds between people and dolphins, snatches of conversation about respect for the river's animals.

The experience is partly music—Barclay uses sound samples to build rhythm, tonal structure, and melody—but it also feels like architecture as she shapes aural spaces that lack obvious pulse or narrative. Unmediated witness is also present in the portions of the experience where a live hydrophone feeds directly to our ears.

The monotone silver sheet of the lake surface acquires a new character. Like a closed door behind which we can hear lively conversations, the water seems no longer still and dull, but full of personality and possibility. This is the power of sensory connection: we understand in our bodies what the mind acting alone finds hard to apprehend. Before walking with Barclay's composition, I knew that water was full of life and motion. But, in a way, I could not grasp these abstractions. The sound in the headphones directly connects my senses, emotions, and mind to the energies in water, not just to ideas about water.

Unexpectedly, water sounds also change my experience of other senses. I feel a sudden enthusiasm for wavelets and immerse my hands at the water's edge, feeling the pulse on my skin. Hearing a mix of snapping shrimp and insects, I wonder about salinity and taste a drop of water. It is brackish, the union of inland wetlands with seepage from the ocean. The sight and sound of a child rushing into the water and throwing sand into wet piles merge with the less familiar sounds in the headphones, making me puzzle at humanity's playful fascination with water. From sandcastles to sailing dinghies to ocean cruises, we seem to crave contact. On a point that juts into the lake, the wind gusts, and I delight in the convergence of its scouring action on my skin and the rough, stormy textures of the sound in my ears at that moment. The aromas of sodden vegetation strike with particular force. Somehow listening wakes up my nose.

We are familiar with the synesthetic and emotional effects of sound in the everyday human realm. The right music can make food taste better, warm our skin, open us to the effects of touch, awaken and relax our muscles, and heighten our sense of belonging in our bodies and communities. Barclay's work brought those sensory, affective connections to an unfamiliar place, expanding our empathy and imagination into the water.

Of the human voices in the piece, one stood out to me, a recounting of the cooperative bond between the Gubbi Gubbi and dolphins. Before colonial invasion broke this connection, local people would call to dolphins by, in the words of nineteenth-century European observers, "jobbing with their spears into the sand under the water, making a queer noise" or using spears to make a "peculiar splashing in the water." The dolphins heard and understood these sounds, and swam close to join the hunting team. By circling then moving inward toward shore, dolphins corralled the fish. People, wading in the water, then speared or netted the trapped quarry. The dolphins got their share, often fearlessly taking fish proffered on spear tips.

Humans and dolphins each have sophisticated vocal cultures. Their societies thrive through sound-mediated reciprocity and coordinated action. These two great animal cultures, triumphs of mammalian evolution, used sound to knit their intelligences into cooperative action. Only recently have some human cultures forgotten that we belong within a speaking, listening, and intelligent world, one where we can converse with other beings for mutual benefit. The first step back to this knowledge is, perhaps, better listening, along with renewed respect for the cultures of other humans and nonhuman beings.

More than twenty thousand people have experienced Barclay's River Listening Sound Walk, either as I did, in small groups, or through self-guided experiences via a smartphone app. Started here on the Noosa River, the project now includes three other sites in Australia and rivers in Europe, North America, and the Asia-Pacific.

Barclay's mastery of the technologies of recording and composition, along with her ability to offer engaging community experiences, is sonic wizardry, raising the hidden energies in the water into human attention. The results can transform people in unexpected ways. Many local farmers are skeptical of city-based artists and scientists coming to "listen to the river," a place the farmers have known through work and recreation, sometimes for decades, without any need of seemingly esoteric art. But dropping a hydrophone into these familiar places produces jolts of excitement and curiosity. Connecting the hydrophone to a live-feed transmitter deepens the connection. Barclay told me that several farmers now start their days by listening in their kitchens to live feeds from nearby rivers. The fact that the sound is both live and local is important. A recorded track or a feed from some distant locale might be interesting for a while, but the sounds of your home place are of immediate relevance and emotional power. Might readily accessible data from hydrophones and microphones one day become as ubiquitous as the temperature and rainfall readings from weather stations, technological aids to human senses and curiosity?

Among scientists, too, listening to the river can change behavior. Biologists often become inured to the damage they do to their "subjects," walled off by educational curricula that favor vivisection and objectification over affective and sensory connection. In my own early education in biology, I was asked hundreds of times to apply the scalpel or a lethal dose of ethanol to animals from rats to fruit flies to snails, but not once was I challenged to converse with these beings that Darwin taught us were blood kin. In surveys of rivers, field biologists routinely kill the animals they have sampled with electric shocks or nets. After listening to the river through her equipment, Barclay told me that many scientists say, "Well, maybe we'll put them back alive this time." Listening to the many sounds of fish opens human imagination. We hear them not as numbers on a spreadsheet but as communicative creatures in whose voices we hear selfhood and agency. This is a sensory lesson in kinship.

Sound-recording technologies, then, open our ears to the lives of other beings. For aquatic creatures, hydrophones break what is mostly an impenetrable sensory barrier. On land, too, sounds captured by microphones and shared with listeners can reveal hidden stories and encourage connection to place. From "nature sound" albums, to websites that teach us to notice and understand the voices of our nonhuman neighbors, to apps that guide listeners through curated aural experiences of notable sites, recording technology opens our ears, and thus our imagination and empathy, to the beauty and the travails of the world. By freezing ephemeral sound waves on magnetic tape or in a microchip, we bring them partly under our control. We can then share, rework, puzzle over, measure, and celebrate sound's many qualities.

Too much control, though, can distance us from the places and lives we seek to hear. Barclay told me of students whose work integrated the latest aquatic recording devices with sophisticated analysis software. Their work demonstrated great technological proficiency. Yet not one of them had listened to their "study soundscapes" with unaided ears or from raw electronic recordings. Like passive acoustic monitoring in rain forests, microphones and computer software in the hands of artists and scientists do not necessarily displace embodied listening. But their powers can sometimes make us forget the testimony of our own bodies.

Leah Barclay's work seems especially noteworthy to me because it uses technology to reembody listeners in their senses and relocate them in landscapes and water. She builds on the work of pioneers such as Annea Lockwood and Pauline Oliveros, whose music calls us to listen more fully to the places around us, especially to the voices of the beyond-human world. This contrasts with the philosophy underlying so much technological evocation of "nature" where screens and loudspeakers transport us to exciting locales and action-filled narratives, yet do little to open our senses to the stories of our home places. Indeed, after the excitement of a documentary film, the edited highlights of thousands of hours of filming and sound recording, the creatures we live among can seem disappointingly dull. Escape from the mundane has its place, of course, and art should sometimes lift us into other places and times. But the discovery of the rhythms and stories of home is vital too. These are the foundations not only of delight but of wise ethical discernment.

River Listening is not polemical—it contains no gunning outboard engines or throbs of offshore container ships—but instead offers open-ended invitations to listen and extend human sensory attention into aquatic realms. This expanded sensory and imaginative connection is much needed. Beyond the mouth of the Noosa River, along a coast rich in sea life, including the breeding grounds of whales and the edge of the Great Barrier Reef, shipping traffic is increasing by nearly 5 percent per year. Several large new inland mines have recently been approved in Queensland that will export their coal and minerals by ship. Each one of these vessels will haze water with noise. As is true along all shipping routes, the devastating effects of this noise on sea life remain hidden from us. As sensory beings, we are disoriented without direct experience of the consequences of our actions. For a species that transports about 90 percent of our goods by water, our disconnection from aquatic sounds is ruinous to moral clarity and right action. Never have human guides to the underwater world of sound been more needed.

The rain held off. The sun is out. For a November morning in New York City, this is glorious weather. Here on the grounds of the New York Botanical Garden (NYBG), trees are on the hinge of late summer and autumn. The low sun gleams from ginkgo leaves now almost entirely turned gold. Larger beeches, maples, and oaks are bronzed and sulfured. Saplings, though, retain late summer's green, no doubt stealing an extra fortnight of photosynthesis as their frost-exposed elders retire. The mellow aroma and crunch of freshly fallen maple leaves rise from underfoot.

Along walkways through the garden, pedestrian traffic flows inward, toward the forested ridge that forms the spine around which more formal collections of the gardens are arranged. We gather at a small table where a leaf-strewn path breaks away from the wider avenue, the entryway to the forest. We have come to listen to an afternoon performance, one that will mingle human voices with those of nonhuman animals and trees. For the next hour, choral groups, loudspeakers, apps on visitor cell phones, and small wooden "robot" instruments animate the loop path through the forest. Visitors move within this promenade of sound, each at their own pace, going back and forth as they please, creating their own sonic narratives.

The performance, *Chorus of the Forest*, is the work of NYBG's 2019 composer in residence, Angélica Negrón. She composed the piece for this site, bringing her musical ideas into relationship with the sounds of the woodland. As I walk the pathways, I pass through overlapping domes of sound, each one centered on a choral group or a cluster of loudspeakers. In the spaces between, the domes merge into one another and the ambient sounds of forest and city.

Near the start of the loop trail, a speaker near a box of electronics delivers crackling sounds mixed with shifting pure tones. These are stimulated by electrodes that run from a rhododendron's living green leaves. A few steps down the path, wooden automata swing clappers against small sheets of wood and metal bells. Made by sound artist Nick Yulman, these devices have the form of small trees, their trunks and side branches made from recycled sawed lumber. As I walk on, I hear amplified clicks and rasps of insects chewing wood, wind and ice playing against leaves, and the thrum of vibrations inside tree trunks layered into much slower, purer tones. These are sounds I recorded from trees and shared with Negrón, which she then interpreted, mixed, and sculpted with sound-editing software. Another electronic enhancement comes later along the loop walk when visitors dial a number to play the sounds of white-throated sparrows and other birds through their phones.

At six different stations along the pathway, choral groups sing her compositions. Close up, we hear the words and musical details. At a distance, the forest adds its signature, a gentle blur and glow of reverberation. Each piece evokes a different dimension of human relationship to forests. In "Awaken," for example, the Young New Yorkers' Chorus lifts into song dozens of verbs about forest interconnection, words that Negrón drew from books and social media conversations. Other pieces are inspired by poems and stories that explore trees, ecological justice, and human resilience. In all, more than one hundred singers are here, including several local school choirs. At two places along the walk, singers line both sides of the path and the stone bridge over the Bronx River, creating a sonic avenue through which visitors walk. As I pass through these spaces, bathed in harmonized human song, voices seemed to rise within my chest, a joyful sympathetic vibration.

This is a work of convergence. The second-by-second physiology of plants, recorded on electronic sensors, merges with percussive sounds from Yulman's creations and my tree recordings, and reveals the materiality and inner lives of wood. This music offers both a contrast and a complement to that of wooden instruments like violins or pianos, which also draw on the physicality of trees but in a form more highly mediated by human intent. The blend of human song with tree and bird sounds creates contrasts in musical forms. The emotional power of human voices is direct and clear; nonhuman sounds are foreign tongues, harder for our human senses to comprehend.

Unifying all these elements of the composition are the sounds of the site itself. A light wind stirs a sandy hiss from dry maple leaves in the canopy. Near the river, water churns over a short weir. Squirrels rustle through leaf litter. Traffic sound and the occasional siren from the roads that circle the gardens arrive in unpredictable waves, buffeted by wind. Visitors talk as they move between choir stations, laugh when the bird sounds leap from their phones, or stand and whisper as they gaze up into the canopy or at one of the wooden automata.

I'm delighted to hear this convergence of musical evocations of the forest. But what strikes me most in the event is the balance between control and openness. Unlike in a concert hall where great efforts go into excluding "outside" sound, human creativity here exists in active relationship to the site and the moving bodies of listeners. The composer has a central voice but one with only partial control. Human creativity exists within the other energies of the place, including wind, traffic, chatty visitors, birds, and the inner lives of plants. This embeddedness aims to elevate our attention to these uncontrolled sounds. Angélica Negrón said of the project, her hands adding air quotes, "My big hope is that when people walk out of the forest and the sound of the piece 'stops'—so the piece is 'done'—they notice that it's still going on, all the time, around them." For the more than three thousand people who experienced the piece, this is music as invitation to listen. It is also music that invites community. We do not sit in the dark isolated from others. We shed our earbuds and headphones before entering the forest. No rules forbid talk or laughter. I came alone but shared short conversations about the experience with a dozen other visitors, a rare occurrence in public spaces in the city or after a concert at Lincoln Center or another recital hall.

Composer John Luther Adams has also noted the convivial effect of music played in unstructured spaces, with audience members free to move. Reflecting on *Inuksuit*, a piece for percussion usually performed in spaces like the forests of Vermont, he wrote, "When I originally composed *Inuksuit*, I wasn't prepared for the strong sense of community the piece seems to create." When music is placed in relationship with the nonhuman world, human community is intensified too.

By inviting us to listen beyond the rigidly defined boundaries of typical performance spaces, these pieces allow us to better hear and connect with one another. Once one wall is breached, others follow. In this opening, we reinhabit our nature. Most of us now live in places where we must block out sound to retain any hope of focus or well-being. We do this sometimes with technology—noise-canceling headphones, closed doors, or sound-proofed walls—but mostly by acts of the will, withdrawing attention from traffic, whir of computers, sigh of air coming from heating or cooling units, chatter and bang of neighbors and coworkers, rumble of jets overhead, construction noises across the street, and the sounds of birds and insects through cracked windows. Most of these sounds contain no information immediately relevant to our work or social lives. But for our ancestors, attention to sound was the source of food and knowledge about local conditions, just as it remains for people today who live and work in close relationship to the nonhuman world. This is the original function of hearing, to bring the stories around us into human awareness. To shut off listening is, in these circumstances, like an industrialized human turning off the internet and TV: you lose connection to news and to networks that link you to others. People who straddle the industrialized and ecological worlds deliberately switch between modes of listening. When I leave the city for places dominated by nonhuman beings, I repeatedly ask myself to open up. Listen, touch, smell, look, then repeat again and again. Only then can I hope to connect to and properly inhabit the forest, prairie, or seashore. When done with others, this opening necessarily brings us into closer human community too. On reentry to the built environment, I re-wall the senses, steeling myself against the incoming surge and tightening the filter on what gets my attention. This includes mostly not interacting with other humans. To greet them as I would in the forest would be not only exhausting but out of step with the social dynamics of city life. Works such as Angélica Negrón's *Chorus of the Forest* offer invitations to lower the sensory barriers we must sometimes necessarily erect. She fashioned this inducement out of the delight and power of human voices and the intriguing strangeness of plant sounds, experiences rich in their musical forms and reorientations of our senses.

Musician and philosopher David Rothenberg takes the invitation further, beyond the boundaries of the human. His performances with insects, birds, and whales ask other species to participate. We humans are not the only species with keen ears and voices eager to connect. In Rothenberg's hands, clarinets become experiments in cross-species connection and sonic innovation. Unlike the eighteenth- and nineteenth-century players of the *merline* and other bird organs used for training captives, Rothenberg's birds are free-living, and the creative process is interactive, ceding some control to the other singer. Instead of layering prerecorded nonhuman animal sounds into a musical performance, as many contemporary ecologically minded musicians do, Rothenberg goes to living animals and offers them an opportunity for sonic dialogue, for creative reciprocity.

In my conversations with him and in his writings, Rothenberg emphasizes the primary importance of listening. His musical roots are in improvisational jazz, where close attention to the sounds of other players is vital. To listen and play with another human player is hard. To do the same with an animal whose lineage separated from ours tens or hundreds of millions of years ago brings our ears to the edge of a vast chasm of sensory and aesthetic experience. Therein lies much of the power of his work. This is experimental biology and philosophy of sensory experience.

Rothenberg's most recent major project involved playing with nightingales over a span of five years in the city parks of Berlin. He did this sometimes alone with the birds, but also with other people, from violinists and oud players, to vocalists, to electronic musicians. Hearing the interplay between these human sound makers and the birds, experiences captured in the film *Nightingales in Berlin*, I am struck by contrasts of pacing. We must sound to the bird as humpback whales seem to us: creatures for whom time is slowed and aural attention is greatly elongated. The nightingale song comprises bursts of trills, whistles, and gurgles whose details are too fast for our sluggish brains to grasp. Rothenberg asks of the birds and his fellow musicians, "What can be done together? Can you ask questions through music?" Are the nightingales riffing with the humans? To my ears, listening from outside the back-and-forth interaction between human players and birds, it is hard to tell. The birds' songs are complex, like insanely fast electronic music, continually remixed. Discerning responses to humans amid this sonic craziness is beyond me. But for Rothenberg, "the nightingale dances musically in and around samples and transpositions of himself." Can two species with rich vocal cultures—nightingales and humans—engage in creative musical dialogue? Rothenberg explores these questions through participation. He says, "My biggest hope with the project is that it should not end up being strange, but rather familiar. All music education, anyone who studies music . . . should have to reckon with the music of other musicians on this planet, other animals."

Rothenberg honors the rich evolutionary diversity of sound and, for birds and whales, he takes seriously the sophistication of their vocal learning and cognition. Humans, birds, and whales are three pinnacles of sonic culture. To put them into active relationship with one another is an act of respect and kinship, profoundly Darwinian and ecological in its approach. Yet to play music with birds in a city park also seems more than a little odd in the context of industrialized, technological human culture. His work, then, reveals our everyday estrangement from the living Earth. We live among other species with elaborate vocal cultures, yet we seldom reach out to experience what might lie at an intersection of sonic cultures. Rothenberg's playing also uncovers and highlights the great diversity of animal aesthetics. Each species has its own preferences for timbre, pacing, and style, varieties brought into vivid contrast with our own through active, embodied dialogue. Scientists understand, through theory and experiment, that these diverse aesthetics are engines of genetic and cultural evolution. Rothenberg's musical work provides a complement to science, investigating aesthetics from the inside, in ways impossible through the objective, but distant, insights of replicated scientific inquiry. Just as an understanding of human music is deepened through the perspectives of players and singers, cross-species participation might also help us to fathom the music of other species.

After Angélica Negrón's piece ended, I leaned on the wooden fence that defines the pathway and enjoyed a sense of calm after the rush of activity and people had passed. A hermit thrush, likely newly arrived from more northerly forests, snatched a tiny spider from among loops of speaker cable in the freshly fallen maple leaves. The bird flew to the crossbeam of the wooden fence next to me, then gave a loud, low *tchup*. Like the human voices that sounded from this same spot an hour ago, the thrush's sound had a pleasing fatness and resonance. Deciduous forests have an acoustic warmth similar to that of concert halls. Sound waves bounce back from tree trunks and leaves, giving a lively sense of immediacy and a warming touch of reverberation. In our concert halls, we re-create the acoustic properties of woodlands, the sonic homes of our primate ancestors for tens of millions of years. The music we heard this afternoon connects us, perhaps, to some of the aesthetic origins of more conventional performance spaces.

But the link here between sound and times past is deeper than the human or primate lineage. It is fitting that a botanical garden should host a celebration of sound. The first trees and shrubs, four hundred million years ago, caused insects to crawl upward, then to evolve wings. This led to Earth's first animal songs. Later, flowering plants fueled the evolutionary explosion that wrapped Earth in the sounds of birds, most insects, and mammals. In this garden, land animal sound has come home.

In the Deep Past and Future

On a moonless night on an escarpment south of Santa Fe, I am astonished by the brightness of the gleam above. With no city light pollution, few clouds, and little dust to obscure vision, the night sky in New Mexico is a confusion of bright flecks against a silvery haze. I lift my binoculars. The haze resolves into yet more stars, with stellar clouds behind them in depths whose magnitude frightens me. The chill of the cold, dry air reinforces my unease. Although I'm breathing easily and rooted to the ground by gravity, I feel somehow unmoored. Daylight is a mask. When the veil of a glowing daytime sky falls away, it reveals stars of such abundance and brilliance that our senses and imaginations are unearthed into a huge and humbling cosmos.

From the same mountains, starting in 2000, the Sloan Digital Sky Survey used a mirror two and a half meters in diameter to gather light from the night sky. This surface is about twenty thousand times larger than the retinas of my eyes. The telescope scanned back and forth across the sky for five years, recording with electronic sensors the coordinates of galaxies.

The telescope found order within the smokelike multitude of stars. Galaxies are more likely to be separated by spans of five hundred million light-years than by other distances. This regularity is the wave mark left by the first sounds of the universe, a remnant from the early cosmos scored into the patterns of the sky. On a clear sky, then, we can stare up and see the origins of sound in the universe.

Where were these first sounds born?

Not in the "big bang." The primordial expansion of the universe was encompassed in nothingness: no space, no time, and no matter. But sound exists only in space and time, its waves flowing through matter. No sound could announce the universe's birth.

Nor was sound born in planetary or geologic tremors, watery vibrations, or the stirrings of bacterial cells. These are all sounds traveling through matter made from atoms: gases, liquids, and solids. But sound is older than atoms.

After its birth, in its infant years, the universe—all energy, all matter—was packed so tight that the temperature blazed into billions of degrees. No atom could exist in such heat. Instead, protons and electrons roiled in a hot lava, a plasma. The plasma was a mire so dense that particles of light, photons, were trapped. Inside this furnace, sound was born.

Irregularities in the plasma sent out pulses. Each pulse was a sound wave, a traveling front of high and low pressure, just like the waves of compression in air that we create when we snap our fingers. The waves traveled through the plasma hundreds of thousands of times faster than sound on present-day Earth.

As the universe expanded, the crowding eased, causing the temperature to drop from billions to mere millions of degrees. At about 380,000 years after the universe's origin, the cosmos cooled enough for the plasma to transform into material familiar to us now. Protons and electrons combined, making stable atoms. As the traffic jam of protons eased, light was no longer trapped and fled.

As atoms formed, they were marked by the waves that flowed through the plasma. Each wave crest, a place where the plasma was compressed, became an aggregation of atoms, separated by wave troughs where atoms were sparse. Gravity's convivial imperative then drew clusters of atoms together, building the former crests of waves into ever denser crowds. From these early clumps, stars and galaxies grew. By our earthly clocks, this was an unhurried ingathering. One hundred and eighty million years passed before the first stars blazed. It took another billion years for galaxies to flock the skies. Now, 13.5 billion years later, a telescope on a piney ridge in New Mexico can measure distances between galaxies and find the regular peaks of the ancient sound waves.

The wave marks are also discernible in the light that escaped the plasma. This light energy became cosmic microwave background radiation, a faint glow that now permeates the universe, detectable only with the most sensitive instruments. The glow is not uniform but is rippled with slight peaks and troughs. These patterns, like the spacing of galaxies, were imprinted on the radiation in the moment of its origin in the cooling plasma.

All sound relays what is past—even the voices of everyday conversation are created a few milliseconds before we hear them—but these waves are older than Earth itself. These ancient sounds exist on scales that feel preternatural. Waves larger than galaxies? Ancient microwave energies passing through us undetected? Our earthbound senses have no bodily understanding of such beyondness. Our imaginations, though, feed on the gleanings of science, casting our minds into places and times previously undreamed. The brains that ponder the first sound waves are themselves made from these waves because our own planet and star are, like all planets and stars, descendants of the primordial plasma. And so our bodies—and the thought that emerges within them—are made from the remnants of acoustic waves in the plasma. From inside ancient sound, we listen.

Some sound waves dissipate. But others evoke new arrangements of matter and energy. Stars were seeded by ancient sound waves. Sound has always been a creative force. This creative property of sound is not mystical; it emerges from the physical laws of our universe. The arrangements of stars and cosmic radiation are among the first of these creations, the opening salvos of the universe's rich sonic history.

Thirteen billion years after the plasma cooled, sound met its new creative partner, life on Earth. What followed is a flourishing unrivaled, as far as we know, by any other time or place in the cosmos. From the thrum of bacteria, to the effusion of animal voices, to human music in concert halls, ours is a sonic planet, full of listeners and communicative voices. This extraordinary blooming is partly rooted in times much older than Earth, in the ancient generative capacity of sound itself.

What is sound's future?

Cosmologists disagree about the fate of the universe, but all agree that the present state of matter will not last. Either we collapse back into infinitesimal smallness, expand into cold flatness, or are torn into a thin fog of elementary particles. All leads to silence. Earth will be devoured by the sun long before this final end, taking with it all the diverse songs of earthly life.

If all living sound is doomed, why care about creativity, diversity, and diminishment in our present moment? Ethical nihilism is one response to the fleeting and fated nature of existence. But sound itself suggests another answer. All sonic experience moves from silence, into ephemeral existence, and then back into silence. Silence also gives sound its shape, providing the open space in which sonic form emerges. The songs of a blackbird or the music of an orchestra recapitulate the journey of sound in the cosmos: out of nothing, into brief life, then a return to silence. In this lies their value. The Earth's sounds matter in part because they are ephemeral manifestations of order and narrative. There is a parallel here with the value of each of our personal journeys from nonexistence, to form and movement, into death. Listening gives us an experience of the value of temporary existence unlike any other bodily sense. Sound departs as soon as it arrives, whereas a gaze onto a scene, or a touch on the skin, or the aroma of a flower all linger, at least for a while.

But sound has one more quality that gives it special value. Sound waves are fugitive, yet the energies and patterns they leave are creative. Sound seeded the stars, caused voices to arise from primordial living beings, and made music and language in animals.

Sound, then, has value because it is generative. Waves in ancient plasma, the songs of crickets and whales, the babble of young sparrows and humans, the tones of human breath in mammoth ivory: these are creators. Not as gods, but as the living and physical processes that made the universe.

This is why the diversity of sounds is so glorious. We hear not only the result of creation but the very *act* of creation. We inhabit the generative power of the universe, expressed in the particularity of the moment. By killing and smothering Earth's many voices, we silence and destroy what made us.

In the seemingly straightforward act of listening, we discover not endings but connection and creativity in the present. Our senses and aesthetics arrive from deep time, made of atoms built from ancient sound waves, animated by tiny hairs on cells, and shaped by the long evolution of animals reaching out to one another in sonic eagerness. These legacies disclose the beauty and brokenness of the present time, giving us sensory foundations for joy, belonging, and action.

ACKNOWLEDGMENTS

In these pages, I make the case that the acoustic crisis has four main pressing and intersecting dimensions: the silencing that ensues from loss of ecological habitat and attacks on human rights, especially in tropical forests; the nightmare of industrial sound in the oceans; the inequities of noise pollution in cities; and our frequent failures as individuals and cultures to listen to and celebrate the storied sensory richness of our world. I will donate at least half of my net proceeds from this book to organizations that work to heal and reverse these aggressions, fragmentations, and losses.

Books such as this have single names on their covers, but whatever insight is carried in their pages arrives from a community and not a single individual. My listening, comprehension, and writing are immeasurably deepened by Katie Lehman's companionship, keen and curious ears, empathetic imagination, and brilliant mind. Paul Slovak's extraordinary work as my editor has shaped and refined the book's ideas and the text, and I am very grateful for the many ways he invigorates, clarifies, and supports my work. Paul's colleagues at Viking have done outstanding work on this and my previous books. Alice Martell is the best agent I could hope for, providing wise counsel, effective advocacy, and unstinting encouragement. I also thank Stephanie Finman, at the Martell Agency, for her support and assistance, especially amid the difficulties of the pandemic. Meagan Binkley gave invaluable encouragement and practical help with the preparation of the manuscript. My parents, Jean and George Haskell, not only nurtured and spurred my youthful curiosities but steeped my upbringing in the music of humans and nonhumans, and more recently provided many fruitful leads for the research for this book.

For answering my queries and generously sharing their expertise about evolution and ecology I am very grateful to Olivier Béthoux, Muséum National d'Histoire Naturelle, Paris; Luis Alberto Bezares-Calderón, University of Exeter; Martin Brazeau, Imperial College London; John Clarke, Nicolaus Copernicus University in Toruń; Rex Cocroft, University of Missouri; Allison Daley, University of Lausanne; Sammy De Grave, Oxford University Museum of Natural History; Gregory Edgecombe, Natural History Museum, London; Eric Keen, University of the South; Rudy Lerosey-Aubril, Harvard University; Lauren Mathews, Worcester Polytechnic Institute; Eric Montie, University of South Carolina Beaufort; Eric Parmentier, Université de Liège; Sheila Patek, Duke University; Arthur Popper, University of Maryland; Rebecca Safran, University of Colorado; William Shear, Hampden-Sydney College; Kirsty Wan, University of Exeter; and Michael Webster, Cornell University. Tim Low, biologist and writer, was especially helpful in clarifying my thinking through both his writing and our conversations.

Zuzana Burivalova, University of Wisconsin Madison, and Eddie Game, The Nature Conservancy, shared with me both their time and their extraordinary sound recordings. These data are hosted at the Ecoacoustics Research Group at the Queensland University of Technology. Wendy Erb, Cornell Lab of Ornithology, and Martha Stevenson, World Wildlife Fund, also generously shared their insights about tropical forests, fire, and biological conservation.

Composers, sound artists, and musicians Leah Barclay, University of the Sunshine Coast; Angélica Negrón; and David Rothenberg, New Jersey Institute of Technology, opened my ears and mind in new ways, through both their public work and our conversations. Their work at the intersections of art, science, philosophy, and activism provides joyful and hopeful paths to the future. At the New York Botanical Garden, I thank Hillarie

O'Toole and Thomas Mulhare for commissioning and organizing performances and public discussions of forest sounds, and Annie Novak for the many ways she encouraged and catalyzed my work.

Wulf Hein and Anna Friederike Potengowski did extraordinary work on the experimental reconstruction and playing of the mammoth-ivory flute and were a delight to collaborate with. Nicholas Conard, University of Tübingen, was a welcoming and insightful guide to the Paleolithic caves of southern Germany.

At National Sawdust, I thank Paola Prestini, Garth MacAleavey, and Holly Hunter for their welcome, conversation, and in situ demonstrations. John Meyer, Pierre Germain, Steve Ellison, and Jane Eagleson shared their insights into the work of Meyer Sound Laboratories. Jayson Kerr Dobney helped me to see the many layers of story and interconnection in the Musical Instruments collection at the Metropolitan Museum of Art of New York City. Sherry Sylar of the New York Philharmonic kindly shared with me her perspective on the many relationships among musicians and the materiality of their instruments.

Audiologist Dr. Shawn Denham was a skillful and wise guide to the hair cells of my inner ears.

For stimulating conversations about sound and its many manifestations, and for their welcome during my travels, I thank Joseph Bordley, John Boulton, Sunniva Boulton, Nickole Brown, Dror Burstein, Angus Carlyle, Lang Elliott, Charles Foster, Sue Gould, Peter Greste, John Grim, Lyanda Fern Lynn Haupt, Holly Haworth, Caspar Henderson, Christine Jackman, Jessica Jacobs, James Lees, Adam Loften, Sanford McGee, Paul Miller, Indira Naidoo, Kate Nash, Rhiannon Phillips, Richard Prum, Marcus Sheffer, Richard Smyth, Stephen Sparks, Mitchell Thomashow, Mary Evelyn Tucker, Marianne Tyndall, Emmanuel Vaughan-Lee, Sophy Williams, Peter Wimberger, and Kirk Zigler. I offer special thanks to David Abram, for his inspiring work, our fruitful conversations, and his encouragement during this project.

My publishers and colleagues in Australia, Black Inc., along with the Byron Writers Festival, the Bendigo Writers Festival, the National Library of Australia, and Integrity 20 at Griffith University, generously hosted me in Australia. In Ecuador, the managers and staff at Universidad San Francisco de Quito's Tiputini Biodiversity Station were welcoming colleagues. I also thank Esteban Suárez, Andrés Reyes, Given Harper, and Chris Hebdon for their companionship and many insights in Ecuador.

My mentors during my undergraduate education at the University of Oxford, Andrew Pomiankowski and William Hamilton, showed me the extraordinary power and beauty of evolutionary biology, especially the many ways that aesthetics have shaped and diversified sound and other forms of animal communication. Greg Budney, Russ Charif, and Chris Clark were generous guides as I learned sound recording and analysis techniques as a graduate student at Cornell University, and many colleagues at Cornell in the fields of ecology, evolution, and animal behavior deepened my appreciation of evolution's creative powers.

In the research for this book I made extensive use of the libraries at the University of the South, Sewanee, and at the University of Colorado, Boulder, and offer my thanks to the staff at both for their help, especially during the challenges of the pandemic. The University of the South provided funding for my visit to Germany and granted me a leave of absence to work on this project.

I wrote this book while living on the unceded territory of the Arapaho. I pay my respects to elders past, present, and emerging.

Thank you, reader, for spending time with the sounds, living beings, ideas, and places imperfectly evoked by these words. I offer the book as an invitation to both wonder and action, guided by your own listening.

BIBLIOGRAPHY

Primal Sound and the Ancient Roots of Hearing

Aggio, Raphael Bastos Mereschi, Victor Obolonkin, and Silas Granato Villas-Bôas. "Sonic vibration affects the metabolism of yeast cells growing in liquid culture: a metabolomic study." *Metabolomics* 8 (2012): 670–78.

Cox, Charles D., Navid Bavi, and Boris Martinac. "Bacterial mechanosensors." *Annual Review of Physiology* 80 (2018): 71–93.

Fee, David, and Robin S. Matoza. "An overview of volcano infrasound: from Hawaiian to Plinian, local to global." *Journal of Volcanology and Geothermal Research* 249 (2013): 123–39.

Gordon, Vernita D., and Liyun Wang. "Bacterial mechanosensing: the force will be with you, always." *Journal of Cell Science* 132 (2019): jcs227694.

Johnson, Ward L., Danielle Cook France, Nikki S. Rentz, William T. Cordell, and Fred L. Walls. "Sensing bacterial vibrations and early response to antibiotics with phase noise of a resonant crystal." *Scientific Reports* 7 (2017): 1–12.

Kasas, Sandor, Francesco Simone Ruggeri, Carine Benadiba, Caroline Maillard, Petar Stupar, Hélène Tournu, Giovanni Dietler, and Giovanni Longo. "Detecting nanoscale vibrations as signature of life." *Proceedings of the National Academy of Sciences* 112 (2015): 378–81.

Longo, G., L. Alonso-Sarduy, L. Marques Rio, A. Bizzini, A. Trampuz, J. Notz, G. Dietler, and S. Kasas. "Rapid detection of bacterial resistance to antibiotics using AFM cantilevers as nanomechanical sensors." *Nature Nanotechnology* 8 (2013): 522.

Matsuhashi, Michio, Alla N. Pankrushina, Satoshi Takeuchi, Hideyuki Ohshima, Housaku Miyoi, Katsura Endoh, Ken Murayama et al. "Production of sound waves by bacterial cells and the response of bacterial cells to sound." *Journal of General and Applied Microbiology* 44 (1998): 49–55.

Norris, Vic, and Gerard J. Hyland. "Do bacteria sing?" *Molecular Microbiology* 24 (1997): 879–80.

Pelling, Andrew E., Sadaf Sehati, Edith B. Gralla, Joan S. Valentine, and James K. Gimzewski. "Local nanomechanical motion of the cell wall of *Saccharomyces cerevisiae*." *Science* 305 (2004): 1147–50.

Reguera, Gemma. "When microbial conversations get physical." *Trends in Microbiology* 19 (2011): 105–13.

Sarvaiya, Niral, and Vijay Kothari. "Effect of audible sound in form of music on microbial growth and production of certain important metabolites." *Microbiology* 84 (2015): 227–35.

Unity and Diversity

Avan, Paul, Béla Büki, and Christine Petit. "Auditory distortions: origins and functions." *Physiological Reviews* 93 (2013): 1563–619.

Bass, Andrew H., and Boris P. Chagnaud. "Shared developmental and evolutionary origins for neural basis of vocal–acoustic and pectoral–gestural signaling." *Proceedings of the National Academy of Sciences* 109 (2012): 10677–84.

Bass, Andrew H., Edwin H. Gilland, and Robert Baker. "Evolutionary origins for social vocalization in a vertebrate hindbrain–spinal compartment." *Science* 321 (2008): 417–21.

Bezares-Calderón, Luis Alberto, Jürgen Berger, and Gáspár Jékely. "Diversity of cilia-based mechanosensory systems and their functions in marine animal behaviour." *Philosophical Transactions of the Royal Society B* 375 (2020): 20190376.

Bregman, Micah R., Aniruddh D. Patel, and Timothy Q. Gentner. "Songbirds use spectral shape, not pitch, for sound pattern recognition." *Proceedings of the National Academy of Sciences* 113 (2016): 1666–71.

Brown, Jason M., and George B. Witman. "Cilia and diseases." *Bioscience* 64 (2014): 1126–37.

Bush, Brian M. H., and Michael S. Laverack. "Mechanoreception." In *The Biology of Crustacea*, edited by Harold L. Atwood, and David C. Sandeman, 399–468. New York: Academic Press, 1982.

Ekdale, Eric G. "Form and function of the mammalian inner ear." *Journal of Anatomy* 228 (2016): 324–37.

Fine, Michael L., Karl L. Malloy, Charles King, Steve L. Mitchell, and Timothy M. Cameron. "Movement and sound generation by the toadfish swimbladder." *Journal of Comparative Physiology A* 187 (2001): 371–79.

Fishbein, Adam R., William J. Idsardi, Gregory F. Ball, and Robert J. Dooling. "Sound sequences in birdsong: how much do birds really care?" *Philosophical Transactions of the Royal Society B* 375 (2020): 20190044.

Fritzsch, Bernd, and Hans Straka. "Evolution of vertebrate mechanosensory hair cells and inner ears: toward identifying stimuli that select mutation driven altered morphologies." *Journal of Comparative Physiology A* 200 (2014): 5–18.

Göpfert, Martin C., and R. Matthias Hennig. "Hearing in insects." *Annual Review of Entomology* 61 (2016): 257–76.

Hughes, A. Randall, David A. Mann, and David L. Kimbro. "Predatory fish sounds can alter crab foraging behaviour and influence bivalve abundance." *Proceedings of the Royal Society B* 281 (2014): 20140715.

Jones, Gareth, and Marc W. Holderied. "Bat echolocation calls: adaptation and convergent evolution." *Proceedings of the Royal Society B* 274 (2007): 905–12.

Kastelein, Ronald A., Paulien Bunskoek, Monique Hagedoorn, Whitlow W. L. Au, and Dick de Haan. "Audiogram of a harbor porpoise (*Phocoena phocoena*) measured with narrow-band frequency-modulated signals." *Journal of the Acoustical Society of America* 112 (2002): 334–44.

Kreithen, Melvin L., and Douglas B. Quine. "Infrasound detection by the homing pigeon: a behavioral audiogram." *Journal of Comparative Physiology* 129 (1979): 1–4.

Ma, Leung-Hang, Edwin Gilland, Andrew H. Bass, and Robert Baker. "Ancestry of motor innervation to pectoral fin and forelimb." *Nature Communications* 1 (2010): 1–8.

Page, Jeremy. "Underwater Drones Join Microphones to Listen for Chinese Nuclear Submarines." *Wall Street Journal*, October 24, 2014.

Payne, Katharine B., William R. Langbauer, and Elizabeth M. Thomas. "Infrasonic calls of the Asian elephant (*Elephas maximus*)." *Behavioral Ecology and Sociobiology* 18 (1986): 297–301.

Popper, Arthur N., Michael Salmon, and Kenneth W. Horch. "Acoustic detection and communication by decapod crustaceans." *Journal of Comparative Physiology A* 187 (2001): 83–89.

Ramcharitar, John, Dennis M. Higgs, and Arthur N. Popper. "Sciaenid inner ears: a study in diversity." *Brain, Behavior and Evolution* 58 (2001): 152–62.

Ramcharitar, John Umar, Xiaohong Deng, Darlene Ketten, and Arthur N. Popper. "Form and function in the unique inner ear of a teleost: the silver perch (*Bairdiella chrysoura*)." *Journal of Comparative Neurology* 475 (2004): 531–39.

"'Sonar' and Shrimps in Anti-Submarine War." *The Age* (Melbourne, Australia), April 8, 1946.

Versluis, Michel, Barbara Schmitz, Anna von der Heydt, and Detlef Lohse. "How snapping shrimp snap: through cavitating bubbles." *Science* 289 (2000): 2114–7.

Washausen, Stefan, and Wolfgang Knabe. "Lateral line placodes of aquatic vertebrates are evolutionarily conserved in mammals." *Biology Open* 7 (2018): bio031815.

Sensory Bargains and Biases

Dallos, Peter. "The active cochlea." *Journal of Neuroscience* 12 (1992): 4575–85.

Dallos, Peter, and Bernd Fakler. "Prestin, a new type of motor protein." *Nature Reviews Molecular Cell Biology* 3 (2002): 104–11.

Dańko, Maciej J., Jan Kozłowski, and Ralf Schaible. "Unraveling the non-senescence phenomenon in Hydra." *Journal of Theoretical Biology* 382 (2015): 137–49.

Deutsch, Diana. *Musical Illusions and Phantom Words: How Music and Speech Unlock Mysteries of the Brain*. New York: Oxford University Press, 2019.

Fritzsch, Bernd, and Hans Straka. "Evolution of vertebrate mechanosensory hair cells and inner ears: toward identifying stimuli that select mutation driven altered morphologies." *Journal of Comparative Physiology A* 200 (2014): 5–18.

Graven, Stanley N., and Joy V. Browne. "Auditory development in the fetus and infant." *Newborn and Infant Nursing Reviews* 8 (2008): 187–93.

Hall, James W. "Development of the ear and hearing." *Journal of Perinatology* 20 (2000): S11–S19.

Kemp, David T. "Otoacoustic emissions, their origin in cochlear function, and use." *British Medical Bulletin* 63 (2002): 223–41.

Lasky, Robert E., and Amber L. Williams. "The development of the auditory system from conception to term." *NeoReviews* (2005): e141–e152.

Manley, Geoffrey A. "Cochlear mechanisms from a phylogenetic viewpoint." *Proceedings of the National Academy of Sciences* 97 (2000): 11736–43.

———. "Aural history." *Scientist* 29 (2015): 36–42.

———. "The Cochlea: What It Is, Where It Came from, and What Is Special about It." In *Understanding the Cochlea*, edited by Geoffrey A. Manley, Anthony W. Gummer, Arthur N. Popper, and Richard R. Fay, 17–32. New York: Springer, 2017.

Moon, Christine. "Prenatal Experience with the Maternal Voice." In *Early Vocal Contact and Preterm Infant Brain Development*, edited by Manuela Filippa, Pierre Kuhn, and Björn Westrup, 25–37. New York: Springer, 2017.

Parga, Joanna J., Robert Daland, Kalpashri Kesavan, Paul M. Macey, Lonnie Zeltzer, and Ronald M. Harper. "A description of externally recorded womb sounds in human subjects during gestation." *PLOS One* 13 (2018): e0197045.

Pickles, James. *An Introduction to the Physiology of Hearing*. Leiden: Brill, 2013.

Plack, Christopher J. *The Sense of Hearing*, 3rd ed. Oxford and New York: Routledge, 2018.

Robles, Luis, and Mario A. Ruggero. "Mechanics of the mammalian cochlea." *Physiological Reviews* 81 (2001): 1305–52.

Smith, Sherri L., Kenneth J. Gerhardt, Scott K. Griffiths, Xinyan Huang, and Robert M. Abrams. "Intelligibility of sentences recorded from the uterus of a pregnant ewe and from the fetal inner ear." *Audiology and Neurotology* 8 (2003): 347–53.

Wan, Kirsty Y., Sylvia K. Hürlimann, Aidan M. Fenix, Rebecca M. McGillivary, Tatyana Makushok, Evan Burns, Janet Y. Sheung, and Wallace F. Marshall. "Reorganization of complex ciliary flows around regenerating *Stentor coeruleus*." *Philosophical Transactions of the Royal Society B* 375 (2020): 20190167.

Predators, Silence, Wings

Bar-On, Yinon M., Rob Phillips, and Ron Milo. "The biomass distribution on Earth." *Proceedings of the National Academy of Sciences* 115 (2018): 6506–11.

Beraldi-Campesi, Hugo. "Early life on land and the first terrestrial ecosystems." *Ecological Processes* 2 (2013): 1–17.

Betancur-R, Ricardo, Edward O. Wiley, Gloria Arratia, Arturo Acero, Nicolas Bailly, Masaki Miya, Guillaume Lecointre, and Guillermo Orti. "Phylogenetic classification of bony fishes." *BMC Evolutionary Biology* 17 (2017): 162.

Béthoux, Olivier. "Grylloptera—a unique origin of the stridulatory file in katydids, crickets, and their kin (*Archaeorthoptera*)." *Arthropod Systematics & Phylogeny* 70 (2012): 43–68.

Béthoux, Olivier, and André Nel. "Venation pattern and revision of Orthoptera sensu nov. and sister groups. Phylogeny of Palaeozoic and Mesozoic Orthoptera sensu nov." *Zootaxa* 96 (2002): 1–88.

Béthoux, Olivier, André Nel, Jean Lapeyrie, and Georges Gand. "The Permostridulidae fam. n. (Panorthoptera), a new enigmatic insect family from the Upper Permian of France." *European Journal of Entomology* 100 (2003): 581–86.

Bocast, C., R. M. Bruch, and R. P. Koenigs. "Sound production of spawning lake sturgeon (*Acipenser fulvescens* Rafinesque, 1817) in the Lake Winnebago watershed, Wisconsin, USA." *Journal of Applied Ichthyology* 30 (2014): 1186–94.

Brazeau, Martin D., and Per E. Ahlberg. "Tetrapod-like middle ear architecture in a Devonian fish." *Nature* 439 (2006): 318–21.

Brazeau, Martin D., and Matt Friedman. "The origin and early phylogenetic history of jawed vertebrates." *Nature* 520 (2015): 490–97.

Breure, Abraham S. H. "The sound of a snail: two cases of acoustic defence in gastropods." *Journal of Molluscan Studies* 81 (2015): 290–93.

Clack, J. A. "The neurocranium of *Acanthostega gunnari* Jarvik and the evolution of the otic region in tetrapods." *Zoological Journal of the Linnean Society* 122 (1998): 61–97.

Clack, Jennifer A. "Discovery of the earliest-known tetrapod stapes." *Nature* 342 (1989): 425–27.

Clack, Jennifer A., Per E. Ahlberg, S. M. Finney, P. Dominguez Alonso, Jamie Robinson, and Richard A. Ketcham. "A uniquely specialized ear in a very early tetrapod." *Nature* 425 (2003): 65–69.

Clack, Jennifer A., Richard R. Fay, and Arthur N. Popper, eds. *Evolution of the Vertebrate Ear: Evidence from the Fossil Record*. New York: Springer, 2016.

Coombs, Sheryl, Horst Bleckmann, Richard R. Fay, and Arthur N. Popper, eds. *The Lateral Line System*. New York: Springer, 2014.

Daley, Allison C., Jonathan B. Antcliffe, Harriet B. Drage, and Stephen Pates. "Early fossil record of Euarthropoda and the Cambrian Explosion." *Proceedings of the National Academy of Sciences* 115 (2018): 5323–31.

Davranoglou, Leonidas-Romanos, Alice Cicirello, Graham K. Taylor, and Beth Mortimer. "Planthopper bugs use a fast, cyclic elastic recoil mechanism for effective vibrational communication at small body size." *PLOS Biology* 17 (2019): e3000155.

———. Response to "On the evolution of the tymbalian tymbal organ: comment on "Planthopper bugs use a fast, cyclic elastic recoil mechanism for effective vibrational communication at small body size" by Davranoglou et al. 2019." *Cicadina* 18 (2019): 17–26.

Desutter-Grandcolas, Laure, Lauriane Jacquelin, Sylvain Hugel, Renaud Boistel, Romain Garrouste, Michel Henrotay, Ben H. Warren et al. "3-D imaging reveals four extraordinary cases of convergent evolution of acoustic communication in crickets and allies (Insecta)." *Scientific Reports* 7 (2017): 1–8.

Downs, Jason P., Edward B. Daeschler, Farish A. Jenkins, and Neil H. Shubin. "The cranial endoskeleton of Tiktaalik roseae." *Nature* 455 (2008): 925–29.

Dubus, I. G., J. M. Hollis, and C. D. Brown. "Pesticides in rainfall in Europe." *Environmental Pollution* 110 (2000): 331–44.

Dunlop, Jason A., Gerhard Scholtz, and Paul A. Selden. "Water-to-land Transitions." In *Arthropod Biology and Evolution*, edited by Alessandro Minelli, Geoffrey Boxshall, and Giuseppe Fusco, 417–39. Berlin: Springer, 2013.

French, Katherine L., Christian Hallmann, Janet M. Hope, Petra L. Schoon, J. Alex Zumberge, Yosuke Hoshino, Carl A. Peters et al. "Reappraisal of hydrocarbon biomarkers in Archean rocks." *Proceedings of the National Academy of Sciences* 112 (2015): 5915–20.

Galtier, Jean, and Jean Broutin. "Floras from red beds of the Permian Basin of Lodève (Southern France)." *Journal of Iberian Geology* 34 (2008): 57–72.

Goerlitz, Holger R., Stefan Greif, and Björn M. Siemers. "Cues for acoustic detection of prey: insect rustling sounds and the influence of walking substrate." *Journal of Experimental Biology* 211 (2008): 2799–2806.

Goto, Ryutaro, Isao Hirabayashi, and A. Richard Palmer. "Remarkably loud snaps during mouth-fighting by a sponge-dwelling worm." *Current Biology* 29 (2019): R617–R618.

Grimaldi, David, and Michael S. Engel. *Evolution of the Insects*. Cambridge, UK: Cambridge University Press, 2005.

Gu, Jun-Jie, Fernando Montealegre-Z, Daniel Robert, Michael S. Engel, Ge-Xia Qiao, and Dong Ren. "Wing stridulation in a Jurassic katydid (Insecta, Orthoptera) produced low-pitched musical calls to attract females." *Proceedings of the National Academy of Sciences* 109 (2012): 3868–73.

Hochkirch, Axel, Ana Nieto, M. García Criado, Marta Cálix, Yoan Braud, Filippo M. Buzzetti, D. Chobanov et al. *European Red List of Grasshoppers, Crickets and Bush Crickets*. Luxembourg: Publications Office of the European Union, 2016.

Kawahara, Akito Y., and Jesse R. Barber. "Tempo and mode of antibat ultrasound production and sonar jamming in the diverse hawkmoth radiation." *Proceedings of the National Academy of Sciences* 112 (2015): 6407–12.

Ladich, Friedrich, and Andreas Tadler. "Sound production in *Polypterus* (Osteichthyes: Polypteridae)." *Copeia* 4 (1988): 1076–77.

Linz, David M., and Yoshinori Tomoyasu. "Dual evolutionary origin of insect wings supported by an investigation of the abdominal wing serial homologs in *Tribolium*." *Proceedings of the National Academy of Sciences* 115 (2018): E658–E667.

Lopez, Michel, Georges Gand, Jacques Garric, F. Körner, and Jodi Schneider. "The playa environments of the Lodève Permian basin (Languedoc-France)." *Journal of Iberian Geology* 34 (2008): 29–56.

Lozano-Fernandez, Jesus, Robert Carton, Alastair R. Tanner, Mark N. Puttick, Mark Blaxter, Jakob Vinther, Jørgen Olesen, Gonzalo Giribet, Gregory D. Edgecombe, and Davide Pisani. "A molecular palaeobiological exploration of arthropod terrestrialization." *Philosophical Transactions of the Royal Society B* 371 (2016): 20150133.

Masters, W. Mitchell. "Insect disturbance stridulation: its defensive role." *Behavioral Ecology and Sociobiology* 5 (1979): 187–200.

Minter, Nicholas J., Luis A. Buatois, M. Gabriela Mángano, Neil S. Davies, Martin R. Gibling, Robert B. MacNaughton, and Conrad C. Labandeira. "Early bursts of diversification defined the faunal colonization of land." *Nature Ecology & Evolution* 1 (2017): 0175.

Moulds, M. S. "Cicada fossils (Cicadoidea: Tettigarctidae and Cicadidae) with a review of the named fossilised Cicadidae." *Zootaxa* 4438 (2018): 443–70.

Near, Thomas J., Alex Dornburg, Ron I. Eytan, Benjamin P. Keck, W. Leo Smith, Kristen L. Kuhn, Jon A. Moore et al. "Phylogeny and tempo of diversification in the superradiation of spiny-rayed fishes." *Proceedings of the National Academy of Sciences* 110 (2013): 12738–43.

Nedelec, Sophie L., James Campbell, Andrew N. Radford, Stephen D. Simpson, and Nathan D. Merchant. "Particle motion: the missing link in underwater acoustic ecology." *Methods in Ecology and Evolution* 7 (2016): 836–42.

Nel, André, Patrick Roques, Patricia Nel, Alexander A. Prokin, Thierry Bourgoin, Jakub Prokop, Jacek Szwedo et al. "The earliest known holometabolous insects." *Nature* 503 (2013): 257–61.

Parmentier, Eric, and Michael L. Fine. "Fish Sound Production: Insights." In *Vertebrate Sound Production and Acoustic Communication*, edited by Roderick A. Suthers, W. Tecumseh Fitch, Richard R. Fay, and Arthur N. Popper, 19–49. Berlin: Springer, 2016.

Pennisi, Elizabeth. "Carbon dioxide increase may promote 'insect apocalypse.'" *Science* 368 (2020): 459.

Pfeifer, Lily S. "Loess in the Lodeve? Exploring the depositional character of the Permian Salagou Formation, Lodeve Basin (France)." *Geological Society of America Abstracts with Programs* 50 (2018).

Plotnick, Roy E., and Dena M. Smith. "Exceptionally preserved fossil insect ears from the Eocene Green River Formation of Colorado." *Journal of Paleontology* 86 (2012): 19–24.

Prokop, Jakub, André Nel, and Ivan Hoch. "Discovery of the oldest known Pterygota in the Lower Carboniferous of the Upper Silesian Basin in the Czech Republic (Insecta: Archaeorthoptera)." *Geobios* 38 (2005): 383–87.

Prokop, Jakub, Jacek Szwedo, Jean Lapeyrie, Romain Garrouste, and André Nel. "New middle Permian insects from Salagou Formation of the Lodève Basin in southern France (Insecta: Pterygota)." *Annales de la Société Entomologique de France* 51 (2015): 14–51.

Rust, Jes, Andreas Stumpner, and Jochen Gottwald. "Singing and hearing in a Tertiary bushcricket." *Nature* 399 (1999): 650.

Rustán, Juan J., Diego Balseiro, Beatriz Waisfeld, Rodolfo D. Foglia, and N. Emilio Vaccari. "Infaunal molting in Trilobita and escalatory responses against predation." *Geology* 39 (2011): 495–98.

Senter, Phil. "Voices of the past: a review of Paleozoic and Mesozoic animal sounds." *Historical Biology* 20 (2008) 255–87.

Siveter, David J., Mark Williams, and Dieter Waloszek. "A phosphatocopid crustacean with appendages from the Lower Cambrian." *Science* 293 (2001): 479–81.

Song, Hojun, Christiane Amédégnato, Maria Marta Cigliano, Laure Desutter-Grandcolas, Sam W. Heads, Yuan Huang, Daniel Otte, and Michael F. Whiting. "300 million years of diversification: elucidating the patterns of orthopteran evolution based on comprehensive taxon and gene sampling." *Cladistics* 31 (2015): 621–51.

Song, Hojun, Olivier Béthoux, Seunggwan Shin, Alexander Donath, Harald Letsch, Shanlin Liu, Duane D. McKenna et al. "Phylogenomic analysis sheds light on the evolutionary pathways towards acoustic communication in Orthoptera." *Nature Communications* 11 (2020): 1–16.

Stewart, Kenneth W. "Vibrational communication in insects: epitome in the language of stoneflies?" *American Entomologist* 43 (1997): 81–91.

van Klink, Roel, Diana E. Bowler, Konstantin B. Gongalsky, Ann B. Swengel, Alessandro Gentile, and Jonathan M. Chase. "Meta-analysis reveals declines in terrestrial but increases in freshwater insect abundances." *Science* 368 (2020): 417–20.

van Klink, Roel, Diana E. Bowler, Konstantin B. Gongalsky, Ann B. Swengel, Alessandro Gentile, and Jonathan M. Chase. "Erratum for the Report 'Meta-analysis reveals declines in terrestrial but increases in freshwater insect abundances.'" *Science* 370 (2020). DOI: 10.1126/science.abf1915.

Vermeij, Geerat J. "Sound reasons for silence: why do molluscs not communicate acoustically?" *Biological Journal of the Linnean Society* 100 (2010): 485–93.

Welti, Ellen A. R., Karl A. Roeder, Kirsten M. de Beurs, Anthony Joern, and Michael Kaspari. "Nutrient dilution and climate cycles underlie declines in a dominant insect herbivore." *Proceedings of the National Academy of Sciences* 117 (2020): 7271–75.

Wendruff, Andrew J., Loren E. Babcock, Christian S. Wirkner, Joanne Kluessendorf, and Donald G. Mikulic. "A Silurian ancestral scorpion with fossilised internal anatomy illustrating a pathway to arachnid terrestrialisation." *Scientific Reports* 10 (2020): 1–6.

Wessel, Andreas, Roland Mühlethaler, Viktor Hartung, Valerija Kuštor, and Matija Gogala. "The Tymbal: Evolution of a Complex Vibration-producing Organ in the Tymbalia (Hemiptera excl. Sternorrhyncha)." In *Studying Vibrational Communication*, edited by Reginald B. Cocroft, Matija Gogala, Peggy S. M. Hill, and Andreas Wessel, 395–444. Berlin: Springer, 2014.

Wipfler, Benjamin, Harald Letsch, Paul B. Frandsen, Paschalia Kapli, Christoph Mayer, Daniela Bartel, Thomas R. Buckley et al. "Evolutionary history of Polyneoptera and its implications for our understanding of early winged insects." *Proceedings of the National Academy of Sciences* 116 (2019): 3024–29.

Zhang, Xi-guang, David J. Siveter, Dieter Waloszek, and Andreas Maas. "An epipodite-bearing crown-group crustacean from the Lower Cambrian." *Nature* 449 (2007): 595–98.

Zhang, Xi-guang, Andreas Maas, Joachim T. Haug, David J. Siveter, and Dieter Waloszek. "A eucrustacean metanauplius from the Lower Cambrian." *Current Biology* 20 (2010): 1075–79.

Zhang, Yunfeng, Feng Shi, Jiakun Song, Xugang Zhang, and Shiliang Yu. "Hearing characteristics of cephalopods: modeling and environmental impact study." *Integrative Zoology* 10 (2015): 141–51.

Zhang, Zhi-Qiang. "Animal biodiversity: An update of classification and diversity in 2013." *Zootaxa* 3703 (2013): 5–11.

Flowers, Oceans, Milk

Alexander, R. McNeill. "Dinosaur biomechanics." *Proceedings of the Royal Society B* 273 (2006): 1849–55.

Bambach, Richard K. "Energetics in the global marine fauna: a connection between terrestrial diversification and change in the marine biosphere." *Geobios* 32 (1999): 131–44.

Barba-Montoya, Jose, Mario dos Reis, Harald Schneider, Philip C. J. Donoghue, and Ziheng Yang. "Constraining uncertainty in the timescale of angiosperm evolution and the veracity of a Cretaceous Terrestrial Revolution." *New Phytologist* 218 (2018): 819–34.

Barney, Anna, Sandra Martelli, Antoine Serrurier, and James Steele. "Articulatory capacity of Neanderthals, a very recent and human-like fossil hominin." *Philosophical Transactions of the Royal Society B* 367 (2012): 88–102.

Barreda, Viviana D., Luis Palazzesi, and Eduardo B. Olivero. "When flowering plants ruled Antarctica: evidence from Cretaceous pollen grains." *New Phytologist* 223 (2019): 1023–30.

Bateman, Richard M. "Hunting the Snark: the flawed search for mythical Jurassic angiosperms." *Journal of Experimental Botany* 71 (2020): 22–35.

Battison, Leila, and Dallas Taylor. "Tyrannosaurus FX." *Twenty Thousand Hertz*. Podcast. https://www.20k.org/episodes/tyrannosaurusfx.

Bergevin, Christopher, Chandan Narayan, Joy Williams, Natasha Mhatre, Jennifer K. E. Steeves, Joshua GW Bernstein, and Brad Story. "Overtone focusing in biphonic Tuvan throat singing." *eLife* 9 (2020): e50476.

Bowling, Daniel L., Jacob C. Dunn, Jeroen B. Smaers, Maxime Garcia, Asha Sato, Georg Hantke, Stephan Handschuh et al. "Rapid evolution of the primate larynx?" *PLOS Biology* 18 (2020): e3000764.

Boyd, Eric, and John W. Peters. "New insights into the evolutionary history of biological nitrogen fixation." *Frontiers in Microbiology* 4 (2013): 201.

Bracken-Grissom, Heather D., Shane T. Ahyong, Richard D. Wilkinson, Rodney M. Feldmann, Carrie E. Schweitzer, Jesse W. Breinholt, Matthew Bendall et al. "The emergence of lobsters: phylogenetic relationships, morphological evolution and divergence time comparisons of an ancient group (Decapoda: Achelata, Astacidea, Glypheidea, Polychelida)." *Systematic Biology* 63 (2014): 457–79.

Bravi, Sergio, and Alessandro Garassino. "Plattenkalk of the Lower Cretaceous (Albian) of Petina, in the Alburni Mounts (Campania, S Italy) and its decapod crustacean assemblage." *Atti della Societá italiana di Scienze naturali e del Museo civico di Storia naturale in Milano* 138 (1998): 89–118.

Brummitt, Neil A., Steven P. Bachman, Janine Griffiths-Lee, Maiko Lutz, Justin F. Moat, Aljos Farjon, John S. Donaldson et al. "Green plants in the red: a baseline global assessment for the IUCN sampled Red List Index for plants." *PLOS One* 10 (2015): e0135152.

Bush, Andrew M., and Richard K. Bambach. "Paleoecologic megatrends in marine metazoa." *Annual Review of Earth and Planetary Sciences* 39 (2011): 241–69.

Bush, Andrew M., Gene Hunt, and Richard K. Bambach. "Sex and the shifting biodiversity dynamics of marine animals in deep time." *Proceedings of the National Academy of Sciences* 113 (2016): 14073–78.

Chen, Zhuo, and John J. Wiens. "The origins of acoustic communication in vertebrates." *Nature Communications* 11 (2020): 1–8.

Clarke, Julia A., Sankar Chatterjee, Zhiheng Li, Tobias Riede, Federico Agnolin, Franz Goller, Marcelo P. Isasi, Daniel R. Martinioni, Francisco J. Mussel, and Fernando E. Novas. "Fossil evidence of the avian vocal organ from the Mesozoic." *Nature* 538 (2016): 502–5.

Coiro, Mario, James A. Doyle, and Jason Hilton. "How deep is the conflict between molecular and fossil evidence on the age of angiosperms?" *New Phytologist* 223 (2019): 83–99.

Colafrancesco, Kaitlen C., and Marcos Gridi-Papp. "Vocal Sound Production and Acoustic Communication in Amphibians and Reptiles." In *Vertebrate Sound Production and Acoustic Communication*, edited by Roderick A. Suthers, W. Tecumseh Fitch, Richard R. Fay, and Arthur N. Popper, 51–82. Berlin: Springer, 2016.

Conde-Valverde, Mercedes, Ignacio Martínez, Rolf M. Quam, Manuel Rosa, Alex D. Velez, Carlos Lorenzo, Pilar Jarabo, José María Bermúdez de Castro, Eudald Carbonell, and Juan Luis Arsuaga. "Neanderthals and Homo sapiens had similar auditory and speech capacities." *Nature Ecology & Evolution* (2021): 1–7.

Corlett, Richard T. "Plant diversity in a changing world: status, trends, and conservation needs." *Plant Diversity* 38 (2016): 10–16.

Cryan, Jason R., Brian M. Wiegmann, Lewis L. Deitz, and Christopher H. Dietrich. "Phylogeny of the treehoppers (Insecta: Hemiptera: Membracidae): evidence from two nuclear genes." *Molecular Phylogenetics and Evolution* 17 (2000): 317–34.

Dowdy, Nicolas J., and William E. Conner. "Characteristics of tiger moth (Erebidae: Arctiinae) anti-bat sounds can be predicted from tymbal morphology." *Frontiers in Zoology* 16 (2019): 45.

Dunn, Jacob C., Lauren B. Halenar, Thomas G. Davies, Jurgi Cristobal-Azkarate, David Reby, Dan Sykes, Sabine Dengg, W. Tecumseh Fitch, and Leslie A. Knapp. "Evolutionary trade-off between vocal tract and testes dimensions in howler monkeys." *Current Biology* 25 (2015): 2839–44.

Feldmann, Rodney M., Carrie E. Schweitzer, Cory M. Redman, Noel J. Morris, and David J. Ward. "New Late Cretaceous lobsters from the Kyzylkum desert of Uzbekistan." *Journal of Paleontology* 81 (2007): 701–13.

Feng, Yan-Jie, David C. Blackburn, Dan Liang, David M. Hillis, David B. Wake, David C. Cannatella, and Peng Zhang. "Phylogenomics reveals rapid, simultaneous diversification of three major clades of Gondwanan frogs at the Cretaceous–Paleogene boundary." *Proceedings of the National Academy of Sciences* 114 (2017): E5864–E5870.

Field, Daniel J., Antoine Bercovici, Jacob S. Berv, Regan Dunn, David E. Fastovsky, Tyler R. Lyson, Vivi Vajda, and Jacques A. Gauthier. "Early evolution of modern birds structured by global forest collapse at the end-Cretaceous mass extinction." *Current Biology* 28 (2018): 1825–31.

Fine, Michael, and Eric Parmentier. "Mechanisms of Fish Sound Production." In *Sound Communication in Fishes*, edited by Friedrich Ladich, 77–126. Vienna: Springer, 2015.

Fitch, W. Tecumseh. "Empirical approaches to the study of language evolution." *Psychonomic Bulletin & Review* 24 (2017): 3–33.

———. "Production of Vocalizations in Mammals." In *Encyclopedia of Language and Linguistics*, edited by K. Brown, 115–21. Oxford, UK: Elsevier, 2006.

Fitch, W. Tecumseh, Bart De Boer, Neil Mathur, and Asif A. Ghazanfar. "Monkey vocal tracts are speech-ready." *Science Advances* 2 (2016): e1600723.

Frey, Roland, and Alban Gebler. "Mechanisms and evolution of roaring-like vocalization in mammals." *Handbook of Behavioral Neuroscience* 19 (2010): 439–50.

Frey, Roland, and Tobias Riede. "The anatomy of vocal divergence in North American elk and European red deer." *Journal of Morphology* 274 (2013): 307–19.

Fu, Qiang, Jose Bienvenido Diez, Mike Pole, Manuel García Ávila, Zhong-Jian Liu, Hang Chu, Yemao Hou et al. "An unexpected noncarpellate epigynous flower from the Jurassic of China." *eLife* 7 (2018): e38827.

Ghazanfar, Asif A., and Drew Rendall. "Evolution of human vocal production." *Current Biology* 18 (2008): R457–R460.

Griesmann, Maximilian, Yue Chang, Xin Liu, Yue Song, Georg Haberer, Matthew B. Crook, Benjamin Billault-Penneteau et al. "Phylogenomics reveals multiple losses of nitrogen-fixing root nodule symbiosis." *Science* 361 (2018): eaat 1743.

Hoch, Hannelore, Jürgen Deckert, and Andreas Wessel. "Vibrational signalling in a Gondwanan relict insect (Hemiptera: Coleorrhyncha: Peloridiidae)." *Biology Letters* 2 (2006): 222–24.

Hoffmann, Simone, and David W. Krause. "Tongues untied." *Science* 365 (2019): 222–23.

Jézéquel, Youenn, Laurent Chauvaud, and Julien Bonnel. "Spiny lobster sounds can be detectable over kilometres underwater." *Scientific Reports* 10 (2020): 1–11.

Johnson, Kevin P., Christopher H. Dietrich, Frank Friedrich, Rolf G. Beutel, Benjamin Wipfler, Ralph S. Peters, Julie M. Allen et al. "Phylogenomics and the evolution of hemipteroid insects." *Proceedings of the National Academy of Sciences* 115 (2018): 12775–780.

Kaiho, Kunio, Naga Oshima, Kouji Adachi, Yukimasa Adachi, Takuya Mizukami, Megumu Fujibayashi, and Ryosuke Saito. "Global climate change driven by soot at the K-Pg boundary as the cause of the mass extinction." *Scientific Reports* 6 (2016): 28427.

Kawahara, Akito Y., David Plotkin, Marianne Espeland, Karen Meusemann, Emmanuel F. A. Toussaint, Alexander Donath, France Gimnich et al. "Phylogenomics reveals the evolutionary timing and pattern of butterflies and moths." *Proceedings of the National Academy of Sciences* 116 (2019): 22657–663.

Kikuchi, Mumi, Tomonari Akamatsu, and Tomohiro Takase. "Passive acoustic monitoring of Japanese spiny lobster stridulating sounds." *Fisheries Science* 81 (2015): 229–34.

Labandeira, Conrad C. "A compendium of fossil insect families." *Milwaukee Public Museum Contributions in Biology and Geology* 88 (1994): 1–71.

Lefèvre, Christophe M., Julie A. Sharp, and Kevin R. Nicholas. "Evolution of lactation: ancient origin and extreme adaptations of the lactation system." *Annual Review of Genomics and Human Genetics* 11 (2010): 219–38.

Li, Hong-Lei, Wei Wang, Peter E. Mortimer, Rui-Qi Li, De-Zhu Li, Kevin D. Hyde, Jian-Chu Xu, Douglas E. Soltis, and Zhi-Duan Chen. "Large-scale phylogenetic analyses reveal multiple gains of actinorhizal nitrogen-fixing symbioses in angiosperms associated with climate change." *Scientific Reports* 5 (2015): 14023.

Li, Hong-Tao, Ting-Shuang Yi, Lian-Ming Gao, Peng-Fei Ma, Ting Zhang, Jun-Bo Yang, Matthew A. Gitzendanner et al. "Origin of angiosperms and the puzzle of the Jurassic gap." *Nature Plants* 5 (2019): 461.

Lima, Daniel, Arthur Anker, Matúš Hyžný, Andreas Kroh, and Orangel Aguilera. "First evidence of fossil snapping shrimps (Alpheidae) in the Neotropical region, with a checklist of the fossil caridean shrimps from the Cenozoic." *Journal of South American Earth Sciences* (2020): 102795.

Lürling, Miquel, and Marten Scheffer. "Info-disruption: pollution and the transfer of chemical information between organisms." *Trends in Ecology & Evolution* 22 (2007): 374–79.

Lyons, Shelby L., Allison T. Karp, Timothy J. Bralower, Kliti Grice, Bettina Schaefer, Sean P. S. Gulick, Joanna V. Morgan, and Katherine H. Freeman. "Organic matter from the Chicxulub crater exacerbated the K–Pg impact winter." *Proceedings of the National Academy of Sciences* 117 (2020): 25327–34.

Martínez, Ignacio, Juan Luis Arsuaga, Rolf Quam, José Miguel Carretero, Ana Gracia, and Laura Rodríguez. "Human hyoid bones from the middle Pleistocene site of the Sima de los Huesos (Sierra de Atapuerca, Spain)." *Journal of Human Evolution* 54 (2008): 118–24.

McCauley, Douglas J., Malin L. Pinsky, Stephen R. Palumbi, James A. Estes, Francis H. Joyce, and Robert R. Warner. "Marine defaunation: animal loss in the global ocean." *Science* 347 (2015): 1255641.

McKenna, Duane D., Seunggwan Shin, Dirk Ahrens, Michael Balke, Cristian Beza-Beza, Dave J. Clarke, Alexander Donath et al. "The evolution and genomic basis of beetle diversity." *Proceedings of the National Academy of Sciences* 116 (2019): 24729–37.

Mugleston, Joseph D., Michael Naegle, Hojun Song, and Michael F. Whiting. "A comprehensive phylogeny of Tettigoniidae (Orthoptera: Ensifera) reveals extensive ecomorph convergence and widespread taxonomic incongruence." *Insect Systematics and Diversity* 2 (2018): 1–27.

Müller, Johannes, Constanze Bickelmann, and Gabriela Sobral. "The evolution and fossil history of sensory perception in amniote vertebrates." *Annual Review of Earth and Planetary Sciences* 46 (2018): 495–519.

Nakano, Ryo, Takuma Takanashi, and Annemarie Surlykke. "Moth hearing and sound communication." *Journal of Comparative Physiology A* 201 (2015): 111–21.

Near, Thomas J., Alex Dornburg, Ron I. Eytan, Benjamin P. Keck, W. Leo Smith, Kristen L. Kuhn, Jon A. Moore et al. "Phylogeny and tempo of diversification in the superradiation of spiny-rayed fishes." *Proceedings of the National Academy of Sciences* 110 (2013): 12738–43.

Nishimura, Takeshi, Akichika Mikami, Juri Suzuki, and Tetsuro Matsuzawa. "Descent of the hyoid in chimpanzees: evolution of face flattening and speech." *Journal of Human Evolution* 51 (2006): 244–54.

Novack-Gottshall, Philip M. "Love, not war, drove the Mesozoic marine revolution." *Proceedings of the National Academy of Sciences* 113 (2016): 14471–73.

O'Brien, Charlotte L., Stuart A. Robinson, Richard D. Pancost, Jaap S. Sinninghe Damsté, Stefan Schouten, Daniel J. Lunt, Heiko Alsenz et al. "Cretaceous sea-surface temperature evolution: constraints from TEX86 and planktonic foraminiferal oxygen isotopes." *Earth-Science Reviews* 172 (2017): 224–47.

O'Connor, Lauren K., Stuart A. Robinson, B. David A. Naafs, Hugh C. Jenkyns, Sam Henson, Madeleine Clarke, and Richard D. Pancost. "Late Cretaceous temperature evolution of the southern high latitudes: a TEX86 perspective." *Paleoceanography and Paleoclimatology* 34 (2019): 436–54.

Patek, Sheila N. "Squeaking with a sliding joint: mechanics and motor control of sound production in palinurid lobsters." *Journal of Experimental Biology* 205 (2002): 2375–85.

Patek, S. N., and J. E. Baio. "The acoustic mechanics of stick–slip friction in the California spiny lobster (Panulirus interruptus)." *Journal of Experimental Biology* 210 (2007): 3538–46.

Pereira, Graciela, and Helga Josupeit. "The world lobster market." *Globefish Research Programme* 123 (2017).

Perrone-Bertolotti, Marcela, Jan Kujala, Juan R. Vidal, Carlos M. Hamame, Tomas Ossandon, Olivier Bertrand, Lorella Minotti, Philippe Kahane, Karim Jerbi, and Jean-Philippe Lachaux. "How silent is silent reading? Intracerebral evidence for top-down activation of temporal voice areas during reading." *Journal of Neuroscience* 32 (2012): 17554–62.

Pickrell, John. "How the earliest mammals thrived alongside dinosaurs." *Nature* 574 (2019): 468–72.

Rai, A. N., E. Söderbäck, and B. Bergman. "Tansley Review No. 116: Cyanobacterium–plant symbioses." *New Phytologist* 147 (2000): 449–81.

Ramírez-Chaves, Héctor E., Vera Weisbecker, Stephen Wroe, and Matthew J. Phillips. "Resolving the evolution of the mammalian middle ear using Bayesian inference." *Frontiers in Zoology* 13 (2016): 39.

Reidenberg, Joy S., and Jeffrey T. Laitman. "Anatomy of the hyoid apparatus in odontoceli (toothed whales): specializations of their skeleton and musculature compared with those of terrestrial mammals." *Anatomical Record* 240 (1994): 598–624.

Rice, Aaron N., Stacy C. Farina, Andrea J. Makowski, Ingrid M. Kaatz, Philip S. Lobel, William E. Bemis, and Andrew Bass. "Evolution and Ecology in Widespread Acoustic Signaling Behavior Across Fishes." https://www.biorxiv.org/content/biorxiv/early/2020/09/14/2020.09.14.296335.full.pdf.

Riede, Tobias, Heather L. Borgard, and Bret Pasch. "Laryngeal airway reconstruction indicates that rodent ultrasonic vocalizations are produced by an edge-tone mechanism." *Royal Society Open Science* 4 (2017): 170976.

Riede, Tobias, Chad M. Eliason, Edward H. Miller, Franz Goller, and Julia A. Clarke. "Coos, booms, and hoots: the evolution of closed-mouth vocal behavior in birds." *Evolution* 70 (2016): 1734–46.

Ruiz, Michael J., and David Wilken. "Tuvan throat singing and harmonics." *Physics Education* 53 (2018): 035011.

Shcherbakov, Dmitri E. "The earliest leafhoppers (Hemiptera: Karajassidae n. fam.) from the Jurassic of Karatau." *Neues Jahrbuch für Geologie und Paläontologie* 1 (1992): 39–51.

Soltis, Douglas E., Pamela S. Soltis, David R. Morgan, Susan M. Swensen, Beth C. Mullin, Julie M. Dowd, and Peter G. Martin. "Chloroplast gene sequence data suggest a single origin of the predisposition for symbiotic nitrogen fixation in angiosperms." *Proceedings of the National Academy of Sciences* 92 (1995): 2647–51.

Stüeken, Eva E., Michael A. Kipp, Matthew C. Koehler, and Roger Buick. "The evolution of Earth's biogeochemical nitrogen cycle." *Earth-Science Reviews* 160 (2016): 220–39.

Takemoto, Hironori. "Morphological analyses and 3D modeling of the tongue musculature of the chimpanzee (Pan troglodytes)." *American Journal of Primatology* 70 (2008): 966–75.

Vajda, Vivi, and Antoine Bercovici. "The global vegetation pattern across the Cretaceous–Paleogene mass extinction interval: a template for other extinction events." *Global and Planetary Change* 122 (2014): 29–49.

Vega, Francisco J., Rodney M. Feldmann, Pedro García-Barrera, Harry Filkorn, Francis Pimentel, and Javier Avendano. "Maastrichtian Crustacea (Brachyura: Decapoda) from the Ocozocuautla Formation in Chiapas, southeast Mexico." *Journal of Paleontology* 75 (2001): 319–29.

Vermeij, Geerat J. "The Mesozoic marine revolution: evidence from snails, predators and grazers." *Paleobiology* (1977): 245–58.

Veselka, Nina, David D. McErlain, David W. Holdsworth, Judith L. Eger, Rethy K. Chhem, Matthew J. Mason, Kirsty L. Brain, Paul A. Faure, and M. Brock Fenton. "A bony connection signals laryngeal echolocation in bats." *Nature* 463 (2010): 939–42.

Webb, Thomas J., and Beth L. Mindel. "Global patterns of extinction risk in marine and non-marine systems." *Current Biology* 25 (2015): 506–11.

Wing, Scott L., Leo J. Hickey, and Carl C. Swisher. "Implications of an exceptional fossil flora for Late Cretaceous vegetation." *Nature* 363 (1993): 342–44.

Zhou, Chang-Fu, Bhart-Anjan S. Bhullar, April I. Neander, Thomas Martin, and Zhe-Xi Luo. "New Jurassic mammaliaform sheds light on early evolution of mammal-like hyoid bones." *Science* 365 (2019): 276–79.

Air, Water, Wood

Amoser, Sonja, and Friedrich Ladich. "Are hearing sensitivities of freshwater fish adapted to the ambient noise in their habitats?" *Journal of Experimental Biology* 208 (2005): 3533–42.

Bass, Andrew H., and Christopher W. Clark. "The Physical Acoustics of Underwater Sound Communication." In *Acoustic Communication*, edited by A. M. Simmons, A. N. Popper, and R. R. Fay, 15–64. New York: Springer, 2003.

Blasi, Damián E., Steven Moran, Scott R. Moisik, Paul Widmer, Dan Dediu, and Balthasar Bickel. "Human sound systems are shaped by post-Neolithic changes in bite configuration." *Science* 363 (2019): eaav3218.

Charlton, Benjamin D., Megan A. Owen, and Ronald R. Swaisgood. "Coevolution of vocal signal characteristics and hearing sensitivity in forest mammals." *Nature Communications* 10 (2019): 1–7.

Čokl, Andrej, Janez Prešern, Meta Virant-Doberlet, Glen J. Bagwell, and Jocelyn G. Millar. "Vibratory signals of the harlequin bug and their transmission through plants." *Physiological Entomology* 29 (2004): 372–80.

Conner, William E. "Adaptive Sounds and Silences: Acoustic Anti-predator Strategies in Insects." In *Insect Hearing and Acoustic Communication*, edited by Berthold Hedwig, 65–79. Berlin: Springer, 2014.

Derryberry, Elizabeth Perrault, Nathalie Seddon, Santiago Claramunt, Joseph Andrew Tobias, Adam Baker, Alexandre Aleixo, and Robb Thomas Brumfield. "Correlated evolution of beak morphology and song in the neotropical woodcreeper radiation." *Evolution* 66 (2012): 2784–97.

Feighny, J. A., K. E. Williamson, and J. A. Clarke. "North American elk bugle vocalizations: male and female bugle call structure and context." *Journal of Mammalogy* 87 (2006): 1072–77.

Greenfield, Michael D. "Interspecific acoustic interactions among katydids *Neoconocephalus*: inhibition-induced shifts in diel periodicity." *Animal Behaviour* 36 (1988): 684–95.

Heffner, Rickye S. "Primate hearing from a mammalian perspective." *The Anatomical Record Part A: Discoveries in Molecular, Cellular, and Evolutionary Biology: An Official Publication of the American Association of Anatomists* 281 (2004): 1111–22.

Hill, Peggy S. M. "How do animals use substrate-borne vibrations as an information source?" *Naturwissenschaften* 96 (2009): 1355–71.

Hua, Xia, Simon J. Greenhill, Marcel Cardillo, Hilde Schneemann, and Lindell Bromham. "The ecological drivers of variation in global language diversity." *Nature Communications* 10 (2019): 1–10.

Lugli, Marco. "Habitat Acoustics and the Low-Frequency Communication of Shallow Water Fishes." In *Sound Communication in Fishes*, edited by F. Ladich, 175–206. Vienna: Springer, 2015.

Lugli, Marco. "Sounds of shallow water fishes pitch within the quiet window of the habitat ambient noise." *Journal of Comparative Physiology A* 196 (2010): 439–51.

Maddieson, Ian, and Christophe Coupé. "Human spoken language diversity and the acoustic adaptation hypothesis." *Journal of the Acoustical Society of America* 138 (2015): 1838.

McNett, Gabriel D., and Reginald B. Cocroft. "Host shifts favor vibrational signal divergence in Enchenopa binotata treehoppers." *Behavioral Ecology* 19 (2008): 650–56.

Morton, Eugene S. "Ecological sources of selection on avian sounds." *American Naturalist* 109 (1975): 17–34.

Peters, Gustav, and Marcell K. Peters. "Long-distance call evolution in the Felidae: effects of body weight, habitat, and phylogeny." *Biological Journal of the Linnean Society* 101 (2010): 487–500.

Podos, Jeffrey. "Correlated evolution of morphology and vocal signal structure in Darwin's finches." *Nature* 409 (2001): 185–88.

Porter, Cody K., and Julie W. Smith. "Diversification in trophic morphology and a mating signal are coupled in the early stages of sympatric divergence in crossbills." *Biological Journal of the Linnean Society* 129 (2020): 74–87.

Riede, Tobias, Michael J. Owren, and Adam Clark Arcadi. "Nonlinear acoustics in pant hoots of common chimpanzees (*Pan troglodytes*): frequency jumps, subharmonics, biphonation, and deterministic chaos." *American Journal of Primatology* 64 (2004): 277–91.

Riede, Tobias, and Ingo R. Titze. "Vocal fold elasticity of the Rocky Mountain elk (*Cervus elaphus nelsoni*)—producing high fundamental frequency vocalization with a very long vocal fold." *Journal of Experimental Biology* 211 (2008): 2144–54.

Roberts, Seán G. "Robust, causal, and incremental approaches to investigating linguistic adaptation." *Frontiers in Psychology* 9 (2018): 166.

Zapata-Ríos, G., R. E. Suárez, B. V. Utreras, and O. J. Vargas. "Evaluación de amenazas antropogénicas en el Parque Nacional Yasuní y sus implicaciones para la conservación de mamíferos silvestres." *Lyonia* 10 (2006): 47–57.

In the Clamor

Amézquita, Adolfo, Sandra Victoria Flechas, Albertina Pimentel Lima, Herbert Gasser, and Walter Hödl. "Acoustic interference and recognition space within a complex assemblage of dendrobatid frogs." *Proceedings of the National Academy of Sciences* 108 (2011): 17058–63.

Aubin, Thierry, and Pierre Jouventin. "How to vocally identify kin in a crowd: the penguin model." *Advances in the Study of Behavior* 31 (2002): 243–78.

Barringer, Lawrence E., Charles R. Bartlett, and Terry L. Erwin. "Canopy assemblages and species richness of planthoppers (Hemiptera: Fulgoroidea) in the Ecuadorian Amazon." *Insecta Mundi* (2019) 0726: 1–16.

Bass, Margot S., Matt Finer, Clinton N. Jenkins, Holger Kreft, Diego F. Cisneros-Heredia, Shawn F. McCracken, Nigel CA Pitman et al. "Global conservation significance of Ecuador's Yasuní National Park." *PLOS One* 5 (2010): e8767.

Blake, John G., and Bette A. Loiselle. "Enigmatic declines in bird numbers in lowland forest of eastern Ecuador may be a consequence of climate change." *PeerJ* 3 (2015): e1177.

Brumm, Henrik, and Marc Naguib. "Environmental acoustics and the evolution of bird song." *Advances in the Study of Behavior* 40 (2009): 1–33.

Brumm, Henrik, and Hans Slabbekoorn. "Acoustic communication in noise." *Advances in the Study of Behavior* 35 (2005): 151–209.

Carlson, Nora V., Erick Greene, and Christopher N. Templeton. "Nuthatches vary their alarm calls based upon the source of the eavesdropped signals." *Nature Communications* 11 (2020): 1–7.

Colombelli-Négrel, Diane, and Christine Evans. "Superb fairy-wrens respond more to alarm calls from mate and kin compared to unrelated individuals." *Behavioral Ecology* 28 (2017): 1101–12.

Cottingham, John. "'A brute to the brutes?': Descartes' treatment of animals." *Philosophy* 53 (1978): 551–59.

Dalziell, Anastasia H., Alex C. Maisey, Robert D. Magrath, and Justin A. Welbergen. "Male lyrebirds create a complex acoustic illusion of a mobbing flock during courtship and copulation." *Current Biology* (2021). https://doi.org/10.1016/j.cub.2021.02.003.

Evans, Samuel, Carolyn McGettigan, Zarinah K. Agnew, Stuart Rosen, and Sophie K. Scott. "Getting the cocktail party started: masking effects in speech perception." *Journal of Cognitive Neuroscience* 28 (2016): 483–500.

Farrow, Lucy F., Ahmad Barati, and Paul G. McDonald. "Cooperative bird discriminates between individuals based purely on their aerial alarm calls." *Behavioral Ecology* 31 (2020): 440–47.

Flower, Tom P., Matthew Gribble, and Amanda R. Ridley. "Deception by flexible alarm mimicry in an African bird." *Science* 344 (2014): 513–16.

Greene, Erick, and Tom Meagher. "Red squirrels, *Tamiasciurus hudsonicus*, produce predator-class specific alarm calls." *Animal Behaviour* 55 (1998): 511–18.

Hansen, John H. L., Mahesh Kumar Nandwana, and Navid Shokouhi. "Analysis of human scream and its impact on text-independent speaker verification." *Journal of the Acoustical Society of America* 141 (2017): 2957–67.

Hedwig, Berthold, and Daniel Robert. "Auditory Parasitoid Flies Exploiting Acoustic Communication of Insects." In *Insect Hearing and Acoustic Communication*, edited by Hedwig Berthold, 45–63. Berlin: Springer, 2014.

Hulse, Stewart H. "Auditory scene analysis in animal communication." *Advances in the Study of Behavior* 31 (2002): 163–201.

Jain, Manjari, Swati Diwakar, Jimmy Bahuleyan, Rittik Deb, and Rohini Balakrishnan. "A rain forest dusk chorus: cacophony or sounds of silence?" *Evolutionary Ecology* 28 (2014): 1–22.

Krause, Bernard L. "Bioacoustics, habitat ambience in ecological balance." *Whole Earth Review* (57) 14–18.

Krause, Bernard L. "The niche hypothesis: a virtual symphony of animal sounds, the origins of musical expression and the health of habitats." *Soundscape Newsletter* 6 (1993): 6–10.

Lindsay, Jessica, "Why Do Caterpillars Whistle? Acoustic Mimicry of Bird Alarm Calls in the Amorpha juglandis Caterpillar" (2015). University of Montana, Missoula, Undergraduate Theses, Professional Papers, and Capstone Artifacts. https://scholarworks.umt.edu/utpp/60.

Magrath, Robert D., Tonya M. Haff, Pamela M. Fallow, and Andrew N. Radford. "Eavesdropping on heterospecific alarm calls: from mechanisms to consequences." *Biological Reviews* 90 (2015): 560–86.

McLachlan, Jessica R., and Robert D. Magrath. "Speedy revelations: how alarm calls can convey rapid, reliable information about urgent danger." *Proceedings of the Royal Society B* 287 (2020): 20192772.

Price, Tabitha, Philip Wadewitz, Dorothy Cheney, Robert Seyfarth, Kurt Hammerschmidt, and Julia Fischer. "Vervets revisited: a quantitative analysis of alarm call structure and context specificity." *Scientific Reports* 5 (2015): 13220.

Schmidt, Arne K. D., and Rohini Balakrishnan. "Ecology of acoustic signalling and the problem of masking interference in insects." *Journal of Comparative Physiology A* 201 (2015): 133–42.

Schmidt, Arne KD, Klaus Riede, and Heiner Römer. "High background noise shapes selective auditory filters in a tropical cricket." *Journal of Experimental Biology* 214 (2011): 1754–62.

Schmidt, Arne K. D., Heiner Römer, and Klaus Riede. "Spectral niche segregation and community organization in a tropical cricket assemblage." *Behavioral Ecology* 24 (2013): 470–80.

Suarez, Esteban, Manuel Morales, Rubén Cueva, V. Utreras Bucheli, Galo Zapata-Ríos, Eduardo Toral, Javier Torres, Walter Prado, and J. Vargas Olalla. "Oil industry, wild meat trade and roads: indirect effects of oil extraction activities in a protected area in north-eastern Ecuador." *Animal Conservation* 12 (2009): 364–73.

Summers, Kyle, S. E. A. McKeon, J. O. N. Sellars, Mark Keusenkothen, James Morris, David Gloeckner, Corey Pressley, Blake Price, and Holly Snow. "Parasitic exploitation as an engine of diversity." *Biological Reviews* 78 (2003): 639–75.

Swing, Kelly. "Preliminary observations on the natural history of representative treehoppers (Hemiptera, Auchenorrhyncha, Cicadomorpha: Membracidae and Aetalionidae) in the Yasuní Biosphere Reserve, including first reports of 13 genera for Ecuador and the province of Orellana." *Avances en Ciencias e Ingenierías* 4 (2012): B10–B38.

Templeton, Christopher N., Erick Greene, and Kate Davis. "Allometry of alarm calls: black-capped chickadees encode information about predator size." *Science* 308 (2005): 1934–37.

Tobias, Joseph A., Robert Planqué, Dominic L. Cram, and Nathalie Seddon. "Species interactions and the structure of complex communication networks." *Proceedings of the National Academy of Sciences* 111 (2014): 1020–25.

Zuk, Marlene, John T. Rotenberry, and Robin M. Tinghitella. "Silent night: adaptive disappearance of a sexual signal in a parasitized population of field crickets." *Biology Letters* 2 (2006): 521–24.

Sexuality and Beauty

Archetti, Marco. "Evidence from the domestication of apple for the maintenance of autumn colours by coevolution." *Proceedings of the Royal Society B* 276 (2009): 2575–80.

Baker, Myron C., Merrill SA Baker, and Laura M. Tilghman. "Differing effects of isolation on evolution of bird songs: examples from an island-mainland comparison of three species." *Biological Journal of the Linnean Society* 89 (2006): 331–42.

Beasley, V. R., R. Cole, C. Johnson, L. Johnson, C. Lieske, J. Murphy, M. Piwoni, C. Richards, P. Schoff, and A. M. Schotthoefer. "Environmental factors that influence amphibian community structure and health as indicators of ecosystems." Final Report EPA Grant R825867 (2001). https://cfpub.epa.gov/ncer_abstracts/index.cfm/fuseaction/display.highlight/abstract/274/report/F.

Biernaskie, Jay M., Alan Grafen, and Jennifer C. Perry. "The evolution of index signals to avoid the cost of dishonesty." *Proceedings of the Royal Society B* 281 (2014): 20140876.

Boccia, Maddalena, Sonia Barbetti, Laura Piccardi, Cecilia Guariglia, Fabio Ferlazzo, Anna Maria Giannini, and D. W. Zaidel. "Where does brain neural activation in aesthetic responses to visual art occur? Meta-analytic evidence from neuroimaging studies." *Neuroscience & Biobehavioral Reviews* 60 (2016): 65–71.

Butterfield, Brian P., Michael J. Lannoo, and Priya Nanjappa. "*Pseudacris crucifer*. Spring Peeper." AmphibiaWeb. Accessed May 23, 2020. http://amphibiaweb.org.

Conway, Bevil R., and Alexander Rehding. "Neuroaesthetics and the trouble with beauty." *PLOS Biology* 11 (2013): e1001504.

Cresswell, Will. "Song as a pursuit-deterrent signal, and its occurrence relative to other anti-predation behaviours of skylark (*Alauda arvensis*) on attack by merlins (*Falco columbarius*)." *Behavioral Ecology and Sociobiology* 34 (1994): 217–23.

Cummings, Molly E., and John A. Endler. "25 Years of sensory drive: the evidence and its watery bias." *Current Zoology* 64 (2018): 471–84.

Darwin, Charles. *On the Origin of Species by Means of Natural Selection, or the Preservation of Favoured Races in the Struggle for Life*. London: Murray, 1859. http://darwin-online.org.uk/.

Eberhardt, Laurie S. "Oxygen consumption during singing by male Carolina wrens (*Thryothorus ludovicianus*)." *Auk* 111 (1994): 124–30.

Fisher, Ronald A. "The evolution of sexual preference." *Eugenics Review* 7 (1915): 184–92.

Forester, Don C., and Richard Czarnowsky. "Sexual selection in the spring peeper, *Hyla crucifer* (Amphibia, Anura): role of the advertisement call." *Behaviour* 92 (1985): 112–27.

Forester, Don C., and W. Keith Harrison. "The significance of antiphonal vocalisation by the spring peeper, *Pseudacris crucifer* (Amphibia, Anura)." *Behaviour* 103 (1987): 1–15.

Fowler-Finn, Kasey D., and Rafael L. Rodríguez. "The causes of variation in the presence of genetic covariance between sexual traits and preferences." *Biological Reviews* 91 (2016): 498–510.

Grant, Peter R., and B. Rosemary Grant. "The founding of a new population of Darwin's finches." *Evolution* 49 (1995): 229–40.

Gray, David A., and William H. Cade. "Sexual selection and speciation in field crickets." *Proceedings of the National Academy of Sciences* 97 (2000): 14449–54.

Henshaw, Jonathan M., and Adam G. Jones. "Fisher's lost model of runaway sexual selection." *Evolution* 74 (2019): 487–94.

Hill, Brad G., and M. Ross Lein. "The non-song vocal repertoire of the white-crowned sparrow." *Condor* 87 (1985): 327–35.

Humfeld, Sarah C., Vincent T. Marshall, and Mark A. Bee. "Context-dependent plasticity of aggressive signalling in a dynamic social environment." *Animal Behaviour* 78 (2009): 915–24.

Kirkpatrick, Mark. "Sexual selection and the evolution of female choice." *Evolution* 82 (1982): 1–12.

Kruger, M. Charlotte, Carina J. Sabourin, Alexandra T. Levine, and Stephen G. Lomber. "Ultrasonic hearing in cats and other terrestrial mammals." *Acoustics Today* 17 (2021): 18–25.

Kuhelj, Anka, Maarten De Groot, Franja Pajk, Tatjana Simčič, and Meta Virant-Doberlet. "Energetic cost of vibrational signalling in a leafhopper." *Behavioral Ecology and Sociobiology* 69 (2015): 815–28.

Laland, Kevin N. "On the evolutionary consequences of sexual imprinting." *Evolution* 48 (1994): 477–89.

Lande, Russell. "Models of speciation by sexual selection on polygenic traits." *Proceedings of the National Academy of Sciences* 78 (1981): 3721–25.

Lemmon, Emily Moriarty. "Diversification of conspecific signals in sympatry: geographic overlap drives multidimensional reproductive character displacement in frogs." *Evolution* 63 (2009): 1155–70.

Lemmon, Emily Moriarty, and Alan R. Lemmon. "Reinforcement in chorus frogs: lifetime fitness estimates including intrinsic natural selection and sexual selection against hybrids." *Evolution* 64 (2010): 1748–61.

Ligon, Russell A., Christopher D. Diaz, Janelle L. Morano, Jolyon Troscianko, Martin Stevens, Annalyse Moskeland, Timothy G. Laman, and Edwin Scholes III. "Evolution of correlated complexity in the radically different courtship signals of birds-of-paradise." *PLOS Biology* 16 (2018): e2006962.

Lykens, David V., and Don C. Forester. "Age structure in the spring peeper: do males advertise longevity?" *Herpetologica* (1987): 216–23.

Marshall, David C., and Kathy BR Hill. "Versatile aggressive mimicry of cicadas by an Australian predatory katydid." *PLOS One* 4 (2009).

Matsumoto, Yui K., and Kazuo Okanoya. "Mice modulate ultrasonic calling bouts according to sociosexual context." *Royal Society Open Science* 5 (2018): 180378.

Mead, Louise S., and Stevan J. Arnold. "Quantitative genetic models of sexual selection." *Trends in Ecology & Evolution* 19 (2004): 264–71.

Miles, Meredith C., Eric R. Schuppe, R. Miller Ligon IV, and Matthew J. Fuxjager. "Macroevolutionary patterning of woodpecker drums reveals how sexual selection elaborates signals under constraint." *Proceedings of the Royal Society B* 285 (2018): 20172628.

Odom, Karan J., Michelle L. Hall, Katharina Riebel, Kevin E. Omland, and Naomi E. Langmore. "Female song is widespread and ancestral in songbirds." *Nature Communications* 5 (2014): 1–6.

Pašukonis, Andrius, Matthias-Claudio Loretto, and Walter Hödl. "Map-like navigation from distances exceeding routine movements in the three-striped poison frog (*Ameerega trivittata*)." *Journal of Experimental Biology* 221 (2018).

Pašukonis, Andrius, Katharina Trenkwalder, Max Ringler, Eva Ringler, Rosanna Mangione, Jolanda Steininger, Ian Warrington, and Walter Hödl. "The significance of spatial memory for water finding in a tadpole-transporting frog." *Animal Behaviour* 116 (2016): 89–98.

Patricelli, Gail L., Eileen A. Hebets, and Tamra C. Mendelson. "Book review of Prum, RO 2018. The evolution of beauty." *Evolution* 73 (2019): 115–24.

Pomiankowski, Andrew, and Yoh Iwasa. "Evolution of multiple sexual preferences by Fisher's runaway process of sexual selection." *Proceedings of the Royal Society of London*. Series B 253 (1993): 173–81.

Proctor, Heather C. "Sensory exploitation and the evolution of male mating behaviour: a cladistic test using water mites (Acari: Parasitengona)." *Animal Behaviour* 44 (1992): 745–52.

Prokop, Zofia M., and Szymon M. Drobniak. "Genetic variation in male attractiveness: it is time to see the forest for the trees." *Evolution* 70 (2016): 913–21.

Prokop, Zofia M., Łukasz Michalczyk, Szymon M. Drobniak, Magdalena Herdegen, and Jacek Radwan. "Meta-analysis suggests choosy females get sexy sons more than 'good genes.'" *Evolution* 66 (2012): 2665–73.

Prum, Richard O. "Aesthetic evolution by mate choice: Darwin's really dangerous idea." *Philosophical Transactions of the Royal Society B* 367 (2012): 2253–65.

Prum, Richard O. *The Evolution of Beauty*. Doubleday: New York, 2017.

Prum, Richard O. "The Lande–Kirkpatrick mechanism is the null model of evolution by intersexual selection: implications for meaning, honesty, and design in intersexual signals." *Evolution* 64 (2010): 3085–100.

Purnell, Beverly A. "Intersexuality in female moles." *Science* 370 (2020): 182.

Reeder, Amy L., Marilyn O. Ruiz, Allan Pessier, Lauren E. Brown, Jeffrey M. Levengood, Christopher A. Phillips, Matthew B. Wheeler, Richard E. Warner, and Val R. Beasley. "Intersexuality and the cricket frog decline: historic and geographic trends." *Environmental Health Perspectives* 113 (2005): 261–65.

Rendell, Luke, Laurel Fogarty, and Kevin N. Laland. "Runaway cultural niche construction." *Philosophical Transactions of the Royal Society B* 366 (2011): 823–35.

Riebel, Katharina, Karan J. Odom, Naomi E. Langmore, and Michelle L. Hall. "New insights from female bird song: towards an integrated approach to studying male and female communication roles." *Biology Letters* 15 (2019): 20190059.

Rothenberg, David. *Survival of the Beautiful*. New York: Bloomsbury Press, 2011.

Roughgarden, Joan. "Homosexuality and Evolution: A Critical Appraisal." In *On Human Nature*, edited by Michel Tibayrenc and Francisco J. Ayala, 495–516. New York: Academic Press, 2017.

Ryan, Michael J. "Coevolution of sender and receiver: effect on local mate preference in cricket frogs." *Science* 240 (1988): 1786.

Schoffelen, Richard L. M., Johannes M. Segenhout, and Pim Van Dijk. "Mechanics of the exceptional anuran ear." *Journal of Comparative Physiology A* 194 (2008): 417–28.

Short, Stephen, Gongda Yang, Peter Kille, and Alex T. Ford. "A widespread and distinctive form of amphipod intersexuality not induced by known feminising parasites." *Sexual Development* 6 (2012): 320–24.

Skelly, David K., Susan R. Bolden, and Kirstin B. Dion. "Intersex frogs concentrated in suburban and urban landscapes." *EcoHealth* 7 (2010): 374–79.

Solnit, Rebecca. *Recollections of My Nonexistence*. New York: Viking, 2020.

Starnberger, Iris, Doris Preininger, and Walter Hödl. "The anuran vocal sac: a tool for multimodal signalling." *Animal Behaviour* 97 (2014): 281–88.

Stewart, Kathryn. "Contact Zone Dynamics and the Evolution of Reproductive Isolation in a North American Treefrog, the Spring Peeper (*Pseudacris crucifer*)." (PhD diss., Queen's University, 2013).

Taborsky, Michael, and H. Jane Brockmann. "Alternative Reproductive Tactics and Life History Phenotypes." In *Animal Behaviour: Evolution and Mechanisms*, edited by Peter M. Kappeler, 537–86. Berlin: Springer, 2010.

Wilczynski, Walter, Harold H. Zakon, and Eliot A. Brenowitz. "Acoustic communication in spring peepers." *Journal of Comparative Physiology A* 155 (1984): 577–84.

Zamudio, Kelly R., and Lauren M. Chan. "Alternative Reproductive Tactics in Amphibians." In *Alternative Reproductive Tactics: An Integrative Approach*, edited by Rui F. Oliveira, Michael Taborsky, and Jane Brockmann, 300–31. Cambridge: Cambridge University Press, 2008.

Zhang, Fang, Juan Zhao, and Albert S. Feng. "Vocalizations of female frogs contain nonlinear characteristics and individual signatures." *PLOS One* 12 (2017).

Zimmitti, Salvatore J. "Individual variation in morphological, physiological, and biochemical features associated with calling in spring peepers (*Pseudacris crucifer*)." *Physiological and Biochemical Zoology* 72 (1999): 666–76.

Vocal Learning and Culture

Bolhuis, Johan J., Kazuo Okanoya, and Constance Scharff. "Twitter evolution: converging mechanisms in birdsong and human speech." *Nature Reviews Neuroscience* 11 (2010): 747–59.

Brakes, Philippa, Sasha R. X. Dall, Lucy M. Aplin, Stuart Bearhop, Emma L. Carroll, Paolo Ciucci, Vicki Fishlock et al. "Animal cultures matter for conservation." *Science* 363 (2019): 1032–34.

Cavitt, John F., and Carola A. Haas (2020). Brown Thrasher (*Toxostoma rufum*). In *Birds of the World*, edited by A. F. Poole. https://doi.org/10.2173/bow.brnthr.01.

Cheney, Dorothy L., and Robert M. Seyfarth. "Flexible usage and social function in primate vocalizations." *Proceedings of the National Academy of Sciences* 115 (2018): 1974–79.

Chilton, G., M. C. Baker, C. D. Barrentine, and M. A. Cunningham (2020). White-crowned Sparrow (*Zonotrichia leucophrys*). In *Birds of the World*, edited by A. F. Poole and F. B. Gill. https://doi.org /10.2173/bow.whcspa.01.

Crates, Ross, Naomi Langmore, Louis Ranjard, Dejan Stojanovic, Laura Rayner, Dean Ingwersen, and Robert Heinsohn. "Loss of vocal culture and fitness costs in a critically endangered songbird." *Proceedings of the Royal Society B* 288 (2021): 20210225.

Derryberry, Elizabeth P. "Ecology shapes birdsong evolution: variation in morphology and habitat explains variation in white-crowned sparrow song." *American Naturalist* 174 (2009): 24–33.

Ferrigno, Stephen, Samuel J. Cheyette, Steven T. Piantadosi, and Jessica F. Cantlon. "Recursive sequence generation in monkeys, children, US adults, and native Amazonians." *Science Advances* 6 (2020): eaaz1002.

Gentner, Timothy Q., Kimberly M. Fenn, Daniel Margoliash, and Howard C. Nusbaum. "Recursive syntactic pattern learning by songbirds." *Nature* 440 (2006): 1204–7.

Gero, Shane, Hal Whitehead, and Luke Rendell. "Individual, unit and vocal clan level identity cues in sperm whale codas." *Royal Society Open Science* 3 (2016): 150372.

Kroodsma, Donald E. "Vocal Behavior." In *Handbook of Bird Biology*, 2nd ed. Ithaca, NY: Cornell Lab of Ornithology, 2004.

Lachlan, Robert F., Oliver Ratmann, and Stephen Nowicki. "Cultural conformity generates extremely stable traditions in bird song." *Nature Communications* 9 (2018): 1–9.

Lipshutz, Sara E., Isaac A. Overcast, Michael J. Hickerson, Robb T. Brumfield, and Elizabeth P. Derryberry. "Behavioural response to song and genetic divergence in two subspecies of white-crowned sparrows (*Zonotrichia leucophrys*)." *Molecular Ecology* 26 (2017): 3011–27.

Marler, Peter. "A comparative approach to vocal learning: song development in white-crowned sparrows." *Journal of Comparative and Physiological Psychology* 71 (1970): 1.

May, Michael. "Recordings That Made Waves: The Songs That Saved the Whales." National Public Radio, *All Things Considered*. December 26, 2014.

Nelson, Douglas A. "A preference for own-subspecies' song guides vocal learning in a song bird." *Proceedings of the National Academy of Sciences* 97 (2000): 13348–53.

Nelson, Douglas A., Karen I. Hallberg, and Jill A. Soha. "Cultural evolution of Puget sound white-crowned sparrow song dialects." *Ethology* 110 (2004): 879–908.

Nelson, Douglas A., Peter Marler, and Alberto Palleroni. "A comparative approach to vocal learning: intraspecific variation in the learning process." *Animal Behaviour* 50 (1995): 83–97.

Otter, Ken A., Alexandra Mckenna, Stefanie E. LaZerte, and Scott M. Ramsay. "Continent-wide shifts in song dialects of white-throated sparrows." *Current Biology* 30 (2020): 3231–35.

Paxton, Kristina L., Esther Sebastián-González, Justin M. Hite, Lisa H. Crampton, David Kuhn, and Patrick J. Hart. "Loss of cultural song diversity and the convergence of songs in a declining Hawaiian forest bird community." *Royal Society Open Science* 6 (2019): 190719.

Rosenberg, Kenneth V., Adriaan M. Dokter, Peter J. Blancher, John R. Sauer, Adam C. Smith, Paul A. Smith, Jessica C. Stanton et al. "Decline of the North American avifauna." *Science* 366 (2019): 120–24.

Safina, Carl. *Becoming Wild*. New York: Henry Holt, 2020.

Simmons, Andrea Megela, and Darlene R. Ketten. "How a frog hears." *Acoustics Today* 16 (2020): 67–74.

Slabbekoorn, Hans, and Thomas B. Smith. "Bird song, ecology and speciation." *Philosophical Transactions of the Royal Society of London*. Series B 357 (2002): 493–503.

Thornton, Alex, and Tim Clutton-Brock. "Social learning and the development of individual and group behaviour in mammal societies." *Philosophical Transactions of the Royal Society B* 366 (2011): 978–87.

Trainer, Jill M. "Cultural evolution in song dialects of yellow-rumped caciques in Panama." *Ethology* 80 (1989): 190–204.

Tyack, Peter L. "A taxonomy for vocal learning." *Philosophical Transactions of the Royal Society B* 375 (2020): 20180406.

Uy, J. Albert C., Darren E. Irwin, and Michael S. Webster. "Behavioral isolation and incipient speciation in birds." *Annual Review of Ecology, Evolution, and Systematics* 49 (2018): 1–24.

Whitehead, Hal, Kevin N. Laland, Luke Rendell, Rose Thorogood, and Andrew Whiten. "The reach of gene–culture coevolution in animals." *Nature Communications* 10 (2019): 1–10.

Whiten, Andrew. "A second inheritance system: the extension of biology through culture." *Interface Focus* 7 (2017): 20160142.

Wickman, Forrest. "Who Really Said You Should 'Kill Your Darlings'?" *Slate Magazine*, October 18, 2013. https://slate.com/culture/2013/10/kill-your-darlings-writing-advice-what-writer-really -said-to-murder-your-babies.html.

The Imprints of Deep Time

Batista, Romina, Urban Olsson, Tobias Andermann, Alexandre Aleixo, Camila Cherem Ribas, and Alexandre Antonelli. "Phylogenomics and biogeography of the world's thrushes (Aves, *Turdus*): new evidence for a more parsimonious evolutionary history." *Proceedings of the Royal Society B* 287 (2020): 20192400.

Cigliano, María M., Holger Braun, David C. Eades, and Daniel Otte. *Orthoptera Species File*. Version 5.0/5.0. June 22, 2020. http://Orthoptera.SpeciesFile.org.

Curtis, Syndey, and H. E. Taylor. "Olivier Messiaen and the Albert's Lyrebird: from Tamborine Mountain to Éclairs sur l'au-delà.'" In *Olivier Messiaen: The Centenary Papers*, edited by Judith Crispin, 52–79. Newcastle upon Tyne, UK: Cambridge Scholars Publishing, 2010.

Ducker, Sophie. *The Contented Botanist: Letters of W. H. Harvey about Australia and the Pacific*. Melbourne: Miegunyah Press, 1984.

Fuchs, Jérôme, Martin Irestedt, Jon Fjeldså, Arnaud Couloux, Eric Pasquet, and Rauri C. K. Bowie. "Molecular phylogeny of African bush-shrikes and allies: tracing the biogeographic history of an explosive radiation of corvoid birds." *Molecular Phylogenetics and Evolution* 64 (2012): 93–105.

Heads, Sam W., and Léa Leuzinger. "On the placement of the Cretaceous orthopteran *Brauckmannia groeningae* from Brazil, with notes on the relationships of Schizodactylidae (Orthoptera, Ensifera)." *ZooKeys* 77 (2011): 17.

Hill, Kathy B. R., David C. Marshall, Maxwell S. Moulds, and Chris Simon. "Molecular phylogenetics, diversification, and systematics of Tibicen Latreille 1825 and allied cicadas of the tribe Cryptotympanini, with three new genera and emphasis on species from the USA and Canada (Hemiptera: Auchenorrhyncha: Cicadidae)." *Zootaxa* 3985 (2015): 219–51.

Hopper, Stephen D. "OCBIL theory: towards an integrated understanding of the evolution, ecology and conservation of biodiversity on old, climatically buffered, infertile landscapes." *Plant and Soil* 322 (2009): 49–86.

Jønsson, Knud Andreas, Pierre-Henri Fabre, Jonathan D. Kennedy, Ben G. Holt, Michael K. Borregaard, Carsten Rahbek, and Jon Fjeldså. "A supermatrix phylogeny of corvoid passerine birds (Aves: Corvides)." *Molecular Phylogenetics and Evolution* 94 (2016): 87–94.

Kearns, Anna M., Leo Joseph, and Lyn G. Cook. "A multilocus coalescent analysis of the speciational history of the Australo-Papuan butcherbirds and their allies." *Molecular Phylogenetics and Evolution* 66 (2013): 941–52.

Low, Tim. *Where Song Began: Australia's Birds and How They Changed the World*. New Haven, CT: Yale University Press, 2016.

Marshall, David C., Max Moulds, Kathy B. R. Hill, Benjamin W. Price, Elizabeth J. Wade, Christopher L. Owen, Geert Goemans et al. "A molecular phylogeny of the cicadas (Hemiptera: Cicadidae) with a review of tribe and subfamily classification." *Zootaxa* 4424 (2018): 1–64.

Mayr, Gerald. "Old World fossil record of modern-type hummingbirds." *Science* 304 (2004): 861–64.

McGuire, Jimmy A., Christopher C. Witt, J. V. Remsen Jr., Ammon Corl, Daniel L. Rabosky, Douglas L. Altshuler, and Robert Dudley. "Molecular phylogenetics and the diversification of hummingbirds." *Current Biology* 24 (2014): 910–16.

Nicholson, David B., Peter J. Mayhew, and Andrew J. Ross. "Changes to the fossil record of insects through fifteen years of discovery." *PLOS One* 10 (2015): e0128554.

Oliveros, Carl H., Daniel J. Field, Daniel T. Ksepka, F. Keith Barker, Alexandre Aleixo, Michael J. Andersen, Per Alström et al. "Earth history and the passerine superradiation." *Proceedings of the National Academy of Sciences* 116 (2019): 7916–25.

Orians, Gordon H., and Antoni V. Milewski. "Ecology of Australia: the effects of nutrient-poor soils and intense fires." *Biological Reviews* 82 (2007): 393–423.

Ratcliffe, Eleanor, Birgitta Gatersleben, and Paul T. Sowden. "Predicting the perceived restorative potential of bird sounds through acoustics and aesthetics." *Environment and Behavior* 52 (2020): 371–400.

Sætre, G-P., S. Riyahi, Mansour Aliabadian, Jo S. Hermansen, S. Hogner, U. Olsson, M. F. Gonzalez Rojas, S. A. Sæther, C. N. Trier, and T. O. Elgvin. "Single origin of human commensalism in the house sparrow." *Journal of Evolutionary Biology* 25 (2012): 788–96.

Scheffers, Brett R., Brunno F. Oliveira, Ieuan Lamb, and David P. Edwards. "Global wildlife trade across the tree of life." *Science* 366 (2019): 71–76.

Toda, Yasuka, Meng-Ching Ko, Qiaoyi Liang, Eliot T. Miller, Alejandro Rico-Guevara, Tomoya Nakagita, Ayano Sakakibara, Kana Uemura, Timothy Sackton, Takashi Hayakawa, Simon Yung Wa Sin, Yoshiro Ishimaru, Takumi Misaka, Pablo Oteiza, James Crall, Scott V. Edwards, William Buttemer, Shuichi Matsumura, and Maude W. Baldwin. "Early Origin of Sweet Perception in the Songbird Radiation." *Science* 373 (2021): 226–31.

Wang, H., Y. N. Fang, Y. Fang, E. A. Jarzembowski, B. Wang, and H. C. Zhang. "The earliest fossil record of true crickets belonging to the Baissogryllidae (Insecta, Orthoptera, Grylloidea)." *Geological Magazine* 156 (2019): 1440–44.

Whitehouse, Andrew. "Senses of Being: The Atmospheres of Listening to Birds in Britain, Australia and New Zealand." In *Exploring Atmospheres Ethnographically*, edited by Sara Asu Schroer and Susanne Schmitt, 61–75. Abingdon, UK: Routledge, 2018.

Bone, Ivory, Breath

Albouy, Philippe, Lucas Benjamin, Benjamin Morillon, and Robert J. Zatorre. "Distinct sensitivity to spectrotemporal modulation supports brain asymmetry for speech and melody." *Science* 367 (2020): 1043–47.

Aubert, Maxime, Rustan Lebe, Adhi Agus Oktaviana, Muhammad Tang, Basran Burhan, Andi Jusdi, Budianto Hakim et al. "Earliest hunting scene in prehistoric art." *Nature* (2019): 1–4.

Centre Pompidou. *Préhistoire, Une Énigme Moderne*. Exhibition. Paris, France (2019).

Conard, Nicholas., Michael Bolus, Paul Goldberg, and Suzanne C. Münzel. "The Last Neanderthals and First Modern Humans in the Swabian Jura." In *When Neanderthals and Modern Humans Met*, edited by Nicholas Conrad. Tübingen, Germany: Tübingen Publications in Prehistory, 2006.

Conard, Nicholas J., Michael Bolus, and Susanne C. Münzel. "Middle Paleolithic land use, spatial organization and settlement intensity in the Swabian Jura, southwestern Germany." *Quaternary International* 247 (2012): 236–45.

Conard, Nicholas J., Keiko Kitagawa, Petra Krönneck, Madelaine Böhme, and Susanne C. Münzel. "The importance of fish, fowl and small mammals in the Paleolithic diet of the Swabian Jura, southwestern Germany." In *Zooarchaeology and Modern Human Origins*, edited by Jamie Clark, and John D. Speth, 173–90. Dordrecht: Springer, 2013.

Conard, Nicholas J., and Maria Malina. "New evidence for the origins of music from caves of the Swabian Jura." *Orient-archäologie* 22 (2008): 13–22.

Conard, Nicholas J., Maria Malina, and Susanne C. Münzel. "New flutes document the earliest musical tradition in southwestern Germany." *Nature* 460 (2009): 737.

d'Errico, Francesco, Paola Villa, Ana C. Pinto Llona, and Rosa Ruiz Idarraga. "A Middle Palaeolithic origin of music? Using cave-bear bone accumulations to assess the Divje Babe I bone 'flute.'" *Antiquity* 72 (1998): 65–79.

d'Errico, Francesco, Christopher Henshilwood, Graeme Lawson, Marian Vanhaeren, Anne-Marie Tillier, Marie Soressi, Frédérique Bresson et al. "Archaeological evidence for the emergence of language, symbolism, and music—an alternative multidisciplinary perspective." *Journal of World Prehistory* 17 (2003): 1–70.

Dutkiewicz, Ewa, Sibylle Wolf, and Nicholas J. Conard. "Early symbolism in the Ach and the Lone valleys of southwestern Germany." *Quaternary International* 491 (2017): 30–45.

Floss, Harald. "Same as it ever was? The Aurignacian of the Swabian Jura and the origins of Palaeolithic art." *Quaternary International* 491 (2018): 21–29.

Guenther, Mathias. "N//àe ("Talking"): The oral and rhetorical base of San culture." *Journal of Folklore Research* 43 (2006): 241–61.

Güntürkün, Onur, Felix Ströckens, and Sebastian Ocklenburg. "Brain lateralization: a comparative perspective." *Physiological Reviews* 100 (2020): 1019–63.

Hahn, Joachim, and Susanne C. Münzel. "Knochenflöten aus dem Aurignacien des Geißenklösterle bei Blaubeuren, Alb-Donau-Kreis." *Fundberichte aus Baden-Württemberg* 20 (1995): 1–12.

Hardy, Bruce L., Michael Bolus, and Nicholas J. Conard. "Hammer or crescent wrench? Stone-tool form and function in the Aurignacian of southwest Germany." *Journal of Human Evolution* 54 (2008): 648–62.

Heinsohn, Robert, Christina N. Zdenek, Ross B. Cunningham, John A. Endler, and Naomi E. Langmore. "Tool-assisted rhythmic drumming in palm cockatoos shares key elements of human instrumental music." *Science Advances* 3 (2017): e1602399.

Henshilwood, Christopher S., Francesco d'Errico, Karen L. van Niekerk, Laure Dayet, Alain Queffelec, and Luca Pollarolo. "An abstract drawing from the 73,000-year-old levels at Blombos Cave, South Africa." *Nature* 562 (2018): 115.

Higham, Thomas, Laura Basell, Roger Jacobi, Rachel Wood, Christopher Bronk Ramsey, and Nicholas J. Conard. "Testing models for the beginnings of the Aurignacian and the advent of figurative art and music: the radiocarbon chronology of Geißenklösterle." *Journal of Human Evolution* 62 (2012): 664–76.

Jewell, Edward Alden. "Art Museum Opens Prehistoric Show." *New York Times*, April 28, 1937.

Kehoe, Laura. "Mysterious new behaviour found in our closest living relatives." *Conversation*, February 29, 2016.

Killin, Anton. "The origins of music: evidence, theory, and prospects." *Music & Science* 1 (2018): 2059204317751971.

Kühl, Hjalmar S., Ammie K. Kalan, Mimi Arandjelovic, Floris Aubert, Lucy D'Auvergne, Annemarie Goedmakers, Sorrel Jones et al. "Chimpanzee accumulative stone throwing." *Scientific Reports* 6 (2016): 1–8.

Malina, Maria, and Ralf Ehmann. "Elfenbeinspaltung im Aurignacien Zur Herstellungstechnik der Elfenbeinflöte aus dem Geißenklösterle." *Mitteilungen der Gesellschaft für Urgeschichte* 18 (2009): 93–107.

Mehr, Samuel A., Manvir Singh, Dean Knox, Daniel M. Ketter, Daniel Pickens-Jones, Stephanie Atwood, Christopher Lucas et al. "Universality and diversity in human song." *Science* 366 (2019): eaax0868.

Morley, Iain. *The Prehistory of Music: Human Evolution, Archaeology, and the Origins of Musicality.* Oxford, UK: Oxford University Press, 2013.

Münzel, Susanne, Nicholas J. Conrad, Wulf Hein, Frances Gill, Anna Friederike Potengowski. "Interpreting three Upper Palaeolithic wind instruments from Germany and one from France as flutes. (Re)construction, playing techniques and sonic results." *Studien zur Musikarchäologie* X (2016): 225–43.

Münzel, Susanne, Friedrich Seeberger, and Wulf Hein. "The Geißenklösterle Flute—discovery, experiments, reconstruction." *Studien zur Musikarchäologie* III (2002): 107–18.

Museum of Modern Art. *Prehistoric Rock Pictures in Europe and Africa*, 28 April to 30 May 1937. https://www.moma.org/interactives/exhibitions/2016/spelunker/exhibitions/3037/.

Novitskaya, E., C. J. Ruestes, M. M. Porter, V. A. Lubarda, M. A. Meyers, and J. McKittrick. "Reinforcements in avian wing bones: experiments, analysis, and modeling." *Journal of the Mechanical Behavior of Biomedical Materials* 76 (2017): 85–96.

Peretz, Isabelle, Dominique Vuvan, Marie-Élaine Lagrois, and Jorge L. Armony. "Neural overlap in processing music and speech." *Philosophical Transactions of the Royal Society B* 370 (2015): 20140090.

Potengowski, Anna Friederike, and Susanne C. Münzel. "Hörbeispiele, Examples 1–33." *Mitteilungen der Gesellschaft für Urgeschichte*, 2015. https://uni-tuebingen.de/fakultaeten/mathematisch -naturwissenschaftliche-fakultaet/fachbereiche/geowissenschaften/arbeitsgruppen/urgeschichte -naturwissenschaftliche-archaeologie/forschungsbereich/aeltere-urgeschichte-quartaeroekologie /publikationen/gfu-mitteilungen/hoerbeispiele/.

Potengowski, A.F., and S. C. Münzel, 2015. Die musikalische "Vermessung" paläolithischer Blasinstrumente der Schwäbischen Albhand von Rekonstruktionen. Anblastechniken, Tonmaterial und Klangwelt." *Mitteilungen der Gesellschaft für Urgeschichte* 24 (2015): 173–91.

Potengowski, Anna Friederike (bone flutes), and Georg Wieland Wagner (percussion). *The Edge of Time: Palaeolithic Bone Flutes of France and Germany*, compact disc. Edinburgh, UK: Delphian Records, 2017.

Rhodes, Sara E., Reinhard Ziegler, Britt M. Starkovich, and Nicholas J. Conard. "Small mammal taxonomy, taphonomy, and the paleoenvironmental record during the Middle and Upper Paleolithic at Geißenklösterle Cave (Ach Valley, southwestern Germany)." *Quaternary Science Reviews* 185 (2018): 199–221.

Richard, Maïlys, Christophe Falguères, Helene Valladas, Bassam Ghaleb, Edwige Pons-Branchu, Norbert Mercier, Daniel Richter, and Nicholas J. Conard. "New electron spin resonance (ESR) ages from Geißenklösterle Cave: a chronological study of the Middle and early Upper Paleolithic layers." *Journal of Human Evolution* 133 (2019): 133–45.

Riehl, Simone, Elena Marinova, Katleen Deckers, Maria Malina, and Nicholas J. Conard. "Plant use and local vegetation patterns during the second half of the Late Pleistocene in southwestern Germany." *Archaeological and Anthropological Sciences* 7 (2015): 151–67.

Tomlinson, Gary. *A Million Years of Music: The Emergence of Human Modernity.* New York: Zone Books, 2015.

Zhang, Juzhong, Garman Harbottle, Changsui Wang, and Zhaochen Kong. "Oldest playable musical instruments found at Jiahu early Neolithic site in China." *Nature* 401 (1999): 366.

Resonant Spaces

Anderson, Tim. "How CDs Are Remastering the Art of Noise." *Guardian*, January 18, 2007.

Barron, M. "The Royal Festival Hall acoustics revisited." *Applied Acoustics* 24 (1988): 255–73.

Boyden, David D., Peter Walls, Peter Holman, Karel Moens, Robin Stowell, Anthony Barnett, Matt Glaser et al. "Violin." *Grove Music Online.* January 20, 2001. https://www.oxfordmusiconline.com/.

Cooper, Michel, and Robin Pogrebin. "After Years of False Starts, Geffen Hall Is Being Rebuilt. Really." *New York Times.* December 2, 2019.

Díaz-Andreu, M., and T. Mattioli. "Rock Art Music, and Acoustics: A Global Overview." In *The Oxford Handbook of the Archaeology and Anthropology of Rock Art,* edited by Bruno David and Ian J. McNiven, 503–28. Oxford, UK: Oxford University Press, 2017.

Ellison, Steve. "Innovations: Meyer Sound Spacemap Go." *Pro Sound News* (2020). https://www.prosoundnetwork.com/gear-and-technology/innovations-meyer-sound-spacemap-go.

Emmerling, Caey, and Dallas Taylor. "The Loudness Wars." *Twenty Thousand Hertz.* Podcast. https://www.20k.org/episodes/loudnesswars.

Fazenda, Bruno, Chris Scarre, Rupert Till, Raquel Jiménez Pasalodos, Manuel Rojo Guerra, Cristina Tejedor, Roberto Ontañón Peredo et al. "Cave acoustics in prehistory: exploring the association of Palaeolithic visual motifs and acoustic response." *Journal of the Acoustical Society of America* 142 (2017): 1332–49.

Fei, Faye Chunfang. *Chinese Theories of Theater and Performance from Confucius to the Present.* Ann Arbor, MI: University of Michigan Press, 2002.

Giordano, Nicholas. "The invention and evolution of the piano." *Acoustics Today* 12 (2016): 12–19.

Henahan, Donal. "Philharmonic Hall Is Returning." *New York Times,* July 8, 1969.

Hill, Peggy S. M. "Environmental and social influences on calling effort in the prairie mole cricket (Gryllotalpa major)." *Behavioral Ecology* 9 (1998): 101–8.

Kopf, Dan. "How Headphones Are Changing the Sound of Music." *Quartz,* December 18, 2019.

Kozinn, Allan. "More Tinkering with Acoustics at Avery Fisher." *New York Times,* November 16, 1991.

Lardner, Björn, and Maklarin bin Lakim. "Tree-hole frogs exploit resonance effects." *Nature* 420 (2002): 475.

Lawergren, Bo. "Neolithic drums in China." *Studien zur Musik* V (2006): 109–27.

Manniche, Lise. *Music and Musicians in Ancient Egypt.* London: British Museum Press, 1991.

Manoff, Tom. "Do Electronics Have a Place in the Concert Hall? Maybe." *New York Times,* March 31, 1991.

McKinnon, James W. "Hydraulis." *Grove Music Online.* 2001. https://www.oxfordmusiconline.com/.

Michaels, Sean. "Metallica Album Latest Victim in 'Loudness War'?" *Guardian,* September 17, 2008.

Montagu, Jeremy, Howard Mayer Brown, Jaap Frank, and Ardal Powell. "Flute." *Grove Music Online.* 2001. https://www.oxfordmusiconline.com/.

Petrusich, Amanda. "Headphones Everywhere." *New Yorker,* July 12, 2016.

Pike, Alistair W. G., Dirk L. Hoffmann, Marcos García-Diez, Paul B. Pettitt, Jose Alcolea, Rodrigo De Balbin, César Gonzalez-Sainz et al. "U-series dating of Paleolithic art in 11 caves in Spain." *Science* 336 (2012): 1409–13.

Reznikoff, Iégor. "Sound resonance in prehistoric times: a study of Paleolithic painted caves and rocks." *Journal of the Acoustical Society of America* 123 (2008): 3603.

Reznikoff, Iégor, and Michel Dauvois. "La dimension sonore des grottes ornées." *Bulletin de la Société Préhistorique Française* 85 (1988): 238–46.

Ross, Alex. "Wizards of Sound." *New Yorker*, February 16, 2015.

Scarre, Chris. "Painting by resonance." *Nature* 338 (1989): 382.

Sound on Sound magazine. "Jeff Ellis: Engineering Frank Ocean," November 17, 2016. https://www.youtube.com/watch?v=izZMM5eHCtQ.

Tommasini, Anthony. "Defending the operatic voice from technology's wiles." *New York Times*, November 3, 1999.

Velliky, Elizabeth C., Martin Porr, and Nicholas J. Conard. "Ochre and pigment use at Hohle Fels cave: results of the first systematic review of ochre and ochre-related artefacts from the Upper Palaeolithic in Germany." *PLOS One* 13 (2018): e0209874.

Wu, Chih-Wei, Chih-Fang Huang, and Yi-Wen Liu. "Sound analysis and synthesis of Marquis Yi of Zeng's chime-bell set." *Proceedings of Meetings on Acoustics* ICA2013 19 (2013): 035077.

Music, Forest, Body

Anthwal, Neal, Leena Joshi, and Abigail S. Tucker. "Evolution of the mammalian middle ear and jaw: adaptations and novel structures." *Journal of Anatomy* 222 (2013): 147–60.

Ball, Stephen M. J. "Stocks and exploitation of East African blackwood." *Oryx* 38 (2004): 1–7.

Beachey, Richard W. "The East African ivory trade in the nineteenth century." *Journal of African History* (1967): 269–90.

Bennett, Bradley C. "The sound of trees: wood selection in guitars and other chordophones." *Economic Botany* 70 (2016): 49–63.

Chaiklin, Martha. "Ivory in world history—early modern trade in context." *History Compass* 8 (2010): 530–42.

Christensen-Dalsgaard, Jakob, and Catherine E. Carr. "Evolution of a sensory novelty: tympanic ears and the associated neural processing." *Brain Research Bulletin* 75 (2008): 365–70.

Clack, Jennifer A. "Patterns and processes in the early evolution of the tetrapod ear." *Journal of Neurobiology* 53 (2002): 251–64.

Conniff, Richard. "When the music in our parlors brought death to darkest Africa." *Audubon* 89 (1987): 77–92.

Currie, Adrian, and Anton Killin. "Not music, but musics: a case for conceptual pluralism in aesthetics." *Estetika: Central European Journal of Aesthetics* 54 (2017).

Davies, Stephen. "On defining music." *Monist* 95 (2012): 535–55.

Dick, Alastair. "The earlier history of the shawm in India." *Galpin Society Journal* 37 (1984): 80–98.

Fuller, Trevon L., Thomas P. Narins, Janet Nackoney, Timothy C. Bonebrake, Paul Sesink Clee, Katy Morgan, Anthony Tróchez et al. "Assessing the impact of China's timber industry on Congo Basin land use change." *Area* 51 (2019): 340–49.

Godt, Irving. "Music: a practical definition." *Musical Times* 146 (2005): 83–88.

Gracyk, Theodore, and Andrew Kania, eds. *The Routledge Companion to Philosophy and Music*. London: Taylor & Francis Group, 2011.

Hansen, Matthew C., Peter V. Potapov, Rebecca Moore, Matt Hancher, Svetlana A. Turubanova, Alexandra Tyukavina, David Thau et al. "High-resolution global maps of 21st-century forest cover change." *Science* 342 (2013): 850–53.

Jenkins, Martin, Sara Oldfield, and Tiffany Aylett. *International Trade in African Blackwood*. Cambridge, UK: Fauna & Flora International, 2002.

Kania, Andrew. "The Philosophy of Music." In *Stanford Encyclopedia of Philosophy* (Fall 2017 Edition), edited by Edward N. Zalta. Accessed October 16, 2020. https://plato.stanford.edu/archives/fall2017/entries/music/.

Levinson, Jerrold. *Music, Art, and Metaphysics*. Oxford, UK: Oxford University Press, 2011.

Luo, Zhe-Xi. "Developmental patterns in Mesozoic evolution of mammal ears." *Annual Review of Ecology, Evolution, and Systematics* 42 (2011): 355–80.

Mao, Fangyuan, Yaoming Hu, Chuankui Li, Yuanqing Wang, Morgan Hill Chase, Andrew K. Smith, and Jin Meng. "Integrated hearing and chewing modules decoupled in a Cretaceous stem therian mammal." *Science* 367 (2020): 305–8.

Mhatre, Natasha, Robert Malkin, Rittik Deb, Rohini Balakrishnan, and Daniel Robert. "Tree crickets optimize the acoustics of baffles to exaggerate their mate-attraction signal." *eLife* 6 (2017): e32763.

Mpingo Conservation and Development Initiative. Accessed October 12, 2020. http://www .mpingoconservation.org/.

New York Philharmonic. "Program Notes." (2019), January 26, 2019.

New York Philharmonic. "Sheryl Staples on Her Instrument." March 30, 2011. https://www .youtube.com/watch?v=UuWlIa27Fuo.

Nieder, Andreas, Lysann Wagener, and Paul Rinnert. "A neural correlate of sensory consciousness in a corvid bird." *Science* 369 (2020): 1626–29.

Page, Janet K., Geoffrey Burgess, Bruce Haynes, and Michael Finkelman. "Oboe." *Grove Music Online.* 2001. https://www.oxfordmusiconline.com/.

Spatz, H. Ch., H. Beismann, F. Brüchert, A. Emanns, and Th. Speck. "Biomechanics of the giant reed *Arundo donax*." *Philosophical Transactions of the Royal Society of London.* Series B 352 (1997): 1–10.

Thrasher, Alan R. "Sheng." *Grove Music Online.* 2001. https://www.oxfordmusiconline.com/.

Tucker, Abigail S. "Major evolutionary transitions and innovations: the tympanic middle ear." *Philosophical Transactions of the Royal Society B* 372 (2017): 20150483.

United Nations Office on Drugs and Crime. "World Wildlife Crime Report: trafficking in protected species." (2016) Vienna, Austria.

United States Environmental Protection Agency. "Durable goods: product-specific data." Accessed November 12, 2020. https://www.epa.gov/facts-and-figures-about-materials-waste-and-recycling /durable-goods-product-specific-data.

Urban, Daniel J., Neal Anthwal, Zhe-Xi Luo, Jennifer A. Maier, Alexa Sadier, Abigail S. Tucker, and Karen E. Sears. "A new developmental mechanism for the separation of the mammalian middle ear ossicles from the jaw." *Proceedings of the Royal Society B* 284 (2017): 20162416.

Wang, Haibing, Jin Meng, and Yuanqing Wang. "Cretaceous fossil reveals a new pattern in mammalian middle ear evolution." *Nature* 576 (2019): 102–5.

Wegst, Ulrike GK. "Wood for sound." *American Journal of Botany* 93 (2006): 1439–48.

Williams, Keith. "How Lincoln Center Was Built (It Wasn't Pretty)." *New York Times*, December 21, 2017.

World Wildlife Fund. "Timber: Overview." Accessed November 12, 2020. https://www.worldwildlife .org/industries/timber.

Zhu, Annah Lake. "China's rosewood boom: a cultural fix to capital overaccumulation." *Annals of the American Association of Geographers* 110 (2020): 277–96.

Forests

Aliansi Masyarakat Adat Nusantara et al. "Request for consideration of the Situation of Indigenous Peoples in Kalimantan, Indonesia, under the Committee of the Elimination of Racial Discrimination's Urgent Action and Early Warning Procedure." July 2020. https://www .forestpeoples.org/sites/default/files/documents/Early%20Warning%20Urgent%20Action %20Procedure%20CERD%20submission%20Indonesia.pdf.

Astaras, Christos, Joshua M. Linder, Peter Wrege, Robinson Orume, Paul J. Johnson, and David W. Macdonald. "Boots on the ground: the role of passive acoustic monitoring in evaluating anti-poaching patrols." *Environmental Conservation* (2020): 1–4.

Austin, Peter K., and Julia Sallabank, eds. *The Cambridge Handbook of Endangered Languages.* Cambridge, UK: Cambridge University Press, 2011.

Bengtsson, J., J. M. Bullock, B. Egoh, C. Everson, T. Everson, T. O'Connor, P. J. O'Farrell, H. G. Smith, and Regina Lindborg. "Grasslands—more important for ecosystem services than you might think." *Ecosphere* 10 (2019): e02582.

Berry, Nicholas J., Oliver L. Phillips, Simon L. Lewis, Jane K. Hill, David P. Edwards, Noel B. Tawatao, Norhayati Ahmad et al. "The high value of logged tropical forests: lessons from northern Borneo." *Biodiversity and Conservation* 19 (2010): 985–97.

Blackman, Allen, Leonardo Corral, Eirivelthon Santos Lima, and Gregory P. Asner. "Titling indigenous communities protects forests in the Peruvian Amazon." *Proceedings of the National Academy of Sciences* 114 (2017): 4123–28.

Brandt, Jodi S., and Ralf C. Buckley. "A global systematic review of empirical evidence of ecotourism impacts on forests in biodiversity hotspots." *Current Opinion in Environmental Sustainability* 32 (2018): 112–18.

Browning, Ella, Rory Gibb, Paul Glover-Kapfer, and Kate E. Jones. "Passive acoustic monitoring in ecology and conservation." (2017). *WWF Conservation Technology Series* 1(2). WWF-UK, Woking, UK.

Burivalova, Zuzana, Edward T. Game, Bambang Wahyudi, Mohamad Rifqi, Ewan MacDonald, Samuel Cushman, Maria Voigt, Serge Wich, and David S. Wilcove. "Does biodiversity benefit when the logging stops? An analysis of conservation risks and opportunities in active versus inactive logging concessions in Borneo." *Biological Conservation* 241 (2020): 108369.

Burivalova, Zuzana, Michael Towsey, Tim Boucher, Anthony Truskinger, Cosmas Apelis, Paul Roe, and Edward T. Game. "Using soundscapes to detect variable degrees of human influence on tropical forests in Papua New Guinea." *Conservation Biology* 32 (2018): 205–15.

Burivalova, Zuzana, Bambang Wahyudi, Timothy M. Boucher, Peter Ellis, Anthony Truskinger, Michael Towsey, Paul Roe, Delon Marthinus, Bronson Griscom, and Edward T. Game. "Using soundscapes to investigate homogenization of tropical forest diversity in selectively logged forests." *Journal of Applied Ecology* 56 (2019): 2493–504.

Caiger, Paul E., Micah J. Dean, Annamaria I. DeAngelis, Leila T. Hatch, Aaron N. Rice, Jenni A. Stanley, Chris Tholke, Douglas R. Zemeckis, and Sofie M. Van Parijs. "A decade of monitoring Atlantic cod Gadus morhua spawning aggregations in Massachusetts Bay using passive acoustics." *Marine Ecology Progress Series* 635 (2020): 89–103.

Casanova, Vanessa, and Josh McDaniel. "'No sobra y no falta': recruitment networks and guest workers in southeastern US forest industries." *Urban Anthropology and Studies of Cultural Systems and World Economic Development* (2005): 45–84.

Deichmann, Jessica L., Orlando Acevedo-Charry, Leah Barclay, Zuzana Burivalova, Marconi Campos-Cerqueira, Fernando d'Horta, Edward T. Game et al. "It's time to listen: there is much to be learned from the sounds of tropical ecosystems." *Biotropica* 50 (2018): 713–18.

de Oliveira, Gabriel, Jing M. Chen, Scott C. Stark, Erika Berenguer, Paulo Moutinho, Paulo Artaxo, Liana O. Anderson, and Luiz EOC Aragão. "Smoke pollution's impacts in Amazonia." *Science* 369 (2020): 634–35.

Ecosounds. "TNC—Indonesia, East Kalimantan Province." Accessed July 1–August 31, 2020. https://www.ecosounds.org/.

Edwards, David P., Jenny A. Hodgson, Keith C. Hamer, Simon L. Mitchell, Abdul H. Ahmad, Stephen J. Cornell, and David S. Wilcove. "Wildlife-friendly oil palm plantations fail to protect biodiversity effectively." *Conservation Letters* 3 (2010): 236–42.

Edwards, Felicity A., David P. Edwards, Trond H. Larsen, Wayne W. Hsu, Suzan Benedick, Arthur Chung, C. Vun Khen, David S. Wilcove, and Keith C. Hamer. "Does logging and forest conversion to oil palm agriculture alter functional diversity in a biodiversity hotspot?" *Animal Conservation* 17 (2014): 163–73.

Erb, W. M., E. J. Barrow, A. N. Hofner, S. S. Utami-Atmoko, and E. R. Vogel. "Wildfire smoke impacts activity and energetics of wild Bornean orangutans." *Scientific Reports* 8 (2018): 1–8.

Evans, Jonathan P., Kristen K. Cecala, Brett R. Scheffers, Callie A. Oldfield, Nicholas A. Hollingshead, David G. Haskell, and Benjamin A. McKenzie. "Widespread degradation of a vernal pool network in the southeastern United States: challenges to current and future management." *Wetlands* 37 (2017): 1093–1103.

FAO and FILAC. 2021. Forest Governance by Indigenous and Tribal People. An Opportunity for Climate Action in Latin America and the Caribbean. Santiago. https://doi.org/10.4060/cb2953en.

Game, Edward. "The encroaching silence." *Griffith Review* online (2019). https://griffithreview.atavist.com/the-encroaching-silence.

Global Forest Watch. "Indonesia: Land Cover." Accessed August 11, 2020. https://www.globalforestwatch.org/.

Global Forest Watch. "We lost a football pitch of primary rainforest every 6 seconds in 2019." June 2, 2020. https://blog.globalforestwatch.org/data-and-research/global-tree-cover-loss-data-2019.

Global Witness. "Defending Tomorrow." July 2020. https://www.globalwitness.org/documents/19939/Defending_Tomorrow_EN_low_res_-_July_2020.pdf.

Gorenflo, Larry J., Suzanne Romaine, Russell A. Mittermeier, and Kristen Walker-Painemilla. "Co-occurrence of linguistic and biological diversity in biodiversity hotspots and high biodiversity wilderness areas." *Proceedings of the National Academy of Sciences* 109 (2012): 8032–37.

Haskell, David G. "Listening to the Thoughts of the Forest." *Undark* (2017). https://undark.org/2017 /05/07/listening-to-the-thoughts-of-the-forest/.

Haskell, David G., Jonathan P. Evans, and Neil W. Pelkey. "Depauperate avifauna in plantations compared to forests and exurban areas." *PLOS One* 1 (2006): e63.

Hewitt, Gwen, Ann MacLarnon, and Kate E. Jones. "The functions of laryngeal air sacs in primates: a new hypothesis." *Folia Primatologica* 73 (2002): 70–94.

Hill, Andrew P., Peter Prince, Jake L. Snaddon, C. Patrick Doncaster, and Alex Rogers. "AudioMoth: A low-cost acoustic device for monitoring biodiversity and the environment." *HardwareX* 6 (2019): e00073.

Holland, Margaret B., Free De Koning, Manuel Morales, Lisa Naughton-Treves, Brian E. Robinson, and Luis Suárez. "Complex tenure and deforestation: implications for conservation incentives in the Ecuadorian Amazon." *World Development* 55 (2014): 21–36.

Junior, Celso H. L. Silva, Ana CM Pessôa, Nathália S. Carvalho, João BC Reis, Liana O. Anderson, and Luiz E. O. C. Aragão. "The Brazilian Amazon deforestation rate in 2020 is the greatest of the decade." *Nature Ecology & Evolution* 5 (2021): 144–45.

Konopik, Oliver, Ingolf Steffan-Dewenter, and T. Ulmar Grafe. "Effects of logging and oil palm expansion on stream frog communities on Borneo, Southeast Asia." *Biotropica* 47 (2015): 636–43.

Krausmann, Fridolin, Karl-Heinz Erb, Simone Gingrich, Helmut Haberl, Alberte Bondeau, Veronika Gaube, Christian Lauk, Christoph Plutzar, and Timothy D. Searchinger. "Global human appropriation of net primary production doubled in the 20th century." *Proceedings of the National Academy of Sciences* 110 (2013): 10324–29.

Loh, Jonathan, and David Harmon. *Biocultural Diversity: Threatened Species, Endangered Languages.* Zeist, The Netherlands: WWF Netherlands, 2014.

Lohberger, Sandra, Matthias Stängel, Elizabeth C. Atwood, and Florian Siegert. "Spatial evaluation of Indonesia's 2015 fire-affected area and estimated carbon emissions using Sentinel-1." *Global Change Biology* 24 (2018): 644–54.

McDaniel, Josh, and Vanessa Casanova. "Pines in lines: tree planting, H2B guest workers, and rural poverty in Alabama." *Journal of Rural Social Sciences* 19 (2003): 4.

McGrath, Deborah A., Jonathan P. Evans, C. Ken Smith, David G. Haskell, Neil W. Pelkey, Robert R. Gottfried, Charles D. Brockett, Matthew D. Lane, and E. Douglass Williams. "Mapping land-use change and monitoring the impacts of hardwood-to-pine conversion on the Southern Cumberland Plateau in Tennessee." *Earth Interactions* 8 (2004): 1–24.

Mikusiński, Grzegorz, Jakub Witold Bubnicki, Marcin Churski, Dorota Czeszczewik, Wiesław Walankiewicz, and Dries PJ Kuijper. "Is the impact of loggings in the last primeval lowland forest in Europe underestimated? The conservation issues of Białowieża Forest." *Biological Conservation* 227 (2018): 266–74.

National Indigenous Mobilization Network. "Statement in condemnation of draft Law nº 191/20, on the exploration of natural resources on indigenous lands." February 12, 2020. http://apib.info/2020/02 /12/statement-in-condemnation-of-draft-law-no-19120-on-the-exploration-of-natural-resources -on-indigenous-lands/?lang=en.

Natural Resources Defense Council. "NRDC Announces Annual BioGems List of 12 Most Threatened Wildlands in the Americas" (2004). https://www.nrdc.org/media/2004/040226.

Normile, Dennis. "Parched peatlands fuel Indonesia's blazes." *Science* 366 (2019): 18–19.

Oldekop, Johan A., Katharine R. E. Sims, Birendra K. Karna, Mark J. Whittingham, and Arun Agrawal. "Reductions in deforestation and poverty from decentralized forest management in Nepal." *Nature Sustainability* 2 (2019): 421–28.

Open Space Institute. "Protecting the Plateau before it's too late." https://www.openspaceinstitute.org /places/cumberland-plateau.

Scriven, Sarah A., Graeme R. Gillespie, Samsir Laimun, and Benoît Goossens. "Edge effects of oil palm plantations on tropical anuran communities in Borneo." *Biological Conservation* 220 (2018): 37–49.

Sethi, Sarab S., Nick S. Jones, Ben D. Fulcher, Lorenzo Picinali, Dena Jane Clink, Holger Klinck, C. David L. Orme, Peter H. Wrege, and Robert M. Ewers. "Characterizing soundscapes across diverse ecosystems using a universal acoustic feature set." *Proceedings of the National Academy of Sciences* 117 (2020): 17049–55.

Song, Xiao-Peng, Matthew C. Hansen, Stephen V. Stehman, Peter V. Potapov, Alexandra Tyukavina, Eric F. Vermote, and John R. Townshend. "Global land change from 1982 to 2016." *Nature* 560 (2018): 639–43.

Turkewitz, Julie, and Sofía Villamil. "Indigenous Colombians, Facing New Wave of Brutality, Demand Government Action." *New York Times*, October 24, 2020.

V (formerly Eve Ensler). "'The Amazon Is the Entry Door of the World': Why Brazil's Biodiversity Crisis Affects Us All." *Guardian*, August 10, 2020.

Weisse, Mikaela, and Elizabeth Dow Goldman. "We lost a football pitch of primary rainforest every 6 seconds in 2019." *Global Forest Watch*, June 2, 2020. https://blog.globalforestwatch.org/data-and -research/global-tree-cover-loss-data-2019.

Weisse, Mikaela, and Elizabeth Dow Goldman. "The world lost a Belgium-sized area of primary rainforests last year." *World Resources Institute* (2019). https://www.wri.org/blog/2019/04 /world-lost-belgium-sized-area-primary-rainforests-last-year.

Welz, Adam. "Listening to nature: the emerging field of bioacoustics." *Yale Environment 360*, November 5, 2019.

Wiggins, Elizabeth B., Claudia I. Czimczik, Guaciara M. Santos, Yang Chen, Xiaomei Xu, Sandra R. Holden, James T. Randerson, Charles F. Harvey, Fuu Ming Kai, and E. Yu Liya. "Smoke radiocarbon measurements from Indonesian fires provide evidence for burning of millennia-aged peat." *Proceedings of the National Academy of Sciences* 115 (2018): 12419–24.

Wihardandi, Aji. "Dayak Wehea: Kisah Keharmonisan Alam dan Manusia." Mongabay, Indonesia, April 16, 2012. https://www.mongabay.co.id/2012/04/16/dayak-wehea-kisah-keharmonisan -alam-dan-manusia/.

Wijaya, Arief, Tjokorda N. Samadhi, and Reidinar Juliane. "Indonesia is reducing deforestation, but problem areas remain." *World Resources Institute*, July 24, 2019. https://www.wri.org/blog/2019 /07/indonesia-reducing-deforestation-problem-areas-remain.

Yovanda, "Jalan Panjang Hutan Lindung Wehea, Dihantui Pembalakan dan Dikepung Sawit (Bagian 1)." *Mongabay Indonesia*, April 18, 2017. https://www.mongabay.co.id/2017/04/18/jalan-panjang -hutan-lindung-wehea-dihantui-pembalakan-dan-dikepung-sawit/.

Oceans

Andrew, Rex K., Bruce M. Howe, and James A. Mercer. "Long-time trends in ship traffic noise for four sites off the North American West Coast." *Journal of the Acoustical Society of America* 129 (2011): 642–51.

Bernaldo de Quirós, Y., A. Fernandez, R. W. Baird, R. L. Brownell Jr, N. Aguilar de Soto, D. Allen, M. Arbelo et al. "Advances in research on the impacts of anti-submarine sonar on beaked whales." *Proceedings of the Royal Society B* 286 (2019): 20182533.

Best, Peter B. "Increase rates in severely depleted stocks of baleen whales." *ICES Journal of Marine Science* 50 (1993): 169–86.

Branch, Trevor A., Koji Matsuoka, and Tomio Miyashita. "Evidence for increases in Antarctic blue whales based on Bayesian modelling." *Marine Mammal Science* 20 (2004): 726–54.

Brody, Jane. "Scientist at Work: Katy Payne; Picking Up Mammals' Deep Notes." *New York Times*, November 9, 1993.

Buckman, Andrea H., Nik Veldhoen, Graeme Ellis, John K. B. Ford, Caren C. Helbing, and Peter S. Ross. "PCB-associated changes in mRNA expression in killer whales (*Orcinus orca*) from the NE Pacific Ocean." *Environmental Science & Technology* 45 (2011): 10194–202.

Carrigg, David. "Port of Vancouver Hopes Feds Back $2-Billion Expansion Project to Help COVID-19 Recovery." *Vancouver Sun*, April 16, 2020.

Commander, United States Pacific Fleet. "Request for regulations and letters of authorization for the incidental taking of marine mammals resulting from U.S. Navy training and testing activities in the Northwest training and testing study area." December 19, 2019. https://media.fisheries.noaa.gov /dam-migration/navyhstt_2020finalloa_app_oprl_508.pdf.

Cox, Kieran, Lawrence P. Brennan, Travis G. Gerwing, Sarah E. Dudas, and Francis Juanes. "Sound the alarm: a meta-analysis on the effect of aquatic noise on fish behavior and physiology." *Global Change Biology* 24 (2018): 3105–16.

Day, Ryan D., Robert D. McCauley, Quinn P. Fitzgibbon, Klaas Hartmann, and Jayson M. Semmens. "Seismic air guns damage rock lobster mechanosensory organs and impair righting reflex." *Proceedings of the Royal Society B* 286 (2019): 20191424.

Desforges, Jean-Pierre, Ailsa Hall, Bernie McConnell, Aqqalu Rosing-Asvid, Jonathan L. Barber, Andrew Brownlow, Sylvain De Guise et al. "Predicting global killer whale population collapse from PCB pollution." *Science* 361 (2018): 1373–76.

Duncan, Alec J., Linda S. Weilgart, Russell Leaper, Michael Jasny, and Sharon Livermore. "A modelling comparison between received sound levels produced by a marine vibroseis array and those from an airgun array for some typical seismic survey scenarios." *Marine Pollution Bulletin* 119 (2017): 277–88.

Ebdon, Philippa, Leena Riekkola, and Rochelle Constantine. "Testing the efficacy of ship strike mitigation for whales in the Hauraki Gulf, New Zealand." *Ocean & Coastal Management* 184 (2020): 105034.

Erbe, Christine, Sarah A. Marley, Renée P. Schoeman, Joshua N. Smith, Leah E. Trigg, and Clare Beth Embling. "The effects of ship noise on marine mammals—a review." *Frontiers in Marine Science* 6 (2019): 606.

Erisman, Brad E., and Timothy J. Rowell. "A sound worth saving: acoustic characteristics of a massive fish spawning aggregation." *Biology Letters* 13 (2017): 20170656.

Fish, Marie Poland. "Animal sounds in the sea." *Scientific American* 194 (1956): 93–104.

Foote, Andrew D., Michael D. Martin, Marie Louis, George Pacheco, Kelly M. Robertson, Mikkel-Holger S. Sinding, Ana R. Amaral et al. "Killer whale genomes reveal a complex history of recurrent admixture and vicariance." *Molecular Ecology* 28 (2019): 3427–44.

Ford, John K. B. "Vocal traditions among resident killer whales (*Orcinus orca*) in coastal waters of British Columbia." *Canadian Journal of Zoology* (1991): 1454–83.

Ford, John K. B. "Killer Whale: *Orcinus orca*." In *Encyclopedia of Marine Mammals* (3rd ed.), edited by Bernd Würsig, J.G.M. Thewissen, and Kit M. Kovacs, 531–37. London: Academic Press, 2017.

Ford, John K. B. "Dialects." In *Encyclopedia of Marine Mammals* (3rd ed.), edited by Bernd Würsig, J.G.M. Thewissen, and Kit M. Kovacs, 253–54. London: Academic Press, 2018.

"Francis W. Watlington; Recorded Whale Songs." Obituary, *New York Times*, November 24, 1982.

Friends of the San Juans. "Salish Sea vessel traffic projections." Accessed August 28, 2020. https://sanjuans.org/wp-content/uploads/2019/07/SalishSea_VesselTrafficProjections_July27_2019.pdf.

George, Rose. *Ninety Percent of Everything*. New York: Macmillan, 2013.

Giggs, Rebecca. *The World in the Whale*. New York: Simon & Schuster, 2020.

Goldfarb, Ben. "Biologist Marie Fish catalogued the sounds of the ocean for the world to hear." *Smithsonian*, April 2021.

Hildebrand, John A. "Anthropogenic and natural sources of ambient noise in the ocean." *Marine Ecology Progress Series* 395 (2009): 5–20.

International Association of Geophysical Contractors. "Putting Seismic Surveys in Context." Accessed September 1, 2020. https://iagc.org/.

International Association of Geophysical Contractors. "The time is now." Accessed September 1, 2020. http://modernizemmpa.com/.

International Maritime Organization. "Guidelines for the reduction of underwater noise from commercial shipping to address adverse impacts on marine life." MEPC.1/Circ.833 (2014) London, UK.

Jang, Brent. "B.C. Loses Another LNG Project as Woodside Petroleum Axes Grassy Point." *Globe and Mail: Energy and Resources*, March 6, 2018.

Jones, Nicola. "Ocean uproar: saving marine life from a barrage of noise." *Nature* 568 (2019): 158–61.

Kaplan, Maxwell B., and Susan Solomon. "A coming boom in commercial shipping? The potential for rapid growth of noise from commercial ships by 2030." *Marine Policy* 73 (2016): 119–21.

Kavanagh, A. S., M. Nykänen, W. Hunt, N. Richardson, and M. J. Jessopp. "Seismic surveys reduce cetacean sightings across a large marine ecosystem." *Scientific Reports* 9 (2019): 1–10.

Keen, Eric M., Éadin O'Mahony, Chenoah Shine, Erin Falcone, Janie Wray, and Hussein Alidina. "Response to the COSEWIC (2019) reassessment of Pacific Canada fin whales (*Balaenoptera physalus*) to 'Special Concern.'" Manuscript in preparation (2020).

Keen, Eric M., Kylie L. Scales, Brenda K. Rone, Elliott L. Hazen, Erin A. Falcone, and Gregory S. Schorr. "Night and day: diel differences in ship strike risk for fin whales (*Balaenoptera physalus*) in the California current system." *Frontiers in Marine Science* 6 (2019): 730.

Ketten, Darlene R. "Cetacean Ears." In *Hearing by Whales and Dolphins*, edited by Whitlow W. L. Au, Arthur N. Popper, and Richard R. Fay, 43–108. New York: Springer, 2000.

Konrad, Christine M., Timothy R. Frasier, Luke Rendell, Hal Whitehead, and Shane Gero. "Kinship and association do not explain vocal repertoire variation among individual sperm whales or social units." *Animal Behaviour* 145 (2018): 131–40.

Lacy, Robert C., Rob Williams, Erin Ashe, Kenneth C. Balcomb III, Lauren JN Brent, Christopher W. Clark, Darren P. Croft, Deborah A. Giles, Misty MacDuffee, and Paul C. Paquet. "Evaluating anthropogenic threats to endangered killer whales to inform effective recovery plans." *Scientific Reports* 7 (2017): 1–12.

Leaper, R. C., and M. R. Renilson. "A review of practical methods for reducing underwater noise pollution from large commercial vessels." *Transactions of the Royal Institution of Naval Architects* 154, Part A2, *International Journal of Maritime Engineering*, Paper: T2012-2 Transactions (2012).

Leaper, Russell, Martin Renilson, and Conor Ryan. "Reducing underwater noise from large commercial ships: current status and future directions." *Journal of Ocean Technology* 9 (2014): 51–69.

Lotze, Heike K., and Boris Worm. "Historical baselines for large marine animals." *Trends in Ecology & Evolution* 24 (2009): 254–62.

MacGillivray, Alexander O., Zizheng Li, David E. Hannay, Krista B. Trounce, and Orla M. Robinson. "Slowing deep-sea commercial vessels reduces underwater radiated noise." *Journal of the Acoustical Society of America* 146 (2019): 340–51.

Mapes, Lynda V. "Washington state officials slam Navy's changes to military testing program that would harm more orcas." *Seattle Times*, July 29, 2020. https://www.seattletimes.com/seattle-news /environment/washington-state-officials-slam-navys-changes-to-military-testing-program-that -would-harm-more-orcas/.

Mapes, Lynda V., Steve Ringman, Ramon Dompor, and Emily M. Eng, "The Roar Below." *Seattle Times*, May 19, 2019. https://projects.seattletimes.com/2019/hostile-waters-orcas-noise/.

McBarnet, Andrew. "How the seismic map is changing." *Offshore Engineer* (2013). https://www .oedigital.com/news/459029-how-the-seismic-map-is-changing.

McCauley, Robert D., Ryan D. Day, Kerrie M. Swadling, Quinn P. Fitzgibbon, Reg A. Watson, and Jayson M. Semmens. "Widely used marine seismic survey air gun operations negatively impact zooplankton." *Nature Ecology & Evolution* 1 (2017): 0195.

McCoy, Kim, Beatrice Tomasi, and Giovanni Zappa. "JANUS: the genesis, propagation and use of an underwater standard." *Proceedings of Meetings on Acoustics* (2010).

McDonald, Mark A., John A. Hildebrand, and Sean M. Wiggins. "Increases in deep ocean ambient noise in the Northeast Pacific west of San Nicolas Island, California." *Journal of the Acoustical Society of America* 120 (2006): 711–18.

McKenna, Megan F., Donald Ross, Sean M. Wiggins, and John A. Hildebrand. "Underwater radiated noise from modern commercial ships." *Journal of the Acoustical Society of America* 131 (2012): 92–103.

Merchant, Nathan D. "Underwater noise abatement: economic factors and policy options." *Environmental Science & Policy* 92 (2019): 116–23.

Mitson R. B., ed. *Underwater Noise of Research Vessels: Review and Recommendations*, 1995 ICES Cooperative Research Report 209. Copenhagen, Denmark: International Council for the Exploration of the Sea, 1995.

National Marine Fisheries Service. "Puget Sound Salmon Recovery Plan, I (2007). https://repository .library.noaa.gov/view/noaa/16005.

National Marine Fisheries Service. Southern Resident Killer Whales (*Orcinus orca*) 5-Year Review: Summary and Evaluation. (National Marine Fisheries Service West Coast Region, Seattle, 2016) http://www.westcoast.fisheries.noaa.gov/publications/status_reviews/marine_mammals /kw-review-2016.pdf.

NATO. "A new era of digital underwater communications." April 20, 2017. https://www.nato.int/cps /bu/natohq/news_143247.htm.

Nieukirk, Sharon L., David K. Mellinger, Sue E. Moore, Karolin Klinck, Robert P. Dziak, and Jean Goslin. "Sounds from airguns and fin whales recorded in the mid-Atlantic Ocean, 1999–2009." *Journal of the Acoustical Society of America* 131 (2012): 1102–12.

Nowacek, Douglas P., Christopher W. Clark, David Mann, Patrick J. O. Miller, Howard C. Rosenbaum, Jay S. Golden, Michael Jasny, James Kraska, and Brandon L. Southall. "Marine seismic surveys and ocean noise: time for coordinated and prudent planning." *Frontiers in Ecology and the Environment* 13 (2015): 378–86.

Odell, J., D. H. Adams, B. Boutin, W. Collier II, A. Deary, L. N. Havel, J. A. Johnson Jr. et al. "Atlantic Sciaenid Habitats: A Review of Utilization, Threats, and Recommendations for Conservation, Management, and Research." Atlantic States Marine Fisheries Commission Habitat Management Series No. 14 (2017), Arlington, VA.

Ogden, Lesley Evans. "Quieting marine seismic surveys." *BioScience* 64 (2014): 752.

Owen, Brenda, and Alastair Spriggs "For Coast Salish communities, the race to save southern resident orcas is personal." *Canada's National Observer*, September. 17, 2019.

Parsons, Miles J. G., Chandra P. Salgado Kent, Angela Recalde-Salas, and Robert D. McCauley. "Fish choruses off Port Hedland, Western Australia." *Bioacoustics* 26 (2017): 135–52.

Payne, Roger. *Songs of the Humpback Whale*. Vinyl music album. CRM Records, 1970.

Port of Vancouver. "Centerm Expansion Project and South Shore Access Project." Accessed August 28, 2020. https://www.portvancouver.com/projects/terminal-and-facilities/centerm/.

Port of Vancouver. "2020 voluntary vessel slowdown: Haro Strait and Boundary Pass." Accessed August 28, 2020. https://www.portvancouver.com/wp-content/uploads/2020/05/2020-05-15-ECHO -Program-slowdown-fact-sheet.pdf.

Rocha, Robert C., Phillip J. Clapham, and Yulia V. Ivashchenko. "Emptying the oceans: a summary of industrial whaling catches in the 20th century." *Marine Fisheries Review* 76 (2014): 37–48.

Rolland, Rosalind M., Susan E. Parks, Kathleen E. Hunt, Manuel Castellote, Peter J. Corkeron, Douglas P. Nowacek, Samuel K. Wasser, and Scott D. Kraus. "Evidence that ship noise increases stress in right whales." *Proceedings of the Royal Society B* 279 (2012): 2363–68.

Ryan, John. "Washington tribes and Inslee alarmed by Canadian pipeline approval." *KUOW,* June 19, 2019. https://www.kuow.org/stories/washington-tribes-and-inslee-alarmed-by-canadian-pipeline -approval.

Schiffman, Richard. "How ocean noise pollution wreaks havoc on marine life." *Yale Environment 360,* March 31, 2016.

Seely, Elizabeth, Richard W. Osborne, Kari Koski, and Shawn Larson. "Soundwatch: eighteen years of monitoring whale watch vessel activities in the Salish Sea." *PLOS One* 12 (2017): e0189764.

Slabbekoorn, Hans, John Dalen, Dick de Haan, Hendrik V. Winter, Craig Radford, Michael A. Ainslie, Kevin D. Heaney, Tobias van Kooten, Len Thomas, and John Harwood. "Population-level consequences of seismic surveys on fishes: an interdisciplinary challenge." *Fish and Fisheries* 20 (2019): 653–85.

Solan, Martin, Chris Hauton, Jasmin A. Godbold, Christina L. Wood, Timothy G. Leighton, and Paul White. "Anthropogenic sources of underwater sound can modify how sediment-dwelling invertebrates mediate ecosystem properties." *Scientific Reports* 6 (2016): 20540.

Soundwatch Program Annual Contract Report, 2019: *Soundwatch Public Outreach/Boater Education Project.* The Whale Museum. https://cdn.shopify.com/s/files/1/0249/1083/files/2019 _Soundwatch_Program_Annual_Contract_Report.pdf.

Southall, Brandon L., Amy R. Scholik-Schlomer, Leila Hatch, Trisha Bergmann, Michael Jasny, Kathy Metcalf, Lindy Weilgart, and Andrew J. Wright. "Underwater Noise from Large Commercial Ships—International Collaboration for Noise Reduction." *Encyclopedia of Maritime and Offshore Engineering* (2017): 1–9.

Stanley, Jenni A., Sofie M. Van Parijs, and Leila T. Hatch. "Underwater sound from vessel traffic reduces the effective communication range in Atlantic cod and haddock." *Scientific Reports* 7 (2017): 1–12.

Susewind, Kelly, Laura Blackmore, Kaleen Cottingham, Hilary Franz, and Erik Neatherlin. "Comments submitted electronically Re: Taking Marine Mammals Incidental to the U.S. Navy Training and Testing Activities in the Northwest Training and Testing Study Area, NOAA-NMFS-2020-0055." July, 2020. https://www.documentcloud.org/documents/7002861-NMFS-7-16-20.html.

Thomsen, Frank, Dierk Franck, and John K. Ford. "On the communicative significance of whistles in wild killer whales (Orcinus orca)." *Naturwissenschaften* 89 (2002): 404–7.

United States Environmental Protection Agency. "Chinook Salmon." Accessed August 26, 2020. https:// www.epa.gov/salish-sea/chinook-salmon.

Veirs, Scott, Val Veirs, Rob Williams, Michael Jasny, and Jason Wood. "A key to quieter seas: half of ship noise comes from 15% of the fleet." *PeerJ Preprints* 6 (2018): e26525v1.

Veirs, Scott, Val Veirs, and Jason D. Wood. "Ship noise extends to frequencies used for echolocation by endangered killer whales." *PeerJ* 4 (2016): e1657.

Wilcock, William S. D., Kathleen M. Stafford, Rex K. Andrew, and Robert I. Odom. "Sounds in the ocean at 1–100 Hz." *Annual Review of Marine Science* 6 (2014): 117–40.

Wladichuk, Jennifer L., David E. Hannay, Alexander O. MacGillivray, Zizheng Li, and Sheila J. Thornton. "Systematic source level measurements of whale watching vessels and other small boats." *Journal of Ocean Technology* 14 (2019).

Cities

Ayers, B. Drummond. "White Roads Through Black Bedrooms." *New York Times*, December 31, 1967.

Basner, Mathias, Wolfgang Babisch, Adrian Davis, Mark Brink, Charlotte Clark, Sabine Janssen, and Stephen Stansfeld. "Auditory and non-auditory effects of noise on health." *Lancet* 383 (2014): 1325–32.

"Bird Organ." Object Record, Victoria and Albert Museum. May 16, 2001. http://collections.vam .ac.uk/item/O58971/bird-organ-boudin-leonard/.

Caro, Robert. *The Power Broker: Robert Moses and the Fall of New York*. New York: Knopf, 1974.

Casey, Joan A., Rachel Morello-Frosch, Daniel J. Mennitt, Kurt Fristrup, Elizabeth L. Ogburn, and Peter James. "Race/ethnicity, socioeconomic status, residential segregation, and spatial variation in noise exposure in the contiguous United States." *Environmental Health Perspectives* 125 (2017): 077017.

Census Reporter. New York. Accessed September 23, 2020. https://censusreporter.org/profiles /04000US36-new-york/.

Clark, Sierra N., Abosede S. Alli, Michael Brauer, Majid Ezzati, Jill Baumgartner, Mireille B. Toledano, Allison F. Hughes et al. "High-resolution spatiotemporal measurement of air and environmental noise pollution in Sub-Saharan African cities: Pathways to Equitable Health Cities Study protocol for Accra, Ghana." *BMJ open* 10 (2020): e035798.

Commissioners of Prospect Park. First Annual Report (1861). http://home2.nyc.gov/html/records /pdf/govpub/3985annual_report_brooklyn_prospect_park_comm_1861.pdf.

Costantini, David, Timothy J. Greives, Michaela Hau, and Jesko Partecke. "Does urban life change blood oxidative status in birds?" *Journal of Experimental Biology* 217 (2014): 2994–97.

Dalley, Stephanie, editor and translator. *Myths from Mesopotamia. Creation, the Flood, Gilgamesh, and Others*, rev. ed. Oxford, UK: Oxford University Press, 2000.

Derryberry, Elizabeth P., Raymond M. Danner, Julie E. Danner, Graham E. Derryberry, Jennifer N. Phillips, Sara E. Lipshutz, Katherine Gentry, and David A. Luther. "Patterns of song across natural and anthropogenic soundscapes suggest that white-crowned sparrows minimize acoustic masking and maximize signal content." *PLOS One* 11 (2016): e0154456.

Derryberry, Elizabeth P., Jennifer N. Phillips, Graham E. Derryberry, Michael J. Blum, and David Luther. "Singing in a silent spring: birds respond to a half-century soundscape reversion during the COVID-19 shutdown." *Science* 370 (2020): 575–79.

Doman, Mark. "Industry warns noise complaints could see Melbourne's music scene shift to Sydney." *ABC News*, January 4, 2014. https://www.abc.net.au/news/2014-01-05/noise-complaints -threatening-melbourne-live-music/5181126.

Dominoni, Davide M., Stefan Greif, Erwin Nemeth, and Henrik Brumm. "Airport noise predicts song timing of European birds." *Ecology and Evolution* 6 (2016): 6151–59.

Evans, Karl L., Kevin J. Gaston, Alain C. Frantz, Michelle Simeoni, Stuart P. Sharp, Andrew McGowan, Deborah A. Dawson et al. "Independent colonization of multiple urban centres by a formerly forest specialist bird species." *Proceedings of the Royal Society B* 276 (2009): 2403–10.

Evans, Karl L., Kevin J. Gaston, Stuart P. Sharp, Andrew McGowan, Michelle Simeoni, and Ben J. Hatchwell. "Effects of urbanisation on disease prevalence and age structure in blackbird *Turdus merula* populations." *Oikos* 118 (2009): 774–82.

Evans, Karl L., Ben J. Hatchwell, Mark Parnell, and Kevin J. Gaston. "A conceptual framework for the colonisation of urban areas: the blackbird *Turdus merula* as a case study." *Biological Reviews* 85 (2010): 643–67.

Fritsch, Clémentine, Łukasz Jankowiak, and Dariusz Wysocki. "Exposure to Pb impairs breeding success and is associated with longer lifespan in urban European blackbirds." *Scientific Reports* 9 (2019): 1–11.

Guse, Clayton. "Brooklyn's poorest residents get stuck with the MTA's oldest buses." *New York Daily Post*, March 17, 2019.

Halfwerk, Wouter, and Kees van Oers. "Anthropogenic noise impairs foraging for cryptic prey via cross-sensory interference." *Proceedings of the Royal Society B* 287 (2020): 20192951.

Haralabidis, Alexandros S., Konstantina Dimakopoulou, Federica Vigna-Taglianti, Matteo Giampaolo, Alessandro Borgini, Marie-Louise Dudley, Göran Pershagen et al. "Acute effects of night-time noise exposure on blood pressure in populations living near airports." *European Heart Journal* 29 (2008): 658–64.

Hart, Patrick J., Robert Hall, William Ray, Angela Beck, and James Zook. "Cicadas impact bird communication in a noisy tropical rainforest." *Behavioral Ecology* 26 (2015): 839–42.

Heinl, Robert D. "The woman who stopped noises: an account of the successful campaign against unnecessary din in New York City." *Ladies' Home Journal*. April 1908.

Hu, Winnie. "New York Is a Noisy City. One Man Got Revenge." *New York Times*, June 4, 2019

Ibáñez-Álamo, Juan Diego, Javier Pineda-Pampliega, Robert L. Thomson, José I. Aguirre, Alazne Díez-Fernández, Bruno Faivre, Jordi Figuerola, and Simon Verhulst. "Urban blackbirds have shorter telomeres." *Biology Letters* 14 (2018): 20180083.

Injaian, Allison S., Paulina L. Gonzalez-Gomez, Conor C. Taff, Alicia K. Bird, Alexis D. Ziur, Gail L. Patricelli, Mark F. Haussmann, and John C. Wingfield. "Traffic noise exposure alters nestling physiology and telomere attrition through direct, but not maternal, effects in a free-living bird." *General and Comparative Endocrinology* 276 (2019): 14–21.

Jackson, Kenneth, ed. *The Encyclopedia of New York City*, 2nd ed. New Haven, CT: Yale University Press, 2010.

Kunc, Hansjoerg P., and Rouven Schmidt. "The effects of anthropogenic noise on animals: a meta-analysis." *Biology Letters* 15 (2019): 20190649.

Lecocq, Thomas, Stephen P. Hicks, Koen Van Noten, Kasper van Wijk, Paula Koelemeijer, Raphael S. M. De Plaen, Frédérick Massin et al. "Global quieting of high-frequency seismic noise due to COVID-19 pandemic lockdown measures." *Science* 369 (2020): 1338–43.

Legewie, Joscha, and Merlin Schaeffer. "Contested boundaries: explaining where ethnoracial diversity provokes neighborhood conflict." *American Journal of Sociology* 122 (2016): 125–61.

"London market traders in gentrification row as Islington residents complain about street noise." *Telegraph*, October 9, 2016.

López-Barroso, Diana, Marco Catani, Pablo Ripollés, Flavio Dell'Acqua, Antoni Rodríguez-Fornells, and Ruth de Diego-Balaguer. "Word learning is mediated by the left arcuate fasciculus." *Proceedings of the National Academy of Sciences* 110 (2013): 13168–73.

Lühken, Renke, Hanna Jöst, Daniel Cadar, Stephanie Margarete Thomas, Stefan Bosch, Egbert Tannich, Norbert Becker, Ute Ziegler, Lars Lachmann, and Jonas Schmidt-Chanasit. "Distribution of Usutu virus in Germany and its effect on breeding bird populations." *Emerging Infectious Diseases* 23 (2017): 1994.

Luo, Jinhong, Steffen R. Hage, and Cynthia F. Moss. "The Lombard effect: from acoustics to neural mechanisms." *Trends in Neurosciences* 41 (2018): 938–49.

Luther, David, and Luis Baptista. "Urban noise and the cultural evolution of bird songs." *Proceedings of the Royal Society B* 277 (2010): 469–73.

Luther, David A., and Elizabeth P. Derryberry. "Birdsongs keep pace with city life: changes in song over time in an urban songbird affects communication." *Animal Behaviour* 83 (2012): 1059–66.

McDonnell, Evelyn. "It's Time for the Rock & Roll Hall of Fame to Address Its Gender and Racial Imbalances." *Billboard*, November 15, 2019.

Meillère, Alizée, François Brischoux, Paco Bustamante, Bruno Michaud, Charline Parenteau, Coline Marciau, and Frédéric Angelier. "Corticosterone levels in relation to trace element contamination along an urbanization gradient in the common blackbird (*Turdus merula*)." *Science of the Total Environment* 566 (2016): 93–101.

Meillère, Alizée, François Brischoux, Cécile Ribout, and Frédéric Angelier. "Traffic noise exposure affects telomere length in nestling house sparrows." *Biology Letters* 11 (2015): 20150559.

Miller, Vernice D. "Planning, power and politics: a case study of the land use and siting history of the North River Water Pollution Control Plant." *Fordham Urban Law Journal* 21 (1993): 707–22.

Miranda, Ana Catarina, Holger Schielzeth, Tanja Sonntag, and Jesko Partecke. "Urbanization and its effects on personality traits: a result of microevolution or phenotypic plasticity?" *Global Change Biology* 19 (2013): 2634–44.

Mohl, Raymond A. "The interstates and the cities: the US Department of Transportation and the freeway revolt, 1966–1973." *Journal of Policy History* 20 (2008): 193–226.

Møller, Anders Pape, Mario Diaz, Einar Flensted-Jensen, Tomas Grim, Juan Diego Ibáñez-Álamo, Jukka Jokimäki, Raivo Mänd, Gábor Markó, and Piotr Tryjanowski. "High urban population density of birds reflects their timing of urbanization." *Oecologia* 170 (2012): 867–75.

Møller, Anders Pape, Jukka Jokimäki, Piotr Skorka, and Piotr Tryjanowski. "Loss of migration and urbanization in birds: a case study of the blackbird (*Turdus merula*)." *Oecologia* 175 (2014): 1019–27.

Moseley, Dana L., Jennifer N. Phillips, Elizabeth P. Derryberry, and David A. Luther. "Evidence for differing trajectories of songs in urban and rural populations." *Behavioral Ecology* 30 (2019): 1734–42.

Moseley, Dana Lynn, Graham Earnest Derryberry, Jennifer Nicole Phillips, Julie Elizabeth Danner, Raymond Michael Danner, David Andrew Luther, and Elizabeth Perrault Derryberry. "Acoustic adaptation to city noise through vocal learning by a songbird." *Proceedings of the Royal Society B* 285 (2018): 20181356.

Müller, Jakob C., Jesko Partecke, Ben J. Hatchwell, Kevin J. Gaston, and Karl L. Evans. "Candidate gene polymorphisms for behavioural adaptations during urbanization in blackbirds." *Molecular Ecology* 22 (2013): 3629–37.

Neitzel, Richard, Robyn R. M. Gershon, Marina Zeltser, Allison Canton, and Muhammad Akram. "Noise levels associated with New York City's mass transit systems." *American Journal of Public Health* 99 (2009): 1393–99.

Nemeth, Erwin, and Henrik Brumm. "Birds and anthropogenic noise: are urban songs adaptive?" *American Naturalist* 176 (2010): 465–75.

Nemeth, Erwin, Nadia Pieretti, Sue Anne Zollinger, Nicole Geberzahn, Jesko Partecke, Ana Catarina Miranda, and Henrik Brumm. "Bird song and anthropogenic noise: vocal constraints may explain why birds sing higher-frequency songs in cities." *Proceedings of the Royal Society B* 280 (2013): 20122798.

New York City Department of Parks & Recreation. "Riverside Park," Accessed September 24, 2020. https://www.nycgovparks.org/parks/riverside-park/.

New York City Environmental Justice Alliance. "New York City Climate Justice Agenda. Midway to 2030." https://www.nyc-eja.org/wp-content/uploads/2018/04/NYC-Climate-Justice-Agenda -Final-042018-1.pdf.

New York State Joint Commission on Public Ethics 2019 Annual Report. https://jcope.ny.gov/system/files /documents/2020/07/2019_-annual-report-final-web-as-of-7_29_2020.pdf.

New York State Office of the State Comptroller. "Responsiveness to noise complaints related to construction projects" (2016). https://www.osc.state.ny.us/sites/default/files/audits/2018-02 /sga-2017-16n3.pdf.

New York Times. "Anti-noise Bill Passes Aldermen: Only 3 Vote Against Ordinance." April 22, 1936.

New York Times. "Mayor La Guardia's Plea and Proclamation in War on Noise." October 1, 1935.

Nir, Sarah Maslin, "Inside N.Y.C.'s insanely loud car culture." *New York Times*, October 16, 2020.

Oliveira, Maria Joao R., Mariana P. Monteiro, Andreia M. Ribeiro, Duarte Pignatelli, and Artur P. Aguas. "Chronic exposure of rats to occupational textile noise causes cytological changes in adrenal cortex." *Noise and Health* 11 (2009): 118.

Parekh, Trushna. "'They want to live in the Tremé, but they want it for their ways of living': gentrification and neighborhood practice in Tremé, New Orleans." *Urban Geography* 36 (2015): 201–20.

Park, Woon Ju, Kimberly B. Schauder, Ruyuan Zhang, Loisa Bennetto, and Duje Tadin. "High internal noise and poor external noise filtering characterize perception in autism spectrum disorder." *Scientific Reports* 7 (2017): 1–12.

Partecke, Jesko, Eberhard Gwinner, and Staffan Bensch. "Is urbanisation of European blackbirds (*Turdus merula*) associated with genetic differentiation?" *Journal of Ornithology* 147 (2006): 549–52.

Partecke, Jesko, Gergely Hegyi, Patrick S. Fitze, Julien Gasparini, and Hubert Schwabl. "Maternal effects and urbanization: variation of yolk androgens and immunoglobulin in city and forest blackbirds." *Ecology and Evolution* 10 (2020): 2213–24.

Partecke, Jesko, Thomas Van't Hof, and Eberhard Gwinner. "Differences in the timing of reproduction between urban and forest European blackbirds (*Turdus merula*): result of phenotypic flexibility or genetic differences?" *Proceedings of the Royal Society of London*. Series B 271 (2004): 1995–2001.

Partecke, Jesko, Thomas J. Van't Hof, and Eberhard Gwinner. "Underlying physiological control of reproduction in urban and forest-dwelling European blackbirds *Turdus merula*." *Journal of Avian Biology* 36 (2005): 295–305.

Peris, Eulalia et al. "Environmental noise in Europe—2020." (2020) European Environment Agency, Copenhagen, Denmark. https://www.eea.europa.eu/publications/environmental-noise-in-europe.

Phillips, Jennifer N., Catherine Rochefort, Sara Lipshutz, Graham E. Derryberry, David Luther, and Elizabeth P. Derryberry. "Increased attenuation and reverberation are associated with lower maximum frequencies and narrow bandwidth of bird songs in cities." *Journal of Ornithology* (2020): 1–16.

Powell, Michael. "A Tale of Two Cities." *New York Times*, May 6, 2006.

Ransom, Jan. "New Pedestrian Bridge Will Make Riverside Park More Accessible by 2016." *New York Daily News*, December 28, 2014.

Reichmuth, Johannes, and Peter Berster. "Past and Future Developments of the Global Air Traffic." In *Biokerosene*, edited by M. Kaltschmitt and U. Neuling, 13–31. Berlin: Springer, 2018.

Ripmeester, Erwin A. P., Jet S. Kok, Jacco C. van Rijssel, and Hans Slabbekoorn. "Habitat-related birdsong divergence: a multi-level study on the influence of territory density and ambient noise in European blackbirds." *Behavioral Ecology and Sociobiology* 64 (2010): 409–18.

Rosenthal, Brain M., Emma G. Fitzsimmons, and Michael LaForgia. "How Politics and Bad Decisions Starved New York's Subways." *New York Times,* November 18, 2017.

Saccavino, Elisabeth, Jan Krämer, Sebastian Klaus, and Dieter Thomas Tietze. "Does urbanization affect wing pointedness in the Blackbird *Turdus merula?*" *Journal of Ornithology* 159 (2018): 1043–51.

Saiz, Juan-Carlos, and Ana-Belén Blazquez. "Usutu virus: current knowledge and future perspectives." *Virus Adaptation and Treatment* 9 (2017): 27–40.

Schell, Christopher J., Karen Dyson, Tracy L. Fuentes, Simone Des Roches, Nyeema C. Harris, Danica Sterud Miller, Cleo A. Woelfle-Erskine, and Max R. Lambert. "The ecological and evolutionary consequences of systemic racism in urban environments." *Science* 369 (2020).

Schulze, Katrin, Faraneh Vargha-Khadem, and Mortimer Mishkin. "Test of a motor theory of long-term auditory memory." *Proceedings of the National Academy of Sciences* 109 (2012): 7121–25.

Science VS Podcast. "Gentrification: What's really happening?" https://gimletmedia.com/shows/science-vs/39hzkk.

Semuels, Alana. "The role of highways in American poverty." *Atlantic*, March 18, 2016.

Senzaki, Masayuki, Jesse R. Barber, Jennifer N. Phillips, Neil H. Carter, Caren B. Cooper, Mark A. Ditmer, Kurt M. Fristrup et al. "Sensory pollutants alter bird phenology and fitness across a continent." *Nature* 587 (2020): 605–09.

Shah, Ravi R., Jonathan J. Suen, Ilana P. Cellum, Jaclyn B. Spitzer, and Anil K. Lalwani. "The effect of brief subway station noise exposure on commuter hearing." *Laryngoscope Investigative Otolaryngology* 3 (2018): 486–91.

Shah, Ravi R., Jonathan J. Suen, Ilana P. Cellum, Jaclyn B. Spitzer, and Anil K. Lalwani. "The influence of subway station design on noise levels." *Laryngoscope* 127 (2017): 1169–74.

Shannon, Graeme, Megan F. McKenna, Lisa M. Angeloni, Kevin R. Crooks, Kurt M. Fristrup, Emma Brown, Katy A. Warner et al. "A synthesis of two decades of research documenting the effects of noise on wildlife." *Biological Reviews* 91 (2016): 982–1005.

Specter, Michael. "Harlem Groups File Suit to Fight Sewage Odors." *New York Times*, June 22, 1992.

Stremple, Paul. "Brooklyn's Oldest Bus Models Will Be Replaced by Year's End, Says MTA." *Brooklyn Daily Eagle*, March 19, 2019.

Stremple, Paul. "Lowest Income Communities Get Oldest Buses, Sparking Demand for Oversight." *Brooklyn Daily Eagle*, March 18, 2019.

Sze, Julie. *Noxious New York: The Racial Politics of Urban Health and Environmental Justice*. Cambridge, MA: MIT Press, 2007.

Union of Concerned Scientists. "Inequitable exposure to air pollution from vehicles in New York State" (2019). https://www.ucsusa.org/sites/default/files/attach/2019/06/Inequitable-Exposure-to-Vehicle-Pollution-NY.pdf.

Vienneau, Danielle, Christian Schindler, Laura Perez, Nicole Probst-Hensch, and Martin Röösli. "The relationship between transportation noise exposure and ischemic heart disease: a meta-analysis." *Environmental Research* 138 (2015): 372–80.

Vo, Lam Thuy. "They Played Dominoes Outside Their Apartment for Decades. Then the White People Moved In and Police Started Showing Up." *BuzzFeed News*, June 29, 2018. https://www .buzzfeednews.com/article/lamvo/gentrification-complaints-311-new-york.

Walton, Mary. "Elevated railway" (1881). US Patent 237,422.

WE ACT for Environmental Justice. "Mother Clara Hale Bus Depot." https://www.weact.org /campaigns/mother-clara-hale-bus-depot/.

WE ACT for Environmental Justice. "WE ACT calls for retrofitting of North River Waste Treatment Plant." December 15, 2015. https://www.weact.org/2015/12/we-act-calls-for-retrofitting-of -north-river-waste-treatment-plant/.

Zollinger, Sue Anne, and Henrik Brumm. "The Lombard effect." *Current Biology* 21 (2011): R614–R615.

Listening in Community

Barclay, Leah, Toby Gifford, and Simon Linke. "Interdisciplinary approaches to freshwater ecoacoustics." *Freshwater Science* 39 (2020): 356–61.

Bennett, Frank G, Jr. "Legal protection of solar access under Japanese law." UCLA Pac. Basin LJ 5 (1986): 107.

Cantaloupe Music. "Inuksuit by John Luther Adams [online liner notes]." Accessed December 11, 2020. https://cantaloupemusic.com/albums/inuksuit.

Grech, Alana, Laurence McCook, and Adam Smith. "Shipping in the Great Barrier Reef: the miners' highway." *Conversation* (2015). https://theconversation.com/shipping-in-the-great-barrier-reef-the -miners-highway-39251.

Krause, Bernie. *Wild Soundscapes*. Berkeley, CA: Wilderness Press, 2002.

"Ministry compiles list of nation's 100 best-smelling spots." *Japan Times,* October 31, 2001.

Neil, David T. "Cooperative fishing interactions between Aboriginal Australians and dolphins in Eastern Australia." *Anthrozoös* 15 (2002): 3–18.

New York Botanical Garden. "*Chorus of the Forest* by Angélica Negrón." Accessed December 11, 2020. https://www.nybg.org/event/fall-forest-weekends/chorus-of-the-forest/.

Robin, K. "One Hundred Sites of Good Fragrance." (2014) Now Smell This online. Accessed November 17, 2020. http://www.nstperfume.com/2014/04/01/one-hundred-sites-of-good-fragrance/.

Rothenberg, David. *Nightingales in Berlin. Searching for the Perfect Sound*. Chicago: University of Chicago Press, 2019.

Schafer, R. Murray. *The Soundscape: Our Sonic Environment and the Tuning of the World*. Rochester, VT: Destiny Books, 1977.

Soundscape Policy Study Group. "Report on the results of the "100 Soundscapes to Keep." Local Government Questionnaire (in Japanese) (2018). Accessed November 17, 2020. http://mino.eco .coocan.jp/wp/wp-content/uploads/2016/12/20180530report100soundscapesjapan.pdf.

Torigoe, Keiko. "Insights taken from three visited soundscapes in Japan." In *Acoustic Ecology*, Australian Forum for Acoustic Ecology/World Forum for Acoustic Ecology, Melbourne, Australia, 2003.

Torigoe, Keiko. "Recollection and report on incorporation." *Journal of the Soundscape Association of Japan* 20 (2020) 3–4.

Tyler, Royall, tr. *The Tale of the Heike*. New York: Penguin, 2012.

Listening in the Deep Past and Future

Bowman, Judd D., Alan E. E. Rogers, Raul A. Monsalve, Thomas J. Mozdzen, and Nivedita Mahesh. "An absorption profile centered at 78 megahertz in the sky-averaged spectrum." *Nature* 555 (2018): 67.

Eisenstein, Daniel J. 2005. "The acoustic peak primer." Harvard-Smithsonian Center for Astrophysics. Accessed July 31, 2018. https://www.cfa.harvard.edu/~deisenst/acousticpeak/spherical_acoustic.pdf.

Eisenstein, Daniel J. 2005. "Dark energy and cosmic sound." Harvard-Smithsonian Center for Astrophysics. Accessed July 31, 2018. https://www.cfa.harvard.edu/~deisenst/acousticpeak /acoustic.pdf.

Einstein, Daniel J. 2005. "What is the acoustic peak?" Harvard-Smithsonian Center for Astrophysics. Accessed July 31, 2018. https://www.cfa.harvard.edu/~deisenst/acousticpeak/acoustic_physics. html.

Eisenstein, Daniel J., and Charles L. Bennett. "Cosmic sound waves rule." *Physics Today* 61 (2008): 44–50.

Eisenstein, Daniel J., Idit Zehavi, David W. Hogg, Roman Scoccimarro, Michael R. Blanton, Robert C. Nichol, Ryan Scranton et al. "Detection of the baryon acoustic peak in the large-scale correlation function of SDSS luminous red galaxies." *Astrophysical Journal* 633 (2005): 560.

European Space Agency. "Planck science team home." Accessed July 26, 2018. https://www.cosmos .esa.int/web/planck/home.

Follin, Brent, Lloyd Knox, Marius Millea, and Zhen Pan. "First detection of the acoustic oscillation phase shift expected from the cosmic neutrino background." *Physical Review Letters* 115 (2015): 091301.

Gunn, James E., Walter A. Siegmund, Edward J. Mannery, Russell E. Owen, Charles L. Hull, R. French Leger, Larry N. Carey et al. "The 2.5 m telescope of the Sloan digital sky survey." *Astronomical Journal* 131 (2006): 2332.

Siegel, E. "Earliest evidence for stars smashes Hubble's record and points to dark matter." *Forbes*, February 28, 2018. https://www.forbes.com/sites/startswithabang/2018/02/28/earliest-evidence -for-stars-ever-seen-smashes-hubbles-record-and-points-to-dark-matter/#2c56afd01f92.

Siegel, Ethan. "Cosmic neutrinos detected, confirming the big bang's last great prediction." *Forbes*, September 9, 2016. https://www.forbes.com/sites/startswithabang/2016/09/09/cosmic-neutrinos -detected-confirming-the-big-bangs-last-great-prediction/.

INDEX

The Songs of Trees

Stories from Nature's Great Connectors

David George Haskell brings his powers of observation to the biological networks that surround all species. Haskell visits a dozen trees, exploring connections with people, microbes, fungi, and other plants and animals. He takes us to trees in cities, forests, and areas on the front lines of environmental change. In each place he shows how human history, ecology, and well-being are intertwined with the lives of trees. Scientific, lyrical, and contemplative, Haskell reveals the biological connections that underpin all life.

"A love song to trees, an exploration of their biology, and a wonderfully philosophical analysis of [the] role they play in human history and in modern culture." —*Science Friday*

 PENGUIN BOOKS

Ready to find your next great read? Let us help. Visit prh.com/nextread

The Forest Unseen

A Year's Watch in Nature

David George Haskell uses a one-square-meter patch of old-growth Tennessee forest as a window onto the entire natural world. Visiting it almost daily for one year to trace nature's path through the seasons, Haskell spins a brilliant web of biology and ecology, explaining the science that binds together the tiniest microbes and the largest mammals. *The Forest Unseen* is a grand tour of nature in all its profundity. Haskell is a perfect guide into the world that exists beneath our feet and beyond our backyards.

"Injects much-needed vibrancy into the stuffy world of nature writing." –*Outside*, "The Outdoor Books That Shaped the Last Decade"

PENGUIN BOOKS